Lecture Notes in Computer Science   9069

Commenced Publication in 1973
Founding and Former Series Editors:
Gerhard Goos, Juris Hartmanis, and Jan van Leeuwen

T0210480

More information about this series at http://www.springer.com/series/7412

Cheng-Lin Liu · Bin Luo
Walter G. Kropatsch · Jian Cheng (Eds.)

Graph-Based Representations in Pattern Recognition

10th IAPR-TC-15 International Workshop, GbRPR 2015
Beijing, China, May 13–15, 2015
Proceedings

 Springer

Editors

Cheng-Lin Liu
Institute of Automation of CAS
Beijing
China

Walter G. Kropatsch
Vienna University of Technology
Vienna
Austria

Bin Luo
Anhui University
Anhui
China

Jian Cheng
Institute of Automation of CAS
Beijing
China

ISSN 0302-9743
Lecture Notes in Computer Science
ISBN 978-3-319-18223-0
DOI 10.1007/978-3-319-18224-7

ISSN 1611-3349   (electronic)

ISBN 978-3-319-18224-7   (eBook)

Library of Congress Control Number: 2015937201

LNCS Sublibrary: SL6 – Image Processing, Computer Vision, Pattern Recognition, and Graphics

Printed on acid-free paper

Springer International Publishing AG Switzerland is part of Springer Science+Business Media (www.springer.com)

Preface

Welcome to the 10th IAPR-TC15 Workshop on Graph-based Representations in Pattern Recognition (GbR2015) in Beijing, China, 2015! The series of GbR Workshop is sponsored by the TC15 (Graph-based Representations) of IAPR (International Association for Pattern Recognition), started from 1997. The GbR2015 follows the editions of 1997 in Lyon (France), 1999 in Castle of Haindorf (Austria), 2001 in Ischia (Italy), 2003 in York (UK), 2005 in Poitiers (France), 2007 in Alicante (Spain), 2009 in Venice (Italy), 2011 in Münster (Germany), and 2013 in Vienna (Austria).

Graph is a very effective structure for representing structural patterns and has been widely used in pattern recognition, computer vision, machine learning, and data mining. Structural pattern recognition methods usually represent a pattern as a graph modeling the primitives and their interrelationship. A digital image can be viewed as a graph with its pixels as nodes. Many machine learning methods view sample points as the nodes of a graph for modeling their affinity. Social media and biological data are often represented as graphs or networks. Theory and methods of graph data analysis have been studied intensively for graph generation, matching, labeling, segmentation, clustering, classification, and data mining in various fields. The GbR Workshop encourages all related works from such fields.

The GbR2015 received 53 full submissions. The authors are from Asia, Europe, America, and Oceania. Each submission was assigned to three Program Committee (PC) members for review, and 11 sub-reviewers were invited by the PC members. Based on the reviews, 36 papers were accepted for presentation at the workshop and inclusion in the proceedings. The accepted papers cover diverse issues of graph-based methods and applications, with 7 in graph representation, 15 in graph matching, 7 in graph clustering and classification, and 7 in graph-based applications.

In addition to regular presentations, we invited two established researchers to present keynote speeches. Professor Marcello Pelillo of University of Venice (Italy) and Dr. Xing Xie of Microsoft Research Asia (Beijing, China) will present speeches titled "Revealing Structure in Large Graphs: Szemerdi's Regularity Lemma and Its Use in Pattern Recognition" and "Understanding Users by Connecting Large Scale Social Graphs," respectively. We are grateful to the invited speakers, the authors of all submitted papers, the Program Committee members, and reviewers, and the staff of the Organizing Committee. Without their contributions, this workshop would not have been a success. We also thank Springer for publishing the proceedings, especially Mr. Alfred Hofmann and Ms. Anna Kramer for their efforts and patience in collecting and editing the proceedings.

We welcome you to take part in this academic event and hope you find GbR2015 an enjoyable and fruitful workshop.

May 2015

Cheng-Lin Liu
Bin Luo
Walter G. Kropatsch
Jian Cheng

Organization

Sponsors

Institute of Automation of CAS
National Laboratory of Pattern Recognition

Program Co-chairs

Cheng-Lin Liu	Institute of Automation of CAS, China
Bin Luo	Anhui University, China
Walter G. Kropatsch	Vienna University of Technology, Austria

Technical Program Committee

Xiang Bai	Huazhong University of Science and Technology, China
Xiao Bai	Beihang University, China
Luc Brun	GREYC, ENSICAEN, France
Roberto M. Cesar	University of São Paulo, Brazil
Badong Chen	Xi'an Jiaotong University, China
Jian Cheng	Institute of Automation of CAS, China
Ananda Chowdhury	Jadavpur University, India
Donatello Conte	Université François-Rabelais, Tours, France
Guillaume Damiand	LIRIS laboratory, France
Francisco Escolano	Universidad de Alicante, Spain
Volkmar Frinken	Kyushu University, Japan
Pasquale Foggia	University of Salerno, Italy
Edwin R. Hancock	University of York, UK
Qiang Ji	Rensselaer Polytechnic Institute, USA
Xiaoyi Jiang	University of Münster, Germany
Walter G. Kropatsch	Vienna University of Technology, Austria
Cheng-Lin Liu	Institute of Automation of CAS, China
Zhiyong Liu	Institute of Automation of CAS, China
Josep Lladós	Universitat Autònoma de Barcelona, Spain
Bin Luo	Anhui University, China
Jean-Marc Ogier	Université de La Rochelle, France
Shinichiro Omachi	Tohoku University, Japan
Marcello Pelillo	University of Venice, Italy

Peng Ren   China University of Petroleum, China
Kaspar Riesen   University of Applied Sciences and Arts,
                                      Switzerland
Francesc Serratosa   Universitat Rovira i Virgili, Spain
Ali Shokoufandeh   Drexel University, USA
Andrea Torsello   Ca' Foscari University of Venice, Italy
Seiichi Uchida   Kyushu University, Japan
Mario Vento   University of Salerno, Italy
Richard Wilson   University of York, UK
Shuicheng Yan   National University of Singapore, Singapore
Changshui Zhang   Tsinghua University, China

Local Organizing Committee

Jian Cheng   Institute of Automation of CAS, China
Yan-Ming Zhang   Institute of Automation of CAS, China
Ming Li   Institute of Automation of CAS, China

Referees

Xiang Bai   Volkmar Frinken   Peng Ren
Xiao Bai   Edwin R. Hancock   Kaspar Riesen
Luc Brun   Qiang Ji   Francesc Serratosa
Roberto M. Cesar   Xiaoyi Jiang   Ali Shokoufandeh
Badong Chen   Josep Llados   Andrea Torsello
Jian Cheng   Cheng-Lin Liu   Mario Vento
Ananda Chowdhury   Zhi-Yong Liu   Seiichi Uchida
Donatello Conte   Bin Luo   Richard Wilson
Guillaume Damiand   Jean-Marc Ogier   Shuicheng Yan
Francisco Escolano   Shinichiro Omachi   Changshui Zhang
Pasquale Foggia   Marcello Pelillo

Additional Referees

Li Chuan   Xiaodan Liang   Alessia Saggese
Andreas Fischer   Muhammad Muzzamil   Xinggang Wang
Llus-Pere De Las Heras   Luqman   Xu Yang
Yinlin Li   Tomo Miyazaki   Suiwu Zheng

Contents

Graph-Based Representation

Graph Matching

Graph Clustering and Classification

Graph-Based Applications

Graph-Based Representation

Approximation of Graph Edit Distance
in Quadratic Time

Kaspar Riesen[1,3](✉), Miquel Ferrer[1], Andreas Fischer[2],
and Horst Bunke[3]

[1] Institute for Information Systems,
University of Applied Sciences and Arts Northwestern Switzerland,
Riggenbachstrasse 16, 4600, Olten, Switzerland
{kaspar.riesen,miquel.ferrer}@fhnw.ch
[2] DIUF Department, University of Fribourg, Switzerland and iCoSys Institute,
University of Applied Sciences and Arts Western Switzerland, Fribourg, Switzerland
andreas.fischer@unifr.ch
[3] Institute of Computer Science and Applied Mathematics, University of Bern,
Neubrückstrasse 10, 3012, Bern, Switzerland
{riesen,bunke}@iam.unibe.ch

Abstract. The basic idea of a recent graph matching framework is to
reduce the problem of graph edit distance (GED) to an instance of a
linear sum assignment problem (LSAP). The optimal solution for this
simplified GED problem can be computed in cubic time and is eventually used to derive a suboptimal solution for the original GED problem.
Yet, for large scale graphs and/or large scale graph sets the cubic time
complexity remains a severe handicap of this procedure. Therefore, we
propose to use suboptimal algorithms – with quadratic rather than cubic time complexity – for solving the underlying LSAP. In particular, we
introduce several greedy assignment algorithms for approximating GED.
In an experimental evaluation we show that there is great potential for
further speeding up the GED computation. Moreover, we empirically
confirm that the distances obtained by this procedure remain sufficiently
accurate for graph based pattern classification.

1 Introduction

Graph edit distance (GED) is a widely accepted concept for general graph dissimilarity computation [1–5]. Yet, a well known drawback of GED is its computational complexity which is exponential in the number of nodes of the involved
graphs. This means that for large graphs the exact computation of GED is intractable. In recent years, a number of methods addressing the high complexity
of GED computation have been proposed [6–8]. In [9] the authors of the present
paper introduced an algorithmic framework for the approximation of GED. The
basic idea of this approach is to reduce the problem of GED computation to a
linear sum assignment problem (LSAP). For LSAPs quite an arsenal of efficient
(i.e. cubic time) algorithms exist [10].

© Springer International Publishing Switzerland 2015
C.-L. Liu et al. (Eds.): GbRPR 2015, LNCS 9069, pp. 3–12, 2015.
DOI: 10.1007/978-3-319-18224-7_1

The algorithmic procedure described in [9] consists of three major steps. In a first step the graphs to be matched are subdivided into individual nodes including local structural information. Next, in step 2, an LSAP solving algorithm is employed in order to find an optimal assignment of the nodes (plus local structures) of both graphs. Finally, in step 3, an approximate GED, which is globally consistent with the underlying edge structures of both graphs, is derived from the assignment of step 2.

In a recent paper [11] the optimal LSAP algorithm is replaced with a suboptimal greedy algorithm. From the theoretical point of view this approach is very appealing as it makes use of an approximation algorithm for an LSAP which in turn approximates the corresponding GED problem. The present paper introduces three advancements of this novel greedy graph edit distance approximation. All of these refinements are still greedy algorithms with quadratic time complexity. The proposed variants aim at improving the overall quality of the solution for the LSAP (and thus the GED approximation) by means of several heuristics.

Next, in Sect. 2 the original framework for GED approximation [9] is summarized. In Sect. 3 the greedy assignment algorithms are introduced. An experimental evaluation on diverse data sets is carried out in Sect. 4, and in Sect. 5 we draw conclusions and point out possible directions for future work.

2   Bipartite Graph Edit Distance Approximation

By reformulating GED to an instance of an LSAP (as introduced in [9] and denoted with $BP\text{-}GED^1$), three major steps have to be carried out.

First Step: A *cost matrix* $\mathbf{C}$ based on the node sets $V_1 = \{u_1, \ldots, u_n\}$ and $V_2 = \{v_1, \ldots, v_m\}$ of g_1 and g_2, respectively, is established as follows.

$$
\mathbf{C} =
\begin{bmatrix}
c_{11} & c_{12} & \cdots & c_{1m} & c_{1\varepsilon} & \infty & \cdots & \infty \\
c_{21} & c_{22} & \cdots & c_{2m} & \infty & c_{2\varepsilon} & \ddots & \vdots \\
\vdots & \vdots & \ddots & \vdots & \vdots & \ddots & \ddots & \infty \\
c_{n1} & c_{n2} & \cdots & c_{nm} & \infty & \cdots & \infty & c_{n\varepsilon} \\
c_{\varepsilon 1} & \infty & \cdots & \infty & 0 & 0 & \cdots & 0 \\
\infty & c_{\varepsilon 2} & \ddots & \vdots & 0 & 0 & \ddots & \vdots \\
\vdots & \ddots & \ddots & \infty & \vdots & \ddots & \ddots & 0 \\
\infty & \cdots & \infty & c_{\varepsilon m} & 0 & \cdots & 0 & 0
\end{bmatrix}
\tag{1}
$$

Entry c_{ij} thereby denotes the cost of a node substitution $(u_i \to v_j)$, $c_{i\varepsilon}$ denotes the cost of a node deletion $(u_i \to \varepsilon)$, and $c_{\varepsilon j}$ denotes the cost of a node insertion $(\varepsilon \to v_j)$. That is, the left upper corner of $\mathbf{C} = (c_{ij})$ represents the costs of all possible node substitutions, while the diagonals of the right upper and left lower corner represent the costs of all possible node deletions and insertions, respectively (every node can be deleted or inserted at most once and thus any

---

1 *Bipartite Graph Edit Distance* (LSAPs can be formulated by means of *bipartite graphs*).

non-diagonal element is set to ∞ in these parts). Substitutions of the form $(\varepsilon \to \varepsilon)$ should not cause any cost and therefore, any element in the right lower part is set to zero.

Second Step: Next, an LSAP is stated on cost matrix $\mathbf{C} = (c_{ij})$ and eventually solved. The LSAP optimization consists in finding a permutation $(\varphi_1, \ldots, \varphi_{n+m})$ of the integers $(1, 2, \ldots, (n+m))$ that minimizes the overall assignment cost $\sum_{i=1}^{(n+m)} c_{i\varphi_i}$. This permutation corresponds to the assignment

$$\psi = ((u_1 \to v_{\varphi_1}), (u_2 \to v_{\varphi_2}), \ldots, (u_{m+n} \to v_{\varphi_{m+n}}))$$

of the nodes of g_1 to the nodes of g_2. Note that assignment ψ includes node assignments of the form $(u_i \to v_j)$, $(u_i \to \varepsilon)$, $(\varepsilon \to v_j)$, and $(\varepsilon \to \varepsilon)$ (the latter can be dismissed, of course). Hence, the definition of $\mathbf{C} = (c_{ij})$ in Eq. 1 explicitly allows insertions and/or deletions to occur in the optimal assignment.

Assignment ψ does not take any structural constraints of the graphs into account as long as the individual entries in $\mathbf{C} = (c_{ij})$ consider the nodes of both graphs only. In order to integrate knowledge about the graph structure, to each entry c_{ij}, i.e. to each cost of a node edit operation $(u_i \to v_j)$, the minimum sum of edge edit operation costs, implied by the corresponding node operation, is added. This enables the LSAP to consider information about the local, yet not global, edge structure of a graph for optimizing the node assignment.

Third Step: The LSAP optimization finds an assignment ψ in which every node of g_1 and g_2 is either assigned to a unique node of the other graph, deleted or inserted. That is, ψ refers to a consistent node assignment and thus, the edge operations, which are implied by edit operations on their adjacent nodes, can be completely and consistently inferred from ψ. Hence, we get an edit path between the graphs under consideration. Yet, the edit path corresponding to ψ considers the edge structure of g_1 and g_2 in a global and consistent way while the optimal node assignment ψ is able to consider the structural information in an isolated way only (single nodes and their adjacent edges). Hence, the edit path found by this specific framework is not necessarily optimal and thus the distances are – in the best case – equal to, or – in general – larger than the exact graph edit distance.

3   Greedy Assignment Algorithms

3.1   Basic Greedy Assignment

In the second step of BP-GED an assignment of the nodes (plus local structures) of both graphs has to be found. For this task a large number of algorithms exist (see [10] for an exhaustive survey). For optimally solving the LSAP in the existing framework, Munkres' algorithm [12] also referred to as Kuhn-Munkres, or Hungarian algorithm, is deployed. The time complexity of this particular algorithm (as well as the best performing other algorithms for LSAPs) is cubic in the size of the problem, i.e. $O((n+m)^3)$ in our case.

Algorithm 1. Greedy-Assignment($\mathbf{C} = (c_{ij})$)

---

1. $\psi = \{\}$
2. **for** $i = 1, \ldots, (m+n)$ **do**
3. $\varphi_i = \arg\min_{\forall j} c_{ij}$
4.   Remove column φ_i from $\mathbf{C}$
5. $\psi = \psi \cup \{(u_i \rightarrow v_{\varphi_i})\}$
6. **end for**
7. **return** ψ

In [11] a suboptimal (rather than an optimal) algorithm is used for solving the LSAP on cost matrix $\mathbf{C}$. In particular, a greedy algorithm is employed that iterates through $\mathbf{C}$ from top to bottom through all rows and assigns every element to the minimum unused element of the current row. This idea is formalized in Alg. 1. For each row i in the cost matrix $\mathbf{C} = (c_{ij})$ the minimum cost assignment is determined and the corresponding node edit operation $(u_i \rightarrow v_{\varphi_i})$ is added to ψ. By removing column φ_i in $\mathbf{C}$ it is ensured that every column of the cost matrix is considered at most once (i.e. $\forall j$ only refers to available columns in $\mathbf{C}$). Clearly, the complexity of this suboptimal assignment algorithm is only $O((n+m)^2)$. For the remainder of this paper we denote the graph edit distance approximation where the basic node assignment is computed by means of this greedy procedure with *Greedy-GED*.

3.2 Tie Break Strategy

A crucial question in Alg. 1 is how possible ties between two (or more) cost entries in the same row are resolved. In [11] a certain row is always assigned to the first minimum column that occurs from left to right. As a refinement of this coarse heuristic we propose the following procedure (denoted by *Greedy-Tie-GED* from now on).

Assume that in the i-th row the following $t > 1$ cost entries $\{c_{ij_1}, \ldots, c_{ij_t}\}$ offer minimal cost among all available columns. For each of the corresponding columns $\{j_1, \ldots, j_t\}$ we search the minimum cost row that has not yet been assigned. Eventually, row i is assigned to column $\varphi_i \in \{j_1, \ldots, j_t\}$ that holds the highest minimum entry. Formally, column

$$\varphi_i = \arg\max_{j \in \{j_1, \ldots, j_t\}} \left(\min_{\forall k} c_{kj} \right)$$

is assigned to row i.

The intuition behind this strategy is as follows. The minimum cost

$$\min_{\forall k} c_{kj}$$

refers to the best available alternative for column j besides row i ($\forall k$ refers to all unprocessed rows of matrix $\mathbf{C}$). With other words, if we do not select $(u_i \rightarrow v_j)$ as assignment, the alternative assignment costs at least $\min_{\forall k} c_{kj}$. Hence, we should select the assignment among $\{u_i \rightarrow v_{j_1}, \ldots, u_i \rightarrow v_{j_t}\}$ where the best possible alternative would lead to the highest cost.

3.3 Refined Greedy Assignment

A next refinement of the basic greedy algorithm is given in Alg. 2. Similar to Alg. 1 for every row i we search for the minimum cost column φ_i (in case of ties the strategy described in the previous section is applied). In addition to Alg. 1 we also seek for the minimum cost row k for column φ_i.

Algorithm 2. Greedy-Refined-Assignment($\mathbf{C} = (c_{ij})$)

---

1. $\psi = \{\}$
2. **while** row i in $\mathbf{C}$ is available **do**
3. $\varphi_i = \arg\min_{\forall j} c_{ij}$ and $k = \arg\min_{\forall i} c_{i\varphi_i}$
4.   **if** $c_{i\varphi_i} \leq c_{k\varphi_i}$ **then**
5. $\psi = \psi \cup \{(u_i \rightarrow v_{\varphi_i})\}$
6.   Remove row i and column φ_i from $\mathbf{C}$
7.   **else**
8. $\psi = \psi \cup \{(u_k \rightarrow v_{\varphi_i})\}$
9.   Remove row k and column φ_i from $\mathbf{C}$
10.   **end if**
11. **end while**
12. **return** ψ

That is, for row i column φ_i is the best assignment. Yet, considering the assignment process from column φ_i's point of view, row k would be the best choice. We select the assignment which leads to lower cost. That is, if $c_{i\varphi_i} \leq c_{k\varphi_i}$ the assignment $(u_i \rightarrow v_{\varphi_i})$ is added to ψ. This includes the special situation where $i = k$, i.e. column φ_i corresponds to the optimal assignment for row i and vice versa. Otherwise, if $c_{i\varphi_i} > c_{k\varphi_i}$, assignment $(u_k \rightarrow v_{\varphi_i})$ is selected. Regardless of the assignment actually added to ψ, the corresponding row and column are removed from $\mathbf{C}$.

Note that in contrast with Alg. 1 where the i-th assignment added to ψ always considers row i, this algorithm processes the rows of $\mathbf{C}$ not necessarily from top to bottom. However, the complexity of this assignment algorithm remains $O((n+m)^2)$. We denote the graph edit distance approximation where the LSAP is solved by means of this particular greedy assignment algorithm with *Greedy-Refined-GED*.

3.4 Greedy Assignment Regarding Loss

A further refinement of the greedy assignment is given in Alg. 3. Similar to Alg. 2 we first find both the minimum cost column φ_i for row i and the minimum cost row k for column φ_i (again the tie break strategy of Sect. 3.2 is applied). Now three cases have to be considered.

If i equals k, column φ_i is the best available assignment for row i and vice versa (and thus assignment $(u_i \rightarrow v_{\varphi_i})$ can safely be added to ψ). If $i \neq k$ two scenarios have to be distinguished. For both scenarios the second best assignment φ_i' for row i, as well as the best available assignment φ_k for row k (distinct from φ_i) are determined. In the first scenario row i is assigned to its best column φ_i

---

Algorithm 3. Greedy-Loss-Assignment($\mathbf{C} = (c_{ij})$)

---

1. $\psi = \{\}$
2. **while** row i in $\mathbf{C}$ is available **do**
3. $\varphi_i = \arg\min_{\forall j} c_{ij}$ and $k = \arg\min_{\forall i} c_{i\varphi_i}$
4.   **if** $i == k$ **then**
5. $\psi = \psi \cup \{(u_i \to v_{\varphi_i})\}$
6.   Remove row i and column φ_i from $\mathbf{C}$
7.   **else**
8. $\varphi_i' = \arg\min_{\forall j \neq \varphi_i} c_{ij}$ and $\varphi_k = \arg\min_{\forall j \neq \varphi_i} c_{kj}$
9.   **if** $(c_{i\varphi_i} + c_{k\varphi_k}) < (c_{i\varphi_i'} + c_{k\varphi_i})$ **then**
10. $\psi = \psi \cup \{(u_i \to v_{\varphi_i}), (u_k \to v_{\varphi_k})\}$
11.   Remove rows i, k and columns φ_i, φ_k from $\mathbf{C}$
12.   **else**
13. $\psi = \psi \cup \{(u_i \to v_{\varphi_i'}), (u_k \to v_{\varphi_i})\}$
14.   Remove rows i, k and columns φ_i, φ_i' from $\mathbf{C}$
15.   **end if**
16.   **end if**
17. **end while**
18. **return** ψ

---

and thus row k has eventually to be assigned to the best alternative φ_k. The sum of cost of this scenario amounts to $(c_{i\varphi_i} + c_{k\varphi_k})$. In the other scenario, row k is assigned to column φ_i and thus row i has to be assigned to its best possible alternative which is column φ_i'. The sum of cost of this scenario amounts to $(c_{i\varphi_i'} + c_{k\varphi_i})$. We choose the scenario with lower sum of cost and remove the corresponding rows and columns from $\mathbf{C}$.

We denote the graph edit distance approximation using this greedy approach with *Greedy-Loss-GED* (the complexity of this assignment algorithm remains $O((n + m)^2)$.

3.5 Relations to Exact Graph Edit Distance

Note that In contrast with the optimal permutation $(\varphi_1, \ldots, \varphi_{n+m})$ returned by the original framework, the permutations $(\varphi_1', \ldots, \varphi_{n+m}')$ of the proposed greedy algorithms are suboptimal. That is, the sum of assignments costs of all greedy approaches are greater than, or equal to, the minimal sum of assignment cost provided by optimal LSAP solving algorithms. Formally, we have

$$\sum_{i=1}^{(n+m)} c_{i\varphi_i'} \geq \sum_{i=1}^{(n+m)} c_{i\varphi_i}$$

However, for the approximate graph distance values derived by BP-GED and any greedy approach no globally valid order relation exists. That is, the approximate graph edit distance derived from a greedy assignment can be greater than, equal to, or smaller than a distance value returned by BP-GED. Note that approximations derived from both optimal or suboptimal local assignments constitute upper bounds of the true graph edit distance. Hence, the smaller the approximated distance value is, the nearer it is to the exact graph edit distance.

4   Experimental Evaluation

In Table 1 we show different characteristic numbers for BP-GED and all greedy variants for graph edit distance approximation on three different data sets from the IAM graph repository [13]. First we focus on the mean run time for one matching in ms ($\varnothing t$). On the relatively small graphs of the FP data set, the speed-up by Greedy-GED compared to BP-GED is rather small. Yet, on the other two data sets with larger graphs substantial speed-ups can be observed. That is, using Greedy-GED rather than BP-GED on the AIDS data set leads to a decrease of the mean matching time from 3.61ms to 1.21ms. On the MUTA data Greedy-GED is more than seven times faster than the original approximation. We also observe that the enhanced greedy algorithms, viz. Greedy-Tie, Greedy-Refined, and Greedy-Loss, do not need substantially more computation time compared to the plain approximation using Greedy-GED.

The characteristic number $\varnothing o$ measures the overestimation of one particular assignment compared to the optimal assignment. For comparison we take the difference between the optimal sum of costs and the sum of costs of the plain greedy assignment as 100%. Regarding the results in Table 1 we note that our novel tie resolution is not able to substantially reduce this overestimation. Yet, both Greedy-Refined and Greedy-Loss return overall assignment sums which are substantially nearer to the optimal sum than the plain greedy approach. For instance, compared with Greedy-GED the difference to the optimal sum of

Table 1. The mean run time for one matching in ms ($\varnothing t$), the relative overestimation of a greedy sum of assignment costs compared to the optimal sum of assignment costs ($\varnothing o$), the mean relative deviation of greedy GED algorithm variants compared with BP-GED in percentage ($\varnothing e$), and the accuracy of a 1NN classifier

Data		Algorithm				
		BP-GED	Greedy-GED	Greedy-Tie-GED	Greedy-Refined-GED	Greedy-Loss-GED
AIDS	$\varnothing t$ [ms]	3.61	1.21	1.23	1.25	1.24
	$\varnothing o$ [%]	–	+100.0	+99.9	+74.1	+75.0
	$\varnothing e$ [%]	–	+7.1/ − 4.1	+6.5/ − 3.7	+5.8/ − 4.6	+7.0/ − 4.3
	1NN [%]	99.07	98.93	99.20	99.13	98.87
FP	$\varnothing t$ [ms]	0.41	0.30	0.30	0.31	0.31
	$\varnothing o$ [%]	–	+100.0	+99.8	+88.0	+87.0
	$\varnothing e$ [%]	–	+6.6/ − 15.7	+6.5/ − 15.9	+6.0/ − 17.5	+5.7/ − 18.1
	1NN [%]	79.75	77.05	77.20	75.80	76.6
Muta	$\varnothing t$ [ms]	33.89	4.56	4.68	4.88	4.95
	$\varnothing o$ [%]	–	+100.0	+99.9	+2.8	+15.5
	$\varnothing e$ [%]	–	+7.4/ − 3.8	+6.2/ − 4.6	+5.9/ − 5.1	+7.6/ − 4.9
	1NN [%]	70.20	70.10	71.80	71.60	71.10

assignment costs is reduced by about 25% with Greedy-Refined and Greedy-Loss on the AIDS data set. Similar results are observable on the other data sets (note particularly the massive reduction on the Muta data set using Greedy-Refined). The reduction of overestimation of assignment costs can also be seen in the scatter plot in Fig. 1 (a) where the optimal assignment cost (x-axis) is plotted against the greedy assignment costs (y-axis) for Greedy (gray dots) and Greedy-Loss (black dots) on the AIDS data set.

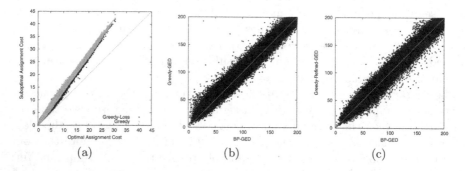

(a)   (b)   (c)

Fig. 1. (a) Optimal assignment cost (x-axis) vs. greedy assignment cost (y-axis) on the AIDS data sets using Greedy (gray dots) and Greedy-Loss (black dots), (b) distances of BP-GED (x-axis) vs. distances of Greedy-GED (y-axis), and (c) distances of BP-GED (x-axis) vs. distances of Greedy-Refined-GED (y-axis)

The characteristic number $\varnothing e$ in Table 1 measures the mean relative over- and underestimation of greedy graph edit distances compared with BP-GED. Note that both means are computed on the sets of distances where a greedy approach actually over- or underestimates the original approximation. We observe that our enhanced greedy assignment algorithms are able to reduce the mean overestimation compared with the plain greedy approach in eight out of nine cases. For instance, the mean overestimation of Greedy-GED amounts to +7.1% on the AIDS data, while the same parameter is reduced to +5.8% with Greedy-Refined-GED. Interestingly, the mean underestimation of Greedy-GED is further increased with our novel enhancements (also in eight out of nine cases). That is, compared with Greedy-GED the refined greedy variants are able to improve the overall distance quality in general. The reduction of overestimation and increase of underestimation by the refined algorithms can also be seen in Fig. 1 (b) and (c) where the distances of BP-GED (x-axis) are plotted against distances returned by Greedy-GED and Greedy-Refined-GED (y-axis), respectively (on the Muta data set).

Not only the mean over- and underestimation, but also the number of matchings, where greedy distances are equal to, smaller than, or greater than, the distances returned by BP-GED are crucially altered using our improved greedy assignments. Fig. 2 shows for every greedy algorithm the relative number of

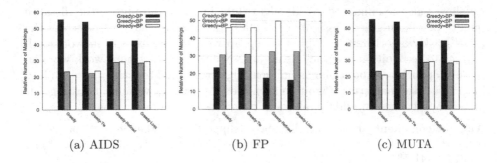

Fig. 2. Relative number of matchings where the greedy algorithms return distances greater than BP-GED (black bars), smaller than than BP-GED (grey bars), or equal to BP-GED (white bars) on all data sets

matchings for these three cases. On the AIDS data set, for instance, the relative number of matchings where the greedy distances overestimate the distances returned by BP-GED are reduced from 55.5% (Greedy-GED) to 42.1% (Greedy-Loss-GED). Likewise, the number of matchings where the greedy approaches lead to equal, or even smaller distances than BP-GED is increased using the proposed refinement algorithms. Similar (or even better) results can be observed on the other data sets. That is, compared with Greedy-GED the novel algorithms decrease the number of matchings which overestimate BP-GED while the number of matchings with equal or even better approximations than BP-GED is increased.

Finally, Table 1 shows the recognition rate of a 1-nearest-neighbor classifier (1NN). The nearest neighbor paradigm is particularly interesting for the present evaluation because it directly uses the distances without any additional classifier training. In comparison with BP-GED we observe that Greedy-GED slightly deteriorates the recognition rates on all data sets. However, the proposed enhanced greedy assignment algorithms improve the recognition rate of a 1NN compared to Greedy-GED in 6 out of 9 cases. Moreover, at least one of the novel greedy algorithms outperforms the recognition rate of Greedy-GED on all data sets. For BP-GED the same observation holds for two of the three data sets.

5   Conclusions and Future Work

In the present paper we propose an extension of graph edit distance approximation algorithms developed previously. While in the original framework the nodes plus local edge structures are assigned to each other in an optimal way, the extension of the present paper uses various suboptimal greedy algorithms for this task. In particular we propose three enhanced versions of a simple greedy assignment algorithm. These novel enhancements allow the graph edit distance approximation in quadratic – rather than cubic – time. The speed up of the approximation is empirically verified on three different graph data sets. Moreover, we show that the enhanced greedy algorithms are able to improve the distance quality of a plain greedy assignment. Finally, we observe that in most cases the novel greedy algorithms are able to keep up with the existing framework with respect to recognition accuracy using a 1NN classifier. In future work we plan to develop further suboptimal assignment algorithms and test their applicability in graph based pattern recognition scenarios. Moreover, we aim at testing our greedy approaches on additional graph data sets and larger graphs

Acknowledgement. This work has been supported by the *Hasler Foundation* Switzerland and the *Swiss National Science Foundation* projects 200021_153249 and PBBEP2_141453.

References

1. Bunke, H., Allermann, G.: Inexact graph matching for structural pattern recognition. Pattern Recognition Letters 1, 245–253 (1983)
2. Sanfeliu, A., Fu, K.: A distance measure between attributed relational graphs for pattern recognition. IEEE Transactions on Systems, Man, and Cybernetics (Part B) 13(3), 353–363 (1983)
3. Gao, X., Xiao, B., Tao, D., Li, X.: A survey of graph edit distance. Pattern Anal. Appl. 13(1), 113–129 (2010)
4. Robles-Kelly, A., Hancock, E.: Graph edit distance from spectral seriation. IEEE Transactions on Pattern Analysis and Machine Intelligence 27(3), 365–378 (2005)
5. Emms, D., Wilson, R., Hancock, E.: Graph edit distance without correspondence from continuous-time quantum walks. In: da Vitoria Lobo, N., Kasparis, T., Roli, F., Kwok, J.T., Georgiopoulos, M., Anagnostopoulos, G.C., Loog, M. (eds.) SSPR&SPR 2008. LNCS, vol. 5342, pp. 5–14. Springer, Heidelberg (2008)
6. Boeres, M., Ribeiro, C., Bloch, I.: A randomized heuristic for scene recognition by graph matching. In: Ribeiro, C.C., Martins, S.L. (eds.) WEA 2004. LNCS, vol. 3059, pp. 100–113. Springer, Heidelberg (2004)
7. Sorlin, S., Solnon, C.: Reactive tabu search for measuring graph similarity. In: Brun, L., Vento, M. (eds.) GbRPR 2005. LNCS, vol. 3434, pp. 172–182. Springer, Heidelberg (2005)
8. Justice, D., Hero, A.: A binary linear programming formulation of the graph edit distance. IEEE Trans. on Pattern Analysis ans Machine Intelligence 28(8), 1200–1214 (2006)
9. Riesen, K., Bunke, H.: Approximate graph edit distance computation by means of bipartite graph matching. Image and Vision Computing 27(4), 950–959 (2009)
10. Burkard, R., Dell'Amico, M., Martello, S.: Assignment Problems. Society for Industrial and Applied Mathematics, Philadelphia (2009)
11. Riesen, K., Ferrer, M., Dornberger, R., Bunke, H.: Greedy graph edit distance (Submitted to MLDM)
12. Munkres, J.: Algorithms for the assignment and transportation problems. Journal of the Society for Industrial and Applied Mathematics 5(1), 32–38 (1957)
13. Riesen, K., Bunke, H.: IAM graph database repository for graph based pattern recognition and machine learning. In: da Vitoria Lobo, N., Kasparis, T., Roli, F., Kwok, J.T., Georgiopoulos, M., Anagnostopoulos, G.C., Loog, M. (eds.) SSPR&SPR 2008. LNCS, vol. 5342, pp. 287–297. Springer, Heidelberg (2008)

Data Graph Formulation as the Minimum-Weight Maximum-Entropy Problem

Samuel de Sousa$^{(\boxtimes)}$ and Walter G. Kropatsch

Pattern Recognition and Image Processing Group,
Vienna University of Technology
Vienna, Austria
{sam,krw}@prip.tuwien.ac.at

Abstract. Consider a point-set coming from an object which was sampled using a digital sensor (depth range, camera, etc). We are interested in finding a graph that would represent that point-set according to some properties. Such a representation would allow us to match two objects (graphs) by exploiting topological properties instead of solely relying on geometrical properties. The Delaunay triangulation is a common out off-the-shelf strategy to triangulate a point-set and it is used by many researchers as the standard way to create the so called data-graph and despite its positive properties, there are also some drawbacks. We are interested in generating a graph with the following properties: the graph is (i) as unique as possible, (ii) and as discriminative as possible regarding the degree distribution. We pose a combinatorial optimization problem (Min-Weight Max-Entropy Problem) to build such a graph by minimizing the total weight cost of the edges and at the same time maximizing the entropy of the degree distribution. Our optimization approach is based on Dynamic Programming (DP) and yields a polynomial time algorithm.

Keywords: Data-graph · Graph realization · Combinatorial optimization

1 Introduction

In many applications it is necessary to create a graph out of an unstructured point-set. Such a point-set could represent the projection of an object onto an image or locations of key features. A common problem in Computer Vision consists of the registration of two or more point-sets. As a result, one would (i) obtain the transformation that maps one set towards the other and (ii) find the pairwise correspondence between points in all sets.

A graph created out of a point-set will be referred here as the data-graph. A common procedure consists of (i) creating the data-graph using the Delaunay triangulation and (ii) performing the registration using an optimization procedure [2, 3, 12]. The Delaunay triangulation [4] is based on the condition that no other point should lie inside the circumcircle of any triangle. It is not clear, though, if the Delaunay triangulation is always the best solution for all possible tasks in Computer Vision when a data-graph is required. There are several

© Springer International Publishing Switzerland 2015
C.-L. Liu et al. (Eds.): GbRPR 2015, LNCS 9069, pp. 13–22, 2015.
DOI: 10.1007/978-3-319-18224-7_2

alternative methods such as Reeb graphs [14], Gabriel Graph [7], and also the Euclidean Minimum Spanning Tree (EMST). Many researchers propose methods of data-graph construction focusing either on aesthetic aspects [13] or designing their own criteria, such as the fan-shaped triangulation of Lian and Zhang [11].

Developing a more unique representation of a graph is not a new concept and a relevant paper in the topic was produced by Dickinson et al. [5] where they focus on a class of graphs which have unique representation of the node labels, a representation ρ of graph $g = (V, E, \alpha, \beta)$ is created where α and β are functions assignings labels to the vertices and edges respectively. They can find if two graphs g' and g'' are isomorphics by comparing their representations $\rho(g')$ and $\rho(g'')$ in $O(N^2)$. Our work differs in nature with the previous work since we are not necessarily interested in isomorphisms between two graphs. Our input is a point-set and we would like to design the best graph to represent it in order to reduce the number of possible isomorphisms between another corresponding data-graph, ideally when two data-graphs are created, the correspondence is solved by our representation. We pose this problem as an optimization problem where we call it the Minimum-Weight Maximum-Entropy Problem which captures the desired behaviour of our graph and we provide an efficient polynomial time algorithm based on Dynamic Programming (DP) to solve it. To the best of our knowledge, we are the first ones to tackle the matching problem in this fashion.

The remainder of this paper is organized as follows: Section 2 formulates the problem being addressed in this paper and the difficulties associated with the minimization of its cost function. Section 3 introduces the *Near Homogeneous Degree Distribution (NHDD)* property which is used for building up the solution. As our solution possesses a recurrence relation, we design a Dynamic Programming algorithm on Section 4. Finally, we draw our conclusions and future work on Section 5.

2   The Minimum-Weight Maximum-Entropy Problem

As the goal of our paper is to design data-graphs which would ease the registration process, the first question to be asked is how we can define such metric, or which property an ideal data graph would have in order to make the registration process as trivial as possible. If we examine a regular graph, i.e. a graph in which all nodes have the same degree, we would notice that all nodes could be matched against all the other nodes, and the registration task would generate many ambiguous solutions. Therefore, if one succeeds to build a graph which is the exact opposite of a regular graph, i.e. a graph whose degree distribution is as diverse as possible, the registration process would be, then, alleviated.

In order to measure the diversity found in the degree distribution, we can calculate the Shannon entropy (H) for a graph $G(V, E)$ as follows:

$$\mathrm{H}(G) = - \sum_{v \in V_{\neq}} p(v) \log_2(p(v)).$$

(1)

where $p(v)$ is the probability of finding a node with a degree of v among all distinct degree values ($V_{\neq}$ is the set of distinct degree values of V). The entropy measures the uncertainty associated with a random variable. As defined in Eq. 1, a high entropy H(G) of a graph $G(V, E)$ would indicate a high "variability" in the distribution of nodes V. The converse is also true, a low entropy means low variability, as in a k-regular graph whose probability $p(k) = 1$ and $log(1) = 0$.

We aim at obtaining a graph with the highest entropy that would let us match the nodes more easily. Nevertheless, even if we are able to obtain a graph whose entropy is as high as possible, there is still the problem of ambiguity. There are multiple solutions with the same entropy value. Therefore, the second question we pose is how to generate a graph as unique as possible. Such question is important due to the fact that it would allow us to match two graphs based only on their degree values. We would like to uniquely identify the nodes unless all points in the graph are equidistant, in this case many possible solutions still exist. To achieve that, we decided also to minimize the total weighted edge cost of our graph. We call this problem *the Minimum-Weight Maximum-Entropy* (MWME):

Definition 1. *The Minimum-Weight Maximum-Entropy (MWME) is the problem of estimating an edge-induced subgraph of a graph whose entropy of the degree distribution is maximum and the total edge weight is minimum.*

Given a point set P, we create a complete graph $K_{|P|}$ using a metric[1] function as the edge weights. Let W denote the weighted edges of $K_{|P|}$. We define a binary vector U that induces an edge-subgraph $G[U]$ composed of all nodes of $K_{|P|}$ and edges $\{W_i | U_i = 1\}$. We search for the vector U which minimizes the cost:

$$
\begin{aligned}
\underset{U}{\text{minimize}} \quad & \sum_{i=1}^{|W|} W_i U_i, \\
\text{subject to} \quad & \text{H}(F) \leq \text{H}(G[U]), \forall F \\
& U \in \{0, 1\}^{|W|}.
\end{aligned}
\tag{2}
$$

under the constraint that the entropy H($G[U]$) of the induced subgraph $G[U]$ is maximum, i.e. for any graph F, the entropy H(F) will be lower or equal to our edge induced subgraph $G[U]$. The second constraint states that the optimization variable is discrete, we either add the edge W_i to our graph $G[U]$ when $U_i = 1$ or we do not add such an edge ($U_i = 0$).

We propose a dynamic programming algorithm that minimizes Eq. 2 by looking deeper into some properties of the desired induced subgraph $G[U]$. It is important to mention that we cannot guarantee unique solutions. The reason for that can be visualized in Fig. 1a. By constructing a graph out of a regular polygon, we could rotate all nodes and the total edge cost would remain the same as well as the entropy. Therefore, there would be many possible solutions.

---

[1] e.g. the Euclidean distance.

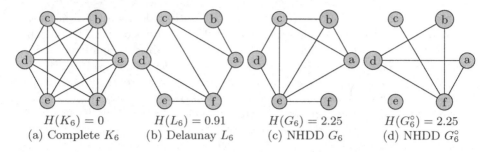

$H(K_6) = 0$ $H(L_6) = 0.91$ $H(G_6) = 2.25$ $H(G_6^\circ) = 2.25$
(a) Complete K_6 (b) Delaunay L_6 (c) NHDD G_6 (d) NHDD G_6°

Fig. 1. Some realizations of a six vertices graph along with their entropy. Graph (a) is a complete graph, Graph (b) is a triangulation and Graphs (c) and (d) fulfill the NHDD property with the highest entropy values.

3   The Near Homogeneous Degree Distribution (NHDD)

In order to minimize the cost function provided in Eq. 2, we will decompose the problem into the entropy maximization part and the edge cost minimization step. In this section we explain how we guarantee the maximum possible entropy in a graph and later how to incorporate our graph with the minimum edge cost.

We need to know, theoretically, how many nodes a simple graph[2] can have with distinct degree values in order to create a graph with the highest variability. This problem is closely related to the graph realization problem [6] which consists of determining if a sequence of degrees is feasible for a graph (called graphic sequence). This problem has been addressed by Erdős-Gallai [6] and Havel-Hakimi [8, 9]. We would like to obtain the maximal variability in a graph, our problem could be considered as generating the graphic sequence with highest entropy. Theorem 1 states the maximum variability in a graph.

Theorem 1. *For every simple graph $G(V, E)$ with $|V| > 2$, there are two nodes with the same degree.*

Proof (Kocay and Kreher [10]). A simple graph does not allow parallel edges and self-loops, therefore, the highest degree is $|V| - 1$. If we create a degree sequence in which all nodes have different degree values (the highest possible variability), this would mean that the nodes would have to be $V = (0, 1, \ldots, |V| - 1)$. If one node has a degree of $|V| - 1$, it means it is connected to all the other nodes, but the degree zero means that one node is not connected to any other. Thus, degree values of 0 and $|V| - 1$ are mutually exclusive and cannot coexist in the same graph, since this leaves only $|V| - 1$ values for $|V|$ nodes, by the pidgeon hole principle, at least two nodes must have the same degree.

Since it is not feasible to create a graph whose nodes have all distinctive degree values, the highest possible variability is $|V| - 1$ (Def. 2).

---

[2] A simple graph is a graph which does not contain parallel edges and self-loops.

Definition 2. *A graph $G(V, E)$ fulfills the Near Homogeneous Degree Distribution (NHDD) property when it contains $|V| - 1$ nodes with different degrees.*

An induced subgraph for our optimization function (Eq. 2) is only feasible if it fulfills the NHDD. We show now that there are two feasible graphs (Def. 3).

Definition 3. G_N *is a connected graph which fulfills the NHDD property and G_N° is a graph which fulfills the NHDD but it contains one isolated node.*

Lemma 1 shows how we can generate the two possible graphs (G_N and G_N°) fulfilling the NHDD definition.

Lemma 1. *For any integer $N \geq 3$, it is possible to generate both G_N and G_N°.*

Proof. We start with a base case of a graph G_N° with $N = 3$, whose degree distribution (the degree values of all nodes in the graph) is $D_3^\circ = (0, 1, 1)$. The NHDD is already fulfilled. Our inductive step states that this holds for any number $N \geq 3$. In order to build a solution for G_{N+1}, there are two possibilities:

a) G_N° has an isolated node. Then, by adding a node and connecting it to all the others, we will obtain a G_{N+1} since the degree of all the other nodes will be increased by one and the graph will be connected. $G_{N+1}^\circ = K_{N+1} - G_{N+1}$.

b) G_N does not have an isolated node. Therefore, we add a new isolated node and we obtain G_{N+1}° in which the NHDD is fulfilled. $G_{N+1} = K_{N+1} - G_{N+1}^\circ$ □

In the proof, we used the complement of a graph. In Lemma 2, we show that the complement of a graph fulfilling NHDD also fulfills the same property. Fig. 1 shows four different induced graphs with 6 nodes and their respective entropy values (Eq. 1). Graph (a) is a complete graph K_6, whose entropy is equal to zero. Graph (b) is created using a Delaunay triangulation (L_6) and its entropy is equal to 0.91. Graphs (c) and (d) are a G_6 and a G_6° respectively. It is clear that both G_N° and G_N are more distinctive than L_6 and K_6 and their entropy is significantly higher. The existence of an isolated node did not affect the entropy as the variability in the number of existing nodes is the same, however, as the nodes have the same distance, those solutions are not unique.

Lemma 2. *The complement graph of G_N also fulfills the NHDD property.*

Proof. The complement graph $\overline{G_N}$ is obtained by taking the difference of $K_N - G_N$. A G_N has distribution equal to $D_N = (1, \ldots, |V| - 1)$. A complete graph K_N has all nodes with degree equal to $|V| - 1$. Therefore, the degree distribution of $K_N - G_N$ will be $D_N^\circ = (|V| - 2, \ldots, 0)$ with one isolated node and maximum degree of $|V| - 2$.

Theorem 2. *For any integer number N, it is always possible to obtain a connected graph that fulfills the NHDD property.*

Proof. This proof comes out naturally as a consequence of Lemmas 1 and 2. For any integer N, we will obtain either G_N or G_N°. In case we obtain G_N°, we take the complement and we can always obtain a connected graph which is *NHDD*.

□

The number of edges of G_N and G_N° can be directly calculated (Theorem 3).

Theorem 3. *The number of edges of graphs G_N and G_N° is equal to* $|E_N| = \frac{1}{2}\left(\frac{N(N-1)}{2} + \lfloor\frac{N}{2}\rfloor\right)$ *and* $|E_N^\circ| = \frac{1}{2}\left(\frac{(N-2)(N-1)}{2} + \lfloor\frac{N-1}{2}\rfloor\right)$.

Proof. We will first prove for graph G_N. The sum of degrees in a G_N is equal to:

$$\sum_{v \in G_N} deg(v) = 1 + \ldots + \left\lfloor\frac{N}{2}\right\rfloor + \left\lfloor\frac{N}{2}\right\rfloor + \ldots + N - 1 \tag{3}$$

which is equivalent to:

$$\sum_{v \in G_N} deg(v) = \sum_{i=1}^{N-1} i + \left\lfloor\frac{N}{2}\right\rfloor = \frac{N(N-1)}{2} + \left\lfloor\frac{N}{2}\right\rfloor \tag{4}$$

The handshake lemma states that $|E_N| = \sum deg(v)/2$, we arrive that the number of edges is:

$$|E_N| = \frac{1}{2}\left(\frac{N(N-1)}{2} + \left\lfloor\frac{N}{2}\right\rfloor\right) \tag{5}$$

For the disconnected case, we conclude that:

$$|E_N^\circ| = \sum_{v \in G_N^\circ} deg(v) = \sum_{i=0}^{N-2} i + \left\lfloor\frac{N-1}{2}\right\rfloor = \frac{1}{2}\left(\frac{(N-2)(N-1)}{2} + \left\lfloor\frac{N-1}{2}\right\rfloor\right) \tag{6}$$

□

4　Optimization

The discussion about the NHDD property was pursued during the attempt to minimize our objective function (Eq. 2). It is known that the uniform distribution is has the maximum entropy ($\log n$) among probability distributions [1], as we cannot obtain the uniform distribution of the degrees in a graph (Theorem 1), we calculate analytically the entropy our of NHDD graphs (Theorem 4).

Theorem 4. *The entropy of $H(G_N) = H(G_N^\circ) = \log_2 N - \frac{2}{N}$.*

Proof. In a probabilistic interpretation of the entropy, the $p(x)$ is the probability of a event x to happen. Our event is the occurrence of a degree x in G_N (or G_N°). There are $N-2$ events with probability $p(x) = 1/N$ and one whose probability is $p(x) = 2/N$. Therefore, the entropy of $H(G_N)$ and $H(G_N^\circ)$ is:

$$H(G_N) = \frac{N \log N}{N} + \frac{2 \log N}{N} - \frac{2 \log N}{N} - \frac{2}{N} = \log N - \frac{2}{N} \tag{7}$$

□

```
   Data: Complete graph $K_N(P, W)$, $N \geq 3$;
 1 begin
 2  |   $dp \leftarrow Matrix(N, N, N, value = \infty)$;
 3  |   // We create $N$ solutions starting with point $i$ (layer).
 4  |   for $i \leftarrow 1$ to $N$ do
 5  |   |   $dp(i, 1, i : N) = 0; join = 0$;
 6  |   |   // $k$ is the capacity of the graph $G_N$ (row).
 7  |   |   for $k \leftarrow 2$ to $N$ do
 8  |   |   |   // $j$ is the node to be added into $G_N$ (column).
 9  |   |   |   for $j \leftarrow 1$ to $N$ do
10  |   |   |   |   if $join$ then
11  |   |   |   |   |   $c_{new} \leftarrow dp(i, k - 1, N) + \sum_{g \in G}^{|G|} W(j, g)$;
12  |   |   |   |   else
13  |   |   |   |   |   $c_{new} \leftarrow dp(i, k - 1, N) + W(j, v_{k-1})$;
14  |   |   |   |   $dp(i, k, j) \leftarrow \min(dp(i, k, j - 1), c_{new})$;
15  |   |   |   $join = \neg join$;
16  |   |   // The new solution $i$ is at least as good as $i - 1$.
17  |   |   $dp(i, n, n) \leftarrow \min(dp(i, n, n), dp(i - 1, n, n))$;
18  |   return $TraceBack(dp)$.
```

Algorithm 1. The n1graph algorithm which generates a G_N graph

Our n1graph algorithm (Alg. 1) starts by defining a Matrix dp (line 2) of dimensions N^3 whose elements are ∞. This matrix will hold the cost to be minimized. We produce N solutions iteratively (line 4), in which the cost of solution i will be most as low as $i - 1$ (line 17). Every iteration of i could be visualized as one layer of dp and the initialization occurs by setting the costs from i to N equal to zero (line 5). Since the weights lie in the edges, the initialization indicates which node is chosen at the moment (i) and it is zero since there are no edges in G_N at capacity 1, the cost is not increased. We have, now, one node (i) in G_N, we start to increase the capacity (k) of our graph G_N (line 7) until all nodes belong to the graph. Whenever the capacity is increased, we are allowed to add one more node (j) and the choose the one with minimum cost.

We alternate between two steps, adding an edge to all the nodes currently in G_N (line 11) and adding an edge to the last node (v_{k-1}) added to G_N (line 13). The algorithm can be visualized as "pushing" the minimum weight towards the right bottom side of the matrix. The optimum for each capacity k is found at (k, N) and the optimum for each layer i is found at $dp(i, N, N)$. Hence, the term $dp(i, k - 1, N)$ can be understood as the best cost for the optimization when the capacity was $k - 1$, that's the total weight we will propagate towards the end. Finally, on line 14, we take the minimum between the cost of adding the current node j and the previous cost of at the capacity k without node j.

Fig. 2 shows a complete K_4 with weighted edges and it highlights the optimum solution for the MWME problem. The highlighted node ⓓ consists of the layer which yielded the optimum solution ($i = 4$). This layer is available on Table 1. On $k = 1$, dp is initialized with node ⓓ $= 0$ (since at this capacity there are no edges in the graph). When capacity is increased to 2, the node ⓐ is added yielding a cost of 4, but within the same capacity there is a lower cost (3) if node ⓒ is added instead. The final cost is displayed at cell $j = d, k = 4$. We trace back on the same row until there is a change in cost (i.e. meaning that a node has been added). Whenever a node is added, we proceed the trace from on the row with a lower capacity $(k - 1, N)$ and continue tracing back until all nodes are added, the reversed node sequence obtained by tracing back is (b, a, c, d).

Theorem 5. *The time complexity of the n1graph algorithm is $\Theta(N^3|E|)$.*

Proof. The three nested loops of i, j, k yield clearly a $\Theta(N^3)$ time. Inside the j loop, two operations will alternate: (i) adding one edge to the last node of G_N and (ii) adding an edge to each node of G_N, the total operations will be: $(1, 2, 1, 4, 1, \ldots)$, which can be split into $\lfloor \frac{N-1}{2} \rfloor$ operations of type (ii) and $\lfloor \frac{N}{2} \rfloor$ operations of type (i). The complexity for those two operations is equivalent to Eq. 5, which is the number of edges in the graph, yielding in total $\Theta(N^3|E|)$. □

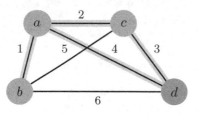

Fig. 2. A K_4 and the induced graph G_4 where d is the reference node i

Table 1. The trace back starts at (d,4) and moves on the same row until a value changes: a node is added to the graph

$k\backslash j$	a	b	c	d
1	∞	∞	∞	0
2	4	4	3 ↰	← 3
3	9 ↰	← 9	← 9	← 9
4	9	10 ↰	← 10	← 10

The TraceBack starts at the right bottom corner (row N), where the minimum is, and goes back on the same line until the cost is changed. This change in cost means that a node was added (cell highlighted in orange). Remember the row means the capacity, when a node of capacity k was added, we only need to go to row $N - 1$ and search for a change of node until we reach the first row. The algorithm produce a sequence in which the nodes were added, e.g. (b,a,c,d) for the Table 1. This sequence is used (as in Lemma 1) to produce G_4.

The matching is performed by bringing each point-set towards this canonical representation. Given point-sets $\mathcal{X}_1, \mathcal{X}_2, \ldots, \mathcal{X}_P$, we build the NHDD graph for each one of them. The matching associates a node with degree $v \in \mathcal{X}_k$ to a node with the same degree $v \in \mathcal{X}_l$. Therefore, the matching is performed via this canonical representation. As each point-set is optimized individually, the scale of one does not affect the scale of the other. The optimization tries to minimize the distances within each point-set individually, therefore, it is able to handle the following two scenarios:

- **Rigid**: The sets are related by a translation and rotation: $\mathcal{X}_k = \mathbf{R}\mathcal{X}_l + \mathbf{t}$. In a rigid transformation, only the length and area are preserved.

– **Similarity**: $\mathcal{X}_l$ and $\mathcal{X}_k$ are not only related by $\mathbf{t}$ and $\mathbf{R}$, but also by an isotropic scaling s: $\mathcal{X}_k = s\mathbf{R}\mathcal{X}_l + \mathbf{t}$. This scale preserves the ratio of lengths.

We took images from the Caltech-256 dataset which contained a single object in the scene. We sampled $N = 50$ points from the sillhouete of the object. A random similarity transformation was applied to the point-sets. Figure 3 shows an example of the matching. Not all edges were displayed to avoid cluttering the scene, but the sets are correctly matched.

Fig. 3. Registration of point-sets under Similarity Transformation: translation + rotation + isotropic scaling

5 Conclusions and Future Work

In this paper we proposed a novel way to match point-sets via a canonical representation in which we minimize the Min-Weight Max Entropy problem. Our approach is based on a Dynamic Programming algorithm yielding a cubic complexity solution. Such a graph turns a registration procedure into a trivial task under, for instance, rigid and similarity transformation since there is a direct mapping between nodes.

As future work, we will extend the algorithm to a more local optimization not to be dependent on all nodes. We will match regions of maximal entropy (i.e. subset of nodes). It would allow us to perform the registration of non-linearly related point-sets by performing piece-wise matchings of a subset of nodes. It will also to be able cope with noise and partially overlapping regions since it is will no longer optimize over the whole graph.

Acknowledgement. The authors would like to thank Rafael Coelho and Emir Demirovic for the discussions on the topic and Giselle Reis for the revision of the manuscript. He also acknowledges research funding from the Vienna PhD School of Informatics.

References

1. Conrad, K.: Probability distributions and maximum entropy (2005), http://www.math.uconn.edu/~kconrad/blurbs/analysis/entropypost.pdf
2. Cross, A.D.J., Hancock, E.R.: Graph matching with a dual-step em algorithm. IEEE Transactions on Pattern Analysis and Machine Intelligence 20(11), 1236–1253 (1998), ISSN 0162-8828

3. de Sousa, S., Kropatsch, W.G.: Graph-based point drift: Graph centrality on the registration of point-sets. Pattern Recognition 48(2), 368–379 (2015), ISSN 0031-3203

4. Delaunay, B.N.: Sur la sphère vide. Bulletin of Academy of Sciences of the USSR, 793–800 (1934)

5. Dickinson, P.J., Bunke, H., Dadej, A., Kraetzl, M.: Matching graphs with unique node labels. Pattern Analysis and Applications 7(3), 243–254 (2004), ISSN 1433-7541

6. Erdös, T., Gallai, P.: Gráfok előírt fokszámú pontokkal. Matematikai Lapok 11, 264–274 (1960)

7. Gabriel, K.R., Sokal, R.R.: A new statistical approach to geographic variation analysis. Systematic Biology 18(3), 259–278 (1969)

8. Hakimi, S.: On realizability of a set of integers as degrees of the vertices of a linear graph. i. Journal of the Society for Industrial and Applied Mathematics 10(3), 496–506 (1962)

9. Havel, V.: Poznámka o existenci konečných graf•367u. Časopis Pro Pěstování Matematiky 080(4), 477–480 (1955)

10. Kocay, W., Kreher, D.L.: Graphs, Algorithms, and Optimization. Discrete Mathematics and Its Applications. Taylor & Francis (2004) ISBN 9780203489055

11. Lian, W., Zhang, L.: Rotation invariant non-rigid shape matching in cluttered scenes. In: Daniilidis, K., Maragos, P., Paragios, N. (eds.) ECCV 2010, Part V. LNCS, vol. 6315, pp. 506–518. Springer, Heidelberg (2010)

12. Luo, B., Hancock, E.R.: Iterative procrustes alignment with the {EM} algorithm. Image and Vision Computing 20(5-6), 377–396 (2002), ISSN 0262-8856

13. Ohrhallinger, S., Mudur, S.: An efficient algorithm for determining an aesthetic shape connecting unorganized 2d points. Computer Graphics Forum 32(8), 72–88 (2013), ISSN 1467-8659

14. Reeb, G.: Sur les points singuliers d'une forme de pfaff complétement intégrable ou d'une fonction numérique. C. R. Acad. Sci. Paris 222, 847–849 (1946)

An Entropic Edge Assortativity Measure

Cheng Ye[✉], Richard C. Wilson, and Edwin R. Hancock

Department of Computer Science, University of York, YO10 5GH, York, UK
{cy666,richard.wilson,edwin.hancock}@york.ac.uk

Abstract. Assortativity or assortative mixing is the tendency of a network's vertices to connect to others with similar characteristics, and has been shown to play a vital role in the structural properties of complex networks. Most of the existing assortativity measures have been developed on the basis of vertex degree information. However, there is a significant amount of additional information residing in the edges in a network, such as the edge directionality and weights. Moreover, the von Neumann entropy has proved to be an efficient entropic complexity level characterization of the structural and functional properties of both undirected and directed networks. Hence, in this paper we aim to combine these two methods and propose a novel edge assortativity measure which quantifies the entropic preference of edges to form connections between similar vertices in undirected and directed graphs. We apply our novel assortativity characterization to both artificial random graphs and real-world networks. The experimental results demonstrate that our measure is effective in characterizing the structural complexity of networks and classifying networks that belong to different complexity classes.

Keywords: Assortative mixing · Von Neumann entropy · Entropic edge assortativity

1 Introduction

Over the past decade there has been a considerable interest in studying the properties of complex networks since they play a crucial role in revealing essential features of the structure, function and dynamics of many large-scale systems in biology, physics and the social sciences. To render such networks tractable, it is imperative to have to hand measures that efficiently reflect their structural, functional and dynamical diversity. An important example is the vertex degree assortativity, which expresses a bias in favor of connections between network vertices with similar degree [8]. Although the vertex degree provides a number of useful characterizations of network structure, significant information also resides in the edges of a network, including the direction of interaction between components and the information conveyed by a random walk on a network. In this paper we present an edge assortativity measure, making use of the von Neumann entropy, which is an effective structural complexity measure designed for complex networks [4][11].

© Springer International Publishing Switzerland 2015
C.-L. Liu et al. (Eds.): GbRPR 2015, LNCS 9069, pp. 23–33, 2015.
DOI: 10.1007/978-3-319-18224-7_3

Assortativity is often formalized as a correlation between the degree distinction of two vertices in a graph. Recently, Foster et al. [3] have pointed out that the classification based on network assortativity is not always efficient for undirected networks. They further achieved that the fundamental feature of edge direction in a network also plays an important role. Thus they propose a set of four directed assortativity measures based on vertex in-degree and out-degree combinations. Recently, computing the importance of edges in a network has attracted considerable interest since many structural and functional features in networks have been shown to reside in the connections between vertices. For instance, the edge betweenness centrality, which is shown to be superior to graph planarization techniques, has been developed as an extension of betweenness centrality from vertices to edges [2].

The von Neumann entropy (or quantum entropy) associated with a density matrix, has proved to provide a highly effective complexity level characterization of a network. Han et al. [4] have taken this work further and have shown how to approximate the calculation of von Neumann entropy in terms of simple degree statistics rather than the normalized Laplacian eigenvalues. Recently, Ye et al. [11] have extended this entropy to the domain of directed graphs. In addition, the distribution of von Neumann entropy associated with edges in a graph, can be encoded as a multi-dimensional histogram, which not only captures the structure of a graph but also reflects its complexity [13] [12].

In this paper, we propose a novel edge assortativity measure based on the edge von Neumann entropy for both undirected and directed graphs. We use this measure to analyze how the entropy is distributed over edges. We show that the measure encodes a number of properties of the intrinsic structural properties of a graph, leading to the possibility of characterizing graphs of different structure.

The remainder of the paper is organized as follows. In Sec. II, we introduce briefly how the von Neumann entropy is defined and computationally simplified for both undirected and directed graphs. In Sec. III, we detail the development of the edge assortativity measure. In Sec. IV, we undertake experiments to demonstrate the usefulness of our method. Finally, in Sec. V we conclude our paper with a summary of our contribution and suggestions for future work.

2   Preliminaries

In this section, we give the definition of the von Neumann entropy for both undirected graph and directed graph, and show the entropy can be simplified in terms of edge entropy contributions.

2.1   Entropy Contribution for Undirected Edges

Suppose $G(V, E)$ is an undirected graph with vertex set V and edge set $E \subseteq V \times V$, then the adjacency matrix A is defined as follows

$$A_{uv} = \begin{cases} 1 & \text{if } (u,v) \in E \\ 0 & \text{otherwise.} \end{cases} \tag{1}$$

The degree of vertex u is $d_u = \sum_{v \in V} A_{uv}$.

According to [9], the normalized Laplacian matrix $\mathcal{L} = D^{-1/2}(D - A)D^{-1/2}$ (D is the degree matrix with the degree of the vertices of the undirected graph along the diagonal and zeros elsewhere) can be interpreted as the density matrix of an undirected graph. As a result, the undirected graph von Neumann entropy can be defined and calculated from the eigenvalues of the normalized Laplacian matrix. With this choice of density matrix, the von Neumann entropy of the undirected graph is the Shannon entropy associated with the normalized Laplacian eigenvalues, i.e., $H_{VN}^U = - \sum_{i=1}^{|V|} (\tilde{\lambda}_i/|V|) \ln(\tilde{\lambda}_i/|V|)$ where $\tilde{\lambda}_i, i = 1, \ldots, |V|$, are the eigenvalues of $\mathcal{L}$.

Commencing from this definition and making use of the quadratic approximation to the Shannon entropy, the von Neumann entropy can be approximated by a quadratic entropy [4] $H_Q^U = \sum_{i=1}^{|V|} \tilde{\lambda}_i/|V|(1 - \tilde{\lambda}_i/|V|)$, which can be expressed in terms of the trace of the normalized Laplacian (is equal to the sum of the normalized Laplacian eigenvalues) and the trace of the squared normalized Laplacian (is equal to the sum of the squares of the normalized Laplacian eigenvalues). Here, the accuracy of the above expression depends on the veracity of the quadratic approximation to the Shannon entropy $-x \ln x \approx x(1 - x)$. This approximation is known to hold well when either $x \to 0$ or $x \to 1$, which guarantees the accuracy of the quadratic entropy since $\tilde{\lambda}_i/|V| \to 0$ when the graph size is very large. Moreover, the trace of normalized Laplacian can be simply computed in terms of vertex degree in a graph, this leads to a more simplified form of the von Neumann entropy of an undirected graph:

$$H_{VN}^U = 1 - \frac{1}{|V|} - \frac{1}{|V|^2} \sum_{(u,v) \in E} \frac{1}{d_u d_v}. \tag{2}$$

This approximation clearly contains two measures of graph structure. The first term measures the effect of graph size and the second term of this formula simply calculates the sum of each edge contribution to the entropy of a graph. This leads to the possibility of defining a normalized local entropic measure for a single edge in the graph.

To this end, we normalize the von Neumann entropy with respect to the total number of edges in the graph in order to obtain the normalized edge entropy contribution, i.e.,

$$I_{uv}^U = \frac{1}{|V||E|d_u d_v}. \tag{3}$$

For an arbitrary graph, this normalized local entropic measure clearly avoids graph size bias and gives the von Neumann entropy contribution associated with each edge in the graph.

2.2   Entropy Contribution for Directed Edges

More recently, Ye et al. [11] have extended the calculation of von Neumann entropy from undirected graphs to directed graphs, using Chung's definition of the normalized Laplacian of a directed graph [1]. The resulting approximate entropy has the following form:

$$H_{VN}^D = 1 - \frac{1}{|V|} - \frac{1}{2|V|^2} \left\{ \sum_{(u,v)\in E} \left(\frac{1}{d_u^{out} d_v^{out}} + \frac{d_u^{in}}{d_v^{in} d_u^{out2}} \right) - \sum_{(u,v)\in E_u} \frac{1}{d_u^{out} d_v^{out}} \right\} \quad (4)$$

or equivalently,

$$H_{VN}^D = 1 - \frac{1}{|V|} - \frac{1}{2|V|^2} \left\{ \sum_{(u,v)\in E} \frac{d_u^{in}}{d_v^{in} d_u^{out2}} + \sum_{(u,v)\in E_b} \frac{1}{d_u^{out} d_v^{out}} \right\}, \quad (5)$$

where $E_u = \{(u,v)|(u,v) \in E$ and $(v,u) \notin E\}$ is the set of unidirectional edges while $E_b = \{(u,v)|(u,v) \in E$ and $(v,u) \in E\}$ is the set of bidirectional edges in the graph.

In particular, when $|E_u| \ll |E_b|$, i.e., few of the edges are unidirectional and the graph is weakly directed (WD), we ignore the summation over E_u in Eq.(4) in order to obtain the approximate von Neumann entropy for WD graphs:

$$H_{VN}^{WD} = 1 - \frac{1}{|V|} - \frac{1}{2|V|^2} \sum_{(u,v)\in E} \left\{ \frac{\frac{d_u^{in}}{d_u^{out}} + \frac{d_v^{in}}{d_v^{out}}}{d_u^{out} d_v^{in}} \right\}. \quad (6)$$

The term $1 - \frac{1}{|V|}$ tends to unity as the graph size becomes large. In the summation, the numerator is given in terms of the sum of the ratios of in-degree and out-degree of the vertices. Since the directed edges cannot start at a sink (a vertex of zero out-degree), the ratios do not become infinite. Moreover, it is natural to realize that in our analysis an undirected graph is equivalent to a WD graph, since their von Neumann entropy expressions Eq.(2) and Eq.(6) are equivalent if we consider each undirected edge as a bidirectional one.

On the other hand, if the cardinality of E_b is very small ($|E_b| \ll |E_u|$), i.e., a graph is strongly directed (SD), this approximate entropy can be simplified one step further by ignoring the summation over E_b in Eq.(5):

$$H_{VN}^{SD} = 1 - \frac{1}{|V|} - \frac{1}{2|V|^2} \sum_{(u,v)\in E} \left\{ \frac{d_u^{in}}{d_v^{in} d_u^{out2}} \right\}. \quad (7)$$

This approximation clearly sums the entropy contribution from each directed edge, which is based on the in and out-degree statistics of the vertices connected by the edge. In other words, by using the same method in the previous subsection, we can compute a normalized local entropy measure for each directed edge in the SD graph. To do this, we remove the term $1 - \frac{1}{|V|}$ and normalize the remaining term with respect to the number of edges in the graph so that we obtain

$$I_{uv}^D = \frac{d_u^{in}}{|V||E|d_v^{in}d_u^{out2}} \tag{8}$$

as the von Neumann entropy contribution for edge $(u,v) \in E$.

3   Entropic Edge Assortativity Measure for Graphs

In this section, we propose a novel assortativity measure for both undirected and directed graphs based on the von Neumann entropy contributions associated with undirected and directed edges. This method provides useful underpinning at the use of entropy in determining graph structure. For instance, a high edge assortativity indicates that edges with large entropy associate preferentially and form some high entropy clusters in a graph. By contrast, a negative assortativity results from edges with high and low entropies that connect to each other.

The traditional assortativity is usually defined as the Pearson correlation coefficient (r) of the degrees of pairs of linked vertices [8]:

$$r = \frac{|E|^{-1}\sum_{(u,v)\in E} d_u d_v - [|E|^{-1}\sum_{(u,v)\in E}\frac{d_u+d_v}{2}]^2}{|E|^{-1}\sum_{(u,v)\in E}\frac{d_u^2+d_v^2}{2} - [|E|^{-1}\sum_{(u,v)\in E}\frac{d_u+d_v}{2}]^2} \in [-1,1]. \tag{9}$$

When $r = 1$, the network is said to be perfectly assortative, when $r = 0$ the network is non-assortative, and when $r = -1$ the network is completely disassortative.

Furthermore, according to Foster et al. [3], in a directed graph, a set of four directed assortativity measures are defined as follows. Let $\alpha, \beta \in \{in, out\}$ be the directionality index for an edge at a vertex (i.e., whether it is incoming or outgoing). Then the directed assortativity measures are

$$r(\alpha,\beta) = \frac{|E|^{-1}\sum_{(u,v)\in E}[(d_u^\alpha - \bar{d_u^\alpha})(d_v^\beta - \bar{d_v^\beta})]}{\sigma^\alpha\sigma^\beta} \tag{10}$$

where $\bar{d_u^\alpha} = |E|^{-1}\sum_{(u,v)\in E} d_u^\alpha$ and $\sigma^\alpha = \sqrt{|E|^{-1}\sum_{(u,v)\in E}(d_u^\alpha - \bar{d_u^\alpha})^2}$; $\bar{d_v^\beta}$ and σ^β are similarly defined.

3.1   Entropic Edge Assortativity Measure for Undirected Graphs

Suppose $G(V,E)$ is an undirected graph, or equivalently, a weakly directed graph, then for an edge $(u,v) \in E$, we define the entropy contribution associated with the end vertex u of this edge S_{uv} as the summation of the entropies on the edges connected with u except the edge (u,v), i.e., $S_{uv} = \sum_{(t,u)\in E, t\neq v} I_{tu}^U$. The entropy contribution associated with another end vertex v is therefore $S_{vu} = \sum_{(v,w)\in E, w\neq u} I_{vw}^U$.

With these to hand, we define the edge assortativity as the Pearson correlation coefficient between all the entropy contributions associated with the two end vertices connected by the edge in the graph $G(V, E)$, with the result that

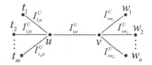

Fig. 1. The illustration of the calculation of quantities S_{uv} and S_{vu} associated with an undirected edge $(u, v) \in E$

$$R^U = \frac{\sum_{(u,v)\in E}(S_{uv} - \bar{S}_{uv})(S_{vu} - \bar{S}_{vu})}{\sigma_u^S \sigma_v^S} \quad (11)$$

where $\bar{S}_{uv} = |E|^{-1}\sum_{(u,v)\in E} S_{uv}$ and $\sigma_u^S = \sqrt{\sum_{(u,v)\in E}(S_{uv} - \bar{S}_{uv})^2}$; $\bar{S}_{vu}$ and σ_v^S are similarly defined. Clearly, this edge assortativity index provides a novel way to understand the entropic preference of edges to form connections between similar vertices in a graph.

3.2 Entropic Edge Assortativity Measure for Directed Graphs

We turn our attention to the domain of directed graphs. Here we mainly focus on the strongly directed graphs. Assume $G(V, E)$ is an SD graph, then for a directed edge starting from vertex u, ending at vertex v, we define the edge assortativity as the Pearson correlation coefficient between the edge entropy contribution H_{uv}^u associated with all the outgoing edges of vertex u (exclude edge (u, v)) and the contribution H_{uv}^v associated with all the incoming connections of vertex v (except edge (u, v)). The reason we use such definition is that this expression conforms to the structure of the approximate von

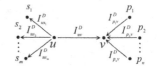

Fig. 2. The illustration of the calculation of quantities H_{uv}^u and H_{uv}^v associated with a directed edge $(u, v) \in E$

Neumann entropy for SD graphs given in Eq.(7). Mathematically, we have $H_{uv}^u = \sum_{(u,s)\in E, s\neq v} I_{us}^D$ and $H_{uv}^v = \sum_{(p,v)\in E, p\neq u} I_{pv}^D$. Therefore the edge assortativity coefficient for SD graphs is given by

$$R^D = \frac{\sum_{(u,v)\in E}(H_{uv}^u - \bar{H}_{uv}^u)(H_{uv}^v - \bar{H}_{uv}^v)}{\sigma_u^H \sigma_v^H} \quad (12)$$

where $\bar{H}_{uv}^u = |E|^{-1}\sum_{(u,v)\in E} H_{uv}^u$ and $\sigma_u^H = \sqrt{\sum_{(u,v)\in E}(H_{uv}^u - \bar{H}_{uv}^u)^2}$; $\bar{H}_{uv}^v$ and σ_v^H are similarly defined. This measure is bounded between -1 and 1: a high coefficient of a graph indicates that most of the directed edges in the graph start from the vertex with outgoing edges that have high entropy contributions, and point to the vertex with incoming edges with high entropy contributions. Conversely, a negative coefficient results from most of the directed edges connect two vertices that have significantly different von Neumann edge entropy contributions.

4   Experiments and Discussion

We have proposed a novel edge assortativity characterization for quantifying the assortative mixing properties for both undirected and directed graphs based on the von Neumann entropy associated with edges. In this section, we explore whether this measure can reveal more useful features of the graph structure than the traditional degree assortativity measures. To this end, we confine our attention to two main tasks. We first apply the edge assortativity measure to some real-world complex networks to show that it can effectively reflect to what extent the vertices are connected preferentially in a network. We then demonstrate one advantage of this novel assortativity characterization, namely that it is more efficient in distinguishing between different classes of complex networks than the traditional measures.

4.1   Experiments and Discussion on Undirected Graphs

We commence by comparing the performance of traditional assortativity coefficients and our novel edge assortativity measure on real-world collaboration networks. These include the Arxiv Astro Physics, Condensed Matter, General Relativity, High Energy Physics and High Energy Physics Theory networks [7]. Table 1 gives the network size, edge number and value of both the degree and edge assortativity measures. From the table it is clear that all the coauthership networks have positive degree assortativity coefficients. This is a reasonable result since productive authors prefer to collaborate. However, the traditional assortativity coefficient has difficultly in distinguishing between CA-HepPh and CA-GrQc networks as their values are similar. The edge assortativity coefficient, on the other hand, is able to characterize these two networks. One of the reasons for this is that the edge assortativity measure can capture not only the degree properties of vertices, but also the underlying entropic structural complexity associated with the edges in a network.

Table 1. Degree assortativity coefficients and edge assortativity measures of real-world undirected complex networks

Datasets	HepTh	HepPh	GrQc	CondMat	AstroPh
Network size	9877	12008	5242	23133	18772
Edge number	51971	237010	28980	186936	396160
Degree assort.	0.2674	0.6322	0.6592	0.1339	0.2051
Edge entropy assort.	0.2012	0.6035	0.3910	0.3435	0.5458

Next we show that the edge assortativity measure is more efficient than the traditional assortativity coefficient in classifying graphs that belong to different random graph models. To do this we first randomly produce a large number of undirected graphs according to one of three models, namely a) the classical Erdős-Rényi model, b) the "small-world" model, and c) the "scale-free" model. The different graphs in the database are generated using a variety of model parameters, e.g. the graph size and the connection probability in the Erdős-Rényi model, the edge rewiring probability in the "small-world" model and the number of added connections at each time step in the "scale-free" model.

Figure 3 shows the mean value of both the degree assortativity coefficients (Eq.(9)) and edge assortativity measures (Eq.(11)) as a function of graph size (standard deviation as an error bar). In the left panel, all three classes of graphs tend to have zero assortative mixing when the graph size becomes very large, and it is difficult to separate the 'small-world" and "scale-free" graphs. Turning our attention to the right panel, the difference in mean edge assortativity coefficients for different models is much larger than the standard deviation of the coefficients for the different models, even when the graph size is large. This suggests that the variance in the edge assortativity measure due to different parameter settings is much smaller than that due to differences in structure. This indicates that different network models have different values of edge assortativity coefficients for a given size. This accords with our expectations since the entropy itself is sensitive to the different graph models.

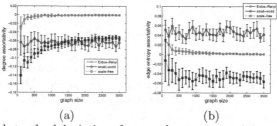

(a)   (b)

Fig. 3. Mean and standard deviation of vertex degree assortativity coefficients (Eq.(9)) and edge assortativity measures (Eq.(11)) for different models of undirected graphs. Red square solid line: Erdős-Rényi; blue circle solid line: "small-world"; black square dotted line: "scale-free".

4.2   Experiments and Discussion on Directed Graphs

For directed graphs, we first provide a comparison of our new directed edge assortativity measure, and the four assortativity coefficients that can be computed from the four combinations of in and out-degree on the two vertices of an edge. We commence with a study on some real-world networks, and these include the Wikipedia vote network, provided by Leskovec et al. [5], the Gnutella peer-to-peer networks from August 5 to 9, 2002, which are a sequence of snapshots of the Gnutella peer-to-peer file sharing network [10] and the Arxiv HEP-TH citation network [6]. Table 2 gives the network size, edge number and the values of in/in-degree, in/out-degree, out/in-degree, out/out-degree and edge assortativity measures. There are a number of observations concerning this data. In the Wikipedia vote network, a person who receives many votes is more likely to vote a person who also obtains a large number of votes, rather than voting for individuals who vote many times. In the file sharing networks, computers that receive a great number of documents preferentially share files with one-another. Computers that send many files are unlikely to share files with computers that receive many documents. For the citation network, important papers are those cited most heavily and this can be reflected accurately by the degree assortativity measures. Although when taken in combination the four types of directed

Table 2. Degree assortativity coefficients and edge assortativity measures of real-world directed complex networks

Datasets	Wiki-Vote	p2p-G05	p2p-G06	p2p-G08	p2p-G09	Arxiv HEP-TH
Network size	7115	8846	8717	6301	8114	27751
Edge number	103689	31839	31525	20777	26013	352807
In/in deg. assort.	0.0051	0.0312	0.0880	0.1079	0.1042	0.0405
In/out deg. assort.	0.0071	-0.0002	0.0322	0.0315	0.0190	0.0055
Out/in deg. assort.	-0.0832	-0.0034	-0.0032	-0.0285	-0.0327	0.0016
Out/out deg. assort.	-0.0161	-0.0017	0.0082	-0.0157	-0.0062	0.0951
Edge entropy assort.	0.0006	0.0053	-0.0092	-0.0038	-0.0055	0.1126

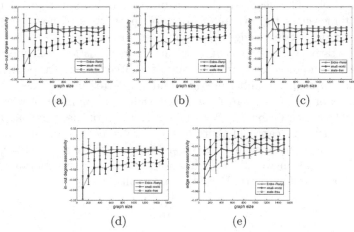

(a) (b) (c)

(d) (e)

Fig. 4. Mean and standard deviation of vertex degree assortativity coefficients (Eq.(10)) and edge assortativity measures (Eq.(12)) for different models of directed graphs. Red square solid line: Erdős-Rényi; blue circle solid line: "small-world"; black square dotted line: "scale-free".

degree assortativity coefficients are useful in characterizing different networks, it is difficult to use a single measure alone to do this. However, when using the novel edge assortativity measure developed for directed graphs, networks with different structures are efficiently characterized.

In Fig. 4 we plot the values of the edge assortativity coefficient, and compare them to the assortativity coefficients obtained with the four different combinations of vertex in and out-degree on an edge (see Eq.(10)). Here we use randomly generated data for three different directed graph models. The figure shows the assortativity measures versus graph size, and shows the mean value and standard deviation. The most important feature in the figure is that although the "scale-free" networks are easily separated, the Erdős-Rényi and "small-world" networks are overlapped significantly, for each of the four degree assortativity coefficients. However, Fig. 4(e) suggests that as the graph size increases, for all three models the mean values of the edge assortativity measures grow slowly and approach zero, with clear separations between them. The result obtained here demonstrates that the edge assortativity measure provides a powerful tool for capturing both the degree properties and the entropic information on edges in a directed network.

5   Conclusions

To conclude, this paper is motivated by the aim of proposing novel measures that quantify the assortative mixing properties for both undirected and directed networks. We commence from the recently developed simplified approximations to the von Neumann entropy for both undirected and directed graphs, which are dependent on the graph size and degree statistics of vertices that are connected. From these approximations we then derive a local measure for quantifying the von Neumann entropy contribution for each edge in the undirected and directed graph respectively. This leads to the possibility of designing a correlation coefficient that measures the average assortative properties of how the entropy contributions that reside in edges are connected in a network, which we name the edge assortativity measure. The resulting expressions for such measures of both undirected and directed graphs are simply related to some graph invariants, including the graph size, number of edges and the vertex degree.

The work reported in this paper can be extended in a number of ways. First, it would be interesting to explore how the distribution of the edge entropy contributions in a network can contribute to the development of novel information theoretic divergence, distance measures and relative entropies. Another interesting line of investigation would be to investigate whether this measure can be applied further to weighted graphs and hypergraphs.

References

1. Chung, F.: Laplacians and the cheeger inequailty for directed graphs. Annals of Combinatorics 9, 1–19 (2005)
2. Cuzzocrea, A., Papadimitriou, A., Katsaros, D., Manolopoulos, Y.: Edge betweenness centrality: A novel algorithm for qos-based topology control over wireless sensor networks. Journal of Network and Computer Applications 35(4), 1210–1217 (2012)
3. Foster, J.G., Foster, D.V., Grassberger, P., Paczuski, M.: Edge direction and the structure of networks. Proceedings of the National Academy of Sciences of the United States of America 107(24), 10815–10820 (2010)
4. Han, L., Escolano, F., Hancock, E.R., Wilson, R.C.: Graph characterizations from von neumann entropy. Pattern Recognition Letters 33, 1958–1967 (2012)
5. Leskovec, J., Huttenlocher, D., Kleinberg, J.: Signed networks in social media. In: CHI (2010)
6. Leskovec, J., Kleinberg, J., Faloutsos, C.: Graphs over time: Densification laws, shrinking diameters and possible explanations. In: ACM SIGKDD International Conference on Knowledge Discovery and Data Mining (2005)
7. Leskovec, J., Kleinberg, J., Faloutsos, C.: Graph evolution: Densification and shrinking diameters. ACM Transactions on Knowledge Discovery from Data (ACM TKDD) 1 (2007)
8. Newman, M.: Assortative mixing in networks. Phys. Rev. Lett. 89(208701) (2002)
9. Passerini, F., Severini, S.: The von neumann entropy of networks. International Journal of Agent Technologies and Systems, 58–67 (2008)

10. Ripeanu, M., Foster, I., Iamnitchi, A.: Mapping the gnutella network: Properties of large-scale peer-to-peer systems and implications for system design. IEEE Internet Computing Journal (2002)
11. Ye, C., Wilson, R.C., Comin, C.H., da F. Costa, L., Hancock, E.R.: Entropy and heterogeneity measures for directed graphs. In: Hancock, E., Pelillo, M. (eds.) SIMBAD 2013. LNCS, vol. 7953, pp. 219–234. Springer, Heidelberg (2013)
12. Ye, C., Wilson, R.C., Hancock, E.R.: Entropic graph embedding via multivariate degree distributions. In: Fränti, P., Brown, G., Loog, M., Escolano, F., Pelillo, M. (eds.) S+SSPR 2014. LNCS, vol. 8621, pp. 163–172. Springer, Heidelberg (2014)
13. Ye, C., Wilson, R.C., Hancock, E.R.: Graph characterization from entropy component analysis. In: 22nd International Conference on Pattern Recognition, ICPR 2014, Stockholm, Sweden, August 24-28, pp. 3845–3850 (2014)

A Subpath Kernel for Learning Hierarchical Image Representations

Yanwei Cui$^{(\boxtimes)}$, Laetitia Chapel, and Sébastien Lefèvre

IRISA (UMR 6074), University Bretagne-Sud, 56000, Vannes, France
{yanwei.cui,laetitia.chapel,sebastien.lefevre}@irisa.fr

Abstract. Tree kernels have demonstrated their ability to deal with hierarchical data, as the intrinsic tree structure often plays a discriminative role. While such kernels have been successfully applied to various domains such as nature language processing and bioinformatics, they mostly concentrate on ordered trees and whose nodes are described by symbolic data. Meanwhile, hierarchical representations have gained increasing interest to describe image content. This is particularly true in remote sensing, where such representations allow for revealing different objects of interest at various scales through a tree structure. However, the induced trees are unordered and the nodes are equipped with numerical features. In this paper, we propose a new structured kernel for hierarchical image representations which is built on the concept of subpath kernel. Experimental results on both artificial and remote sensing datasets show that the proposed kernel manages to deal with the hierarchical nature of the data, leading to better classification rates.

Keywords: Hierarchical representation · Image classification · Structured kernels · Subpath kernel

1 Introduction

Structured data-based learning has become a central topic in machine learning, as such data representations are met in numerous fields. We focus here on tree-based representations, whose typical applications are parse trees in Natural Language Processing [4], XML trees in web mining [8], or even hierarchical image representations [2]. Since tree structure plays an important role in tasks like classification or clustering, similarity measures taking explicitly into account topological characteristics of the tree are sought. Among them, kernels functions are appealing as they allow the use of popular kernel methods [10].

Various kernels have indeed been proposed to cope with a tree as the underlying data structure (see [11] for a review). They mostly rely on a fundamental idea brought by Haussler with its convolution kernels [7], stating that a kernel defined on a complex structure can be formed by kernels computed on its substructures. Most often, those kernels are defined for ordered trees, that is to say trees for which left to right order among nodes or leaves is fixed (mostly because of the specific nature of the data or due to computational complexity issues).

© Springer International Publishing Switzerland 2015
C.-L. Liu et al. (Eds.): GbRPR 2015, LNCS 9069, pp. 34–43, 2015.
DOI: 10.1007/978-3-319-18224-7_4

Examples of such kernels include the subset tree kernel [4] and the subtree kernel [13]. Unordered trees received much less attention, with the subpath kernel [8] being one of the very few existing solutions.

Meanwhile, an emerging paradigm for image classification has advocated the idea of relying on hierarchical representations [2], which are built using series of nested partitions or segmentations, rather than the usual flat representation. This is particularly true in remote sensing, where such representations allow for revealing different objects of interest at various scales through a tree structure. However, the induced trees are unordered and the nodes or regions are associated with numerical features, preventing the use of existing tree kernels.

We propose in this paper a new kernel that arises from the subpath kernel [8]. Based on some existing adaptations to numerical data from the graph kernel literature, the designed kernel is able to cope with unordered trees equipped with numerical data (see Sec. 2). Besides, it considers the complete set of subpaths on tree structures (instead of paths on graphs), leading to an efficient computation scheme. Experimental results are given in Sec. 3. They rely on artificial datasets as well as a real multispectral satellite image. We end the paper with some concluding remarks and directions for future work.

2 Proposed Kernel

We focus here on structured data represented by trees and subpaths. Let us first recall that a tree is a directed and connected acyclic graph with a single root node. A path connecting a node in the tree to one of its descendants is called a subpath. Individual nodes are also included in the set of subpaths.

We build upon the subpath kernel [8] to design a new kernel able to cope with numerical data. Let us recall the principles of the original subpath kernel, that exploits the hierarchical structure by counting all possible common subpaths embedded in two tree structures equipped with symbolic features. Given two trees T and T', the subpath kernel is defined as

$$K(T, T') = \sum_{s \in T, s' \in T'} k(s, s') = \sum_{s \in T, s' \in T'} \delta_{s,s'} \, h(s, T) \, h(s', T') , \qquad (1)$$

where kernels $k(s, s')$ are computed between all subpaths s, s' of tree T, T' respectively. They rely on the number of occurrences of the subpaths in the tree written $h(s, T)$ and $h(s', T')$. The Kronecker delta function $\delta_{s,s'}$ equals 1 iif the two subpaths s, s' are identical. Figure 1 illustrates for a given simple tree its possible subpaths and their occurrences in the tree.

2.1 Adaptation to Numeric Data

The original kernel depicted previously was introduced for data classification in bioinformatics where nodes take symbolic values. Adapting this kernel to numeric data requires one to change the terms $h(s, T)$ and $\delta_{s,s'}$ since strict identity between subpaths (and their respective node features) does not generally occur. We follow

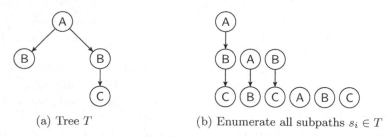

(a) Tree T (b) Enumerate all subpaths $s_i \in T$

Fig. 1. A tree T and all its subpaths s_i, that may have multiple occurrences in T: $h(s_i, T) = 2$ for $s_2 = (A \rightarrow B)$ and $s_5 = (B)$; $h(s_i, T) = 1$ otherwise

here the scheme proposed with graph kernels for image classification [1], but considering subpaths instead of random walks as the substructure component. Indeed, the use of trees allows a complete enumeration of the subpaths that is not achievable with graphs.

We replace $h(s, T)$ and $h(s', T')$ terms in Eq. (1) by a product of some atomic kernel functions $k(n_i, n_i')$ computed between pairs of nodes $n_i \in s$ and $n_i' \in s'$. Various atomic kernels are here available, *e.g.* Gaussian kernel [10] that has often been successfully used in many contexts. As long as these atomic kernels are positive definite, the proposed structured kernel is also positive definite (see [7]). Formulation of the subpath kernel for numeric data is then

$$K(T, T') = \sum_{s \in T, s' \in T'} k(s, s') = \sum_{s \in T, s' \in T'} \delta_{|s|, |s'|} \prod_{n_i \in s, n_i' \in s'} k(n_i, n_i') , \qquad (2)$$

where the Kronecker delta function $\delta_{|s|, |s'|}$ equals 1 iif the two subpaths s, s' have the same length, and nodes n_i, n_i' are scanned in descending order along the two subpaths s, s' (from the root to the leaf). One might notice that if $k(n_i, n_i') = \delta_{n_i, n_i'}$ measures the identicalness of n_i, n_i', Eq. (2) becomes just another form of Eq. (1). The naive complexity of comparing all pairs of subpaths between T and T' is $O(|T|^2 |T'|^2)$ as indicated in [8] (the size of subpath set of a given tree T is in general $|T|^2$, with $|T|$ refers to the number of vertices in the tree).

2.2   Efficient Computation

Besides the proposed adaptation to numeric data, applying kernels on tree-based representations of images also raises a computational issue. Indeed, images are most often made of millions or even billions of pixels, that are put in the tree structure. While such structure aims to reduce the image content through a hierarchical representation, it may still be characterized by a huge number of nodes and edges. So we also need to address this issue to make the proposed kernel relevant when dealing with images.

Let us note that in the original subpath kernel paper [8], some solutions were given to lower the computation time. But they are related to symbolic data and thus cannot be applied here. Inspired from [6], we suggest rather to use dynamic programming for comparing all subpaths. This strategy allows us to break down the complexity of all subpaths comparison into smaller subproblems in a recursive way, and reuse the solution of one subproblem to solve another one. Repeated comparisons are then avoided. More specifically, the comparison is computed recursively between two nodes $n \in T$ and $n' \in T'$:

$$k^p(n, n') = k(n, n') + \left(k(n, n') + 1 \right) \sum_{n_c \in C_T(n), n'_c \in C_{T'}(n')} k^{p-1}(n_c, n'_c) - k(n, n') \sum_{n_c \in C_T(n), n'_c \in C_{T'}(n')} k^{p-2}(n_c, n'_c) ,$$
(3)

with $C_T(n)$ the set of children of n in T, and $k^1(n, n') = k(n, n')$ the atomic kernel measuring the similarity between n and n', $k^0(n, n') = 0$ by convention. We also have p reaching 1 when either n or n' has no child. For more details, let us denote $ST_n, ST_{n'}$ two subtrees rooted at $n \in T$, $n' \in T'$ respectively, (s, s') a pair of subpaths with $s \in ST_n, s' \in ST_{n'}$. Then $k^p(n, n')$ sums the similarity measures of all pairs of subpaths (s, s') with same length and starting at the same height in $ST_n, ST_{n'}$. The similarity is calculated recursively by the first and second term on the right side of equation, together with the third term k^{p-2} introduced to prevent false subpaths that compare non-contiguous alignments.

Given a couple of trees T and T' with respective roots $r(T)$ and $r(T')$, we finally compute all pairs of subpaths embedded in the trees:

$$K(T, T') = k^p(r(T), r(T')) + \sum_{n' \in T', n' \neq r(T')} k^p(r(T), n') + \sum_{n \in T, n \neq r(T)} k^p(n, r(T')) .$$
(4)

Dynamic programming allows us to avoid the explicit computation of all pairs of subpaths between two trees T_1 and T_2. Instead, only all pairs of vertices are considered. The complexity is thus $O(|T_1||T_2|)$ where $|T_i|$ refers to the number of vertices in the tree T_i.

2.3 Additional Improvements

The proposed kernel shares with existing structured kernels two main issues. On the one hand, the value of $K(T, T')$ depends on the size of the trees while some invariance might be sought. This problem has already been tackled in [4] through a normalized kernel, that is computed here as

$$K(T, T') = K(T, T') \left(K(T, T) K(T', T') \right)^{-0.5}.$$
(5)

On the other hand, the structured kernel gives the same importance to every node in the tree. Here nodes represent regions of various size, and larger regions might be given more attention than smaller ones. By weighting the atomic kernel by the relative size A_n (*i.e.*, number of pixels) of a node w.r.t. the root of the tree, and given a parameter $\beta \geq 0$, the updated kernel becomes

$$A_n^\beta A_{n'}^\beta k(n, n') .$$
(6)

3   Experiments

We have conducted two experiments to proceed to a finer understanding of the kernel behavior using an artificial dataset, and to validate the kernel in a realistic context. Before giving in-depth analysis of the results obtained on both datasets, we will present first the common experimental setup.

3.1   Experimental Setup

We evaluate kernels in a classification context, considering a *one-against-one* SVM classifier (using the Java implementation of LibSVM [3]). The proposed subpath kernel is systematically compared with a kernel computed on the root of the tree only (*i.e.* ignoring all remaining nodes of the tree), called rooted kernel in the sequel. Our goal is to assess the importance of the various levels of information contained in the hierarchical representation w.r.t. a raw analysis of the whole data. Let us note that standard tree kernels based on substructures other than paths could hardly be applied here due to their computational complexity [11]). For the proposed kernel, nodes are compared individually using an atomic RBF kernel: each node being described by a feature or vector of N attributes, the kernel is defined for a pair of nodes n, n' with respective features x, x' as

$$k(n, n') = \exp(-\gamma d(x, x')) . \qquad (7)$$

We consider here two types of distances. The first one is an l^2-norm distance

$$d_{\mathcal{G}}(x, x') = \|x - x'\|^2 , \qquad (8)$$

leading to a Gaussian kernel, for which the features x, x' contain the average and variance computed from all information contained in the nodes (that can be accessed from their leaves). The second one is a distance between N-d histograms:

$$d_{\chi^2}(x, x') = \sum_{j=1}^{N} \frac{(x_j - x'_j)^2}{x_j + x'_j} , \qquad (9)$$

where the features x, x' are here histograms of M bins per dimension stacked together. We call χ^2 kernel the resulting atomic kernel.

   Three free parameters are determined by a grid search strategy over potential values: the bandwidth γ of the RBF atomic kernel (Eq. (7)), the SVM regularization parameter C, and the size weight $\beta \in [0, 1]$ (Eq. (6)).

   Accuracies (and standard deviations) of each setup are computed after 100 repetitions of each experiment, choosing randomly 20 data samples from each class as training samples, using the remaining samples for testing.

3.2   Artificial Dataset

In this first experiment, we study the behavior of the proposed tree kernel though 3 different scenarios using an artificial dataset. Unless stated otherwise, we consider $M = 4$ bins by dimension to construct the histogram. We call structure information the way leaves are aggregated and the initial number of leaves.

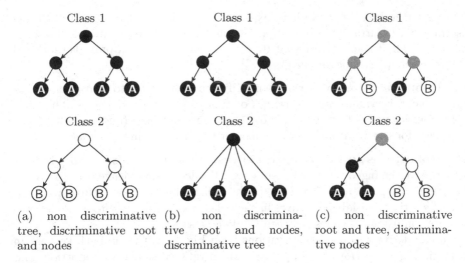

(a) non discriminative tree, discriminative root and nodes

(b) non discriminative root and nodes, discriminative tree

(c) non discriminative root and tree, discriminative nodes

Fig. 2. An illustration of the different scenarios used for experimental evaluation

(a) Only the Root is Discriminative. We generate two types of leaves, A and B, that are described by a 1-D feature generated according to a uniform distribution, with non overlapping intervals. Class 1 trees are composed of leaves of type A only and Class 2 trees of type B only (see Fig. 2a). Number of leaves and node merging parameters are defined randomly to produce various shapes of trees within each class. As shown in Tab. 1a, subpath kernel behaves similarly to rooted kernel: when the structure does not provide additional information, exploiting it in the proposed subpath kernel does not degrade the performances.

(b) Only the Structure is Discriminative. We generate only type A leaves. The two classes of trees can then be discriminated thanks to their structure, *i.e.* with different ranges of related parameters (number of leaves and number of fanouts for each node) for each class, see Fig. 2b. As shown in Tab. 1b, rooted kernel achieves only 50% accuracy, while the subpath kernel is able to discriminate the two classes, thanks to the discriminative structure leading to different subpaths between the two classes. Let us note that when $\gamma = 0$ and $\beta = 0$, the subpath kernel turns into a kernel that computes the product of the number subpaths with common length embedded in the two trees.

(c) Only the Features of the Nodes are Discriminative. We generate both type A and B leaves, and we force type A leaves to merge with type B leaves in Class 1, while in Class 2, type A (resp. B) leaves always merge with type A (resp. B) leaves (see Fig. 2c). Similarly to scenario (a), structure parameters are selected randomly. As shown in Tab. 1c, rooted kernel provides an accuracy about 50%, due to the non discriminative root. Discriminative information contained in the nodes benefits to the proposed subpath kernel, leading to a 100% accuracy. Note that even in the presence of irrelevant features, the subpath kernel still behaves correctly. Indeed, we have experimentally observed that adding 40 non discriminative features to each node (so only one dimension among the 41 is relevant) leads to an accuracy of 97.13 % with Gaussian atomic kernel, and 99.99 % with χ^2 atomic kernel.

(c1) Robustness to Outliers. We modify the scenario (c) to introduce outlier leaves that take values outside ranges of type A and type B leaves. The ratio of such leaves varies from 0% to 100%. We construct here (and here only) the histogram for the χ^2 atomic kernel considering $M = 12$ bins.

(c2) Robustness to Mislabelled Leaves. We update the scenario (c1) with outlier leaves changed to mislabelled leaves. To do so, some leaves of type A are changed into type B, and vice versa. In the binary classification setup considered here, the ratio of mislabelled leaves in each class varies from 0 % to 50 %, leading to more confusing subpaths between the two classes.

We can derive two main observations from Fig. 3: a) the subpath kernel can maintain a good performance up to a certain ratio of structure distortion, and b) both accuracy drops (after 50% of outliers in **c1**, between 20% and 40% mislabelled leaves in **c2**) illustrate that subpath kernel performance is directly related to the discrimination of substructures between two group of complex structured data. Further, one might notice that in both scenarios, subpath using χ^2 atomic kernel always performs better than using Gaussian atomic kernel. Indeed, histograms provide a rich distribution description of leaf attributes.

Table 1. Mean (and standard deviation) of overall accuracies computed over 100 repetitions for the artificial dataset. Best results (with a statistical significance less than 0.01% considering the Wilcoxon signed-rank test for matched samples) between the rooted and subpath kernel – Gaussian or χ^2– are boldfaced.

Scenario	Rooted kernel		Subpath kernel	
	Gaussian	χ^2	Gaussian	χ^2
(a) discriminative root only	**100.0** (0.0)	**100.0** (0.0)	**100.0** (0.0)	**100.0** (0.0)
(b) discriminative structure only	48.9 (4.6)	49.4 (3.4)	**100.0** (0.0)	**100.0** (0.0)
(c) discriminative nodes only	49.5 (3.5)	51.2 (4.6)	**100.0** (0.0)	**100.0** (0.0)

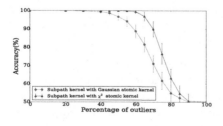 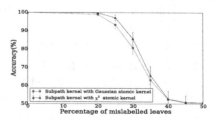

(a) Robustness to outliers   (b) Robustness to mislabelled leaves

Fig. 3. Accuracies for scenarios **(c1)** and **(c2)**

3.3   Satellite Image Dataset

Beyond the evaluation conducted on some artificial datasets, we also perform some experiments on real datasets. Since hierarchical representations are common in remote sensing, we explore the relevance of the proposed kernel in this domain. To do so, we consider a QUICKBIRD satellite image with high spatial resolution (*i.e.*, 2.4 m per pixel) of Strasbourg Illkirch in France, initially proposed and discussed in [9]. We can perform quantitative evaluation of the kernel-based classification procedure thanks to the availability of a ground truth (a partition of the initial image into 400 regions, each of them being associated to one of the 7 classes of interest, see list and distribution in Fig. 5). Figure 4 shows the satellite image and its associated ground truth.

We compute a tree representation of each single region of the ground truth, using a standard open source hierarchical image segmentation method called RHSEG [12]. RHSEG allows us to produce a fine segmentation map containing 3180 regions, that are subsequently aggregated to build coarser layers or segmentation maps with less regions (each iteration contains 300 regions less), as shown in Fig. 4. The coarsest segmentation is nothing but the ground truth. These different layers are stacked within tree structures. The last step consists in deleting the redundant nodes that remain unchanged through different scales. Finally we obtain 400 different trees, where each root represents a ground truth region and other nodes represent its components (subregions) at different scales.

Results. Both rooted and subpath kernels are involved in a supervised classification process. Several statistics are derived: overall accuracy (ratio of correctly classified regions), average accuracy (average of the accuracy measured on each class), and kappa index (percentage of agreement in the test set, corrected by the agreement that could be expected by chance). Results are reported in Tab. 2. We can see that the proposed subpath kernel always outperforms rooted kernel. It is able to exploit the additional spatial features provided by the hierarchical representation of individual regions. A deeper analysis is provided in Fig. 5 where accuracies are provided for each class. Subpath superiority is mainly observed on

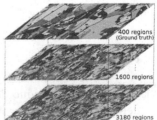

Fig. 4. STRASBOURG dataset. Left: Color composition of a multispectral Quickbird image, ©DigitalGlobe, Inc. Right: examples of RHSEG segmentations at different scales (top: associated ground truth map with 400 regions)

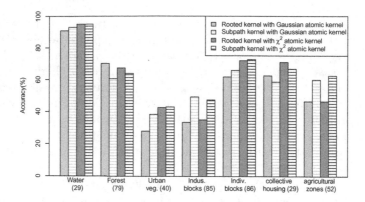

Fig. 5. Classification accuracy for the 7 different classes (number of samples or regions in the ground truth is provided in parentheses)

Table 2. Mean (and standard deviation) of overall accuracies (OA), average accuracies (AA) and Kappa statistics (κ) computed over 100 repetitions for Strasbourg dataset. Computation time (in seconds) is also reported. Best results (with a statistical significance less than 0.01% considering the Wilcoxon signed-rank test for matched samples) are boldfaced.

Method	OA[%]	AA[%]	κ	time
Rooted kernel, with Gaussian atomic kernel	53.1 (3.0)	56.2 (2.9)	0.447 (0.03)	1.4
Subpath kernel, with Gaussian atomic kernel	**58.4** (2.6)	**60.8** (2.9)	**0.498** (0.03)	19.5
Rooted kernel, with χ^2 atomic kernel	57.8 (2.2)	61.3 (2.6)	0.494 (0.03)	2.7
Subpath kernel, with χ^2 atomic kernel	**61.4** (2.8)	**64.4** (2.9)	**0.532** (0.03)	98.8

classes such as urban vegetation, industrial urban blocks, and agricultural zones. For some other classes, performances are weaker. Let us note that the reported results are for the trade-off β parameter providing the best overall accuracies. It may then lead to non adequate values for some classes, as by definition of the proposed kernel, setting $\beta = \infty$ would mimic the rooted kernel.

4   Conclusion

In this paper, we introduced a new structured kernel that is able to cope with unordered trees equipped with numeric data. By doing so, we were able to apply pattern recognition and machine learning techniques to hierarchical image representations that become more and more popular, especially in remote sensing. We built upon a subpath kernel initially designed for bioinformatics data, as well as some graph kernels that relies on random walks. We show by some preliminary experiments the abilities and robustness of the proposed kernel. The encouraging results call for further investigation.

Among future research directions, a comparison with existing kernels in image analysis is planned. Most of them are based on graph kernels. Since trees are a particular class of graphs, various graph kernels may be considered (e.g. [5]).

Acknowledgements. The authors acknowledge the support of the French Agence Nationale de la Recherche (ANR) under reference ANR-13-JS02-0005-01 (Asterix project), and the support of Région Bretagne and Conseil Général du Morbihan (ARIA doctoral project). The authors would also like to thank A. Puissant from LIVE UMR CNRS 7362 (University of Strasbourg) for providing the Strasbourg dataset (Quickbird image and ground truth).

References

1. Aldea, E., Atif, J., Bloch, I.: Image classification using marginalized kernels for graphs. In: Escolano, F., Vento, M. (eds.) GbRPR 2007. LNCS, vol. 4538, pp. 103–113. Springer, Heidelberg (2007)
2. Blaschke, T., et al.: Geographic object-based image analysis–towards a new paradigm. ISPRS J. of Photogrammetry and Remote Sensing 87, 180–191 (2014)
3. Chang, C.C., Lin, C.J.: LIBSVM: A library for support vector machines. ACM Transactions on Intelligent Systems and Technology 2(3), 27:1–27:27 (2011)
4. Collins, M., Duffy, N.: Convolution kernels for natural language. In: Advances in Neural Information Processing Systems, pp. 625–632 (2001)
5. Dupé, F.-X., Brun, L.: Tree covering within a graph kernel framework for shape classification. In: Foggia, P., Sansone, C., Vento, M. (eds.) ICIAP 2009. LNCS, vol. 5716, pp. 278–287. Springer, Heidelberg (2009)
6. Harchaoui, Z., Bach, F.: Image classification with segmentation graph kernels. In: IEEE Conference on Computer Vision and Pattern Recognition, pp. 1–8 (2007)
7. Haussler, D.: Convolution kernels on discrete structures. Tech. rep., Department of Computer Science, University of California at Santa Cruz (1999)
8. Kimura, D., Kuboyama, T., Shibuya, T., Kashima, H.: A subpath kernel for rooted unordered trees. In: Huang, J.Z., Cao, L., Srivastava, J. (eds.) PAKDD 2011, Part I. LNCS, vol. 6634, pp. 62–74. Springer, Heidelberg (2011)
9. Kurtz, C., Passat, N., Gancarski, P., Puissant, A.: Extraction of complex patterns from multiresolution remote sensing images: A hierarchical top-down methodology. Pattern Recognition 45(2), 685–706 (2012)
10. Shawe-Taylor, J., Cristianini, N.: Kernel Methods for Pattern Analysis. Cambridge University Press, Cambridge (2004)
11. Shin, K., Kuboyama, T.: A comprehensive study of tree kernels. In: Nakano, Y., Satoh, K., Bekki, D. (eds.) JSAI-isAI 2013. LNCS, vol. 8417, pp. 337–351. Springer, Heidelberg (2014)
12. Tilton, J.: RHSEG users manual: Including HSWO, HSEG, HSEGExtract, HSEGReader and HSEGViewer, version 1.50 (2010)
13. Vishwanathan, S., Smola, A.J.: Fast kernels for string and tree matching. In: Kernel Methods in Computational Biology, pp. 113–130 (2004)

Coupled-Feature Hypergraph Representation for Feature Selection

Zhihong Zhang[1], Jianbing Xiahou[1], Lu Bai[2(✉)], and Edwin R. Hancock[3]

[1] Software school, Xiamen University, Xiamen, Fujian, China
zhihong@xmu.edu.cn
[2] School of Information,Central University of Finance and Economics, Beijing, China
bailu69@hotmail.com
[3] Department of Computer Science, University of York, York, UK
edwin.hancock@york.ac.uk

Abstract. Real-world objects and their features tend to exhibit multiple relationships rather than simple pairwise ones, and as a result basic graph representation can lead to substantial loss of information. Hypergraph representations, on the other hand, allow vertices to be multiply connected by hyperedges and can hence capture multiple or higher order relationships among features. Due to their effectiveness in representing multiple relationships, in this paper, we draw on recent work on hypergraph clustering to select the relevant feature subset (RFS) from a set of features using high-order (rather than pairwise) similarities. Specifically, we first devise a coupled feature representation to represent the data by utilizing self-coupled and inter-feature coupling relationships, which can be more effective to capture the intrinsic linear and nonlinear information on data structure. Regarding the new data representation, we use a new information theoretic criterion referred to as multivariate mutual information to measure the high-order feature combinations with respect to the class labels. Therefore, we construct a coupled feature hypergraph to model the high-order relations among features. Finally, we locate the relevant feature subset (RFS) from feature hypergraph by maximizing features' average relevance, which has both low redundancy and strong discriminating power. The size of the relevant feature subset (RFS) is determined automatically. Experimental results demonstrate the effectiveness of our feature selection method on a number of standard data-sets.

Keywords: Hypergraph · Coupled feature analysis · Feature selection

1 Introduction

In order to render the analysis of high-dimensional data tractable, it is crucial to identify a smaller subset of features that are informative for classification and clustering. Feature selection is one effective means to remove irrelevant features. Recently, mutual information (MI) has been shown to provide a principled way of measuring the mutual dependence of two variables, and has been used by a

© Springer International Publishing Switzerland 2015
C.-L. Liu et al. (Eds.): GbRPR 2015, LNCS 9069, pp. 44–53, 2015.
DOI: 10.1007/978-3-319-18224-7_5

number of researchers to develop information theoretic feature selection criteria [13,14]. However, one limitation of these MI based methods is that they simply consider pairwise feature dependencies, and do not check for third or higher order dependencies between the candidate features and the existing features. This is because a single feature can be considered irrelevant based on its correlation with the class; but when combined with other features, it may become very relevant.

Another major class of feature selection methods is based on sparsity regularization. The well known Lasso (Least Absolute Shrinkage and Selection Operator) is a penalized least square method with ℓ_1-regularization, which is used to shrink or suppress features to achieve feature selection [15–18]. To our best knowledge, most of existing Lasso-type feature selection methods obtain the final solutions by assuming that the input features are independent with one another and all samples are independent too. However, in real-world data sources, features are more or less strongly interacting and coupled via explicit or implicit relationships. Some researchers also point out that the independence assumption on features often leads to information loss, and several papers have addressed the issue of feature couplings or interactions [7,8].

Despite current research progress, most existing work only captures specific or local aspects of feature interactions in an implicit way rather than systematically taking into account the global relationships among continuous features. To this end, in this work, we propose to find such latent relations among features with a coupled feature representation method [9]. Specifically, we involve both the 'self-coupled' interactions involved in the correlations between features and their own powers, and also 'inter-feature couplings' which measure the correlations between features and the powers of the remaining features. To do this we make use of a hypergraph representation.

Hypergraph representations allow vertices to be multiply connected by hyperedges and can hence capture multiple or higher order relationships between features. Due to their effectiveness in representing multiple relationships, hypergraph based methods have been applied to various practical problems, such as partitioning circuit netlists, clustering [1], clustering categorial data, and image segmentation. For multi-label classification, Sun et al. [2] construct a hypergraph to exploit the correlation information contained in different labels. In this hypergraph, instances correspond to the vertices and each hyperedge includes all instances annotated with a common label. With this hypergraph representation, the higher-order relations among multiple instances sharing the same label can be explored. Following the theory of spectral graph embedding [3], they transform the data into a lower-dimensional space through a linear transformation, which preserves the instance-label relations captured by the hypergraph. The projection is guided by the label information encoded in the hypergraph and a linear Support Vector Machine (SVM) is used to handle the multi-label classification problem. Huang et al. [4] used a hypergraph cut algorithm [1] to solve the unsupervised image categorization problem, where a hypergraph is used to represent the complex relationship among unlabeled images based on shape and appearance features. Specifically, they first extract the region of interest (ROI) of each image, and then construct hyperedges among images based on shape and appearance features in their ROIs. Hyperedges are defined as either a) a group formed by each vertex (image) or b) its k-nearest neighbors (based on shape or appearance descriptors). The weight of each hyperedge is computed as the sum of the pairwise affinities within the hyperedge. In this way, the task of image categorization is transferred into a hypergraph partition problem which can be solved using the hypergraph cut algorithm [6].

One common feature of these existing hypergraph representations is that they exploit domain specific and goal directed representations. Specifically, most of them are confined to uniform hypergraphs where all hyperedges have the same cardinality and do not lend themselves to generalization. The reason for this lies in the difficulty in formulating a nonuniform hypergraph in a mathematically neat way for computation. These has yet to be a widely accepted and consistent way for representing and characterizing nonuniform hypergraphs, and this remains an open problem when exploiting hypergraphs for feature selection.

To address these shortcomings, an effective method for hypergraph construction is needed, such that the ambiguities of relational order can be overcome. In this paper, we improve the hypergraph construction approach presented above by varying the size of the neighborhood, whereby multiple hyperedges are generated for each feature. This makes our approach much more robust than previous hypergraph methods, because we do not need to tune the neighborhood size.

In summary, our method may offer two advantages: Firstly, the proposed model performs feature selection on the new representation of the data. Compared to traditional methods, the coupled representation analyzes the self-coupled and inter-feature couplings, which can be more effective in capturing the intrinsic linear and nonlinear information on data structure; The second advantage is that we develop an nonuniform hypergraph (i.e. the hyperedge cardinality varies) construction approach by varying the size of correlated features. This makes our approach more robust than that in [19], because we do not need to fix the neighborhood size.

2 Coupled Feature Representation

We use $\mathbf{X} \in \Re^{d \times n}$ to denote the data matrix, where n is the number of training samples and d is the number of features. For each dataset, we use $f_1, \cdots, f_d$ to denote the d dimensions of features, and $f_1, \cdots, f_d$ are the corresponding feature vectors, where $f_i \in \Re^n$ and $\mathbf{X} = (f_1, \ldots, f_d)^T$. We also use $x_1, \ldots, x_n$ to denote the n training samples, $\mathbf{X} = (x_1, \ldots, x_n) \in \Re^{d \times n}$ and $\mathbf{y} = (y_1, y_2, \cdots, y_n) \in \Re^{1 \times n}$ be a matrix of corresponding label vectors. Each $x_i = \{f_1^i, f_2^i, \ldots, f_d^i\}^T \in \Re^d$ is a predictive vector of feature measurements for the i-th case and $y_i \in \Re$ is a scalar response for x_i.

For the i-th sample x_i, we extend it to higher dimensions with the incorporation of linear and nonlinear information by means of a matrix expansion as follows:

$$[\langle f_1^i \rangle^1, \langle f_1^i \rangle^2, \langle f_2^i \rangle^1, \langle f_2^i \rangle^2, \cdots, \langle f_d^i \rangle^1, \langle f_d^i \rangle^2]^T$$

where $\langle f_j^i \rangle^l$ indicates the l-th power of the numerical value f_j^i and here we set $l \in \{1, 2\}$ for simplicity.

Utilizing the matrix expansion described above, we first define an self-coupled interaction, which considers the correlations between j-th feature f_j and its own powers as follows:

$$\mathbf{R_a(f_j)} = \begin{pmatrix} \theta_{11}^j & \theta_{12}^j & \cdots & \theta_{1L}^j \\ \theta_{21}^j & \theta_{22}^j & \cdots & \theta_{2L}^j \\ \vdots & \vdots & \ddots & \vdots \\ \theta_{L1}^j & \theta_{L2}^j & \cdots & \theta_{LL}^j \end{pmatrix}$$

where $\theta_{l_1 l_2}^j$ denotes a Person's correlation coefficient between $\langle f_j \rangle^{l_1}$ and $\langle f_j \rangle^{l_2}$, $l_1 = \{1, 2, \cdots, L\}$, $l_2 = \{1, 2, \cdots, L\}$, and L is a maximal power.

Besides the self-coupled interaction, we also define an inter-feature interactions that captures the correlations between the j-th feature f_j and the powers of other features $\{f_{k_c}\}_{k_c \neq j}$ as follows:

$$\mathbf{R_b(f_j)} = \begin{pmatrix} \sigma_{11}^{f_{k_1}} & \cdots & \sigma_{1L}^{f_{k_1}} & \cdots & \sigma_{11}^{f_{k_{d-1}}} & \cdots & \sigma_{1L}^{f_{k_{d-1}}} \\ \sigma_{21}^{f_{k_1}} & \cdots & \sigma_{2L}^{f_{k_1}} & \cdots & \sigma_{21}^{f_{k_{d-1}}} & \cdots & \sigma_{2L}^{f_{k_{d-1}}} \\ \vdots & & \ddots & & & & \vdots \\ \sigma_{L1}^{f_{k_1}} & \cdots & \sigma_{LL}^{f_{k_1}} & \cdots & \sigma_{L1}^{f_{k_{d-1}}} & \cdots & \sigma_{LL}^{f_{k_{d-1}}} \end{pmatrix}$$

where $\{f_{k_c}\}_{k_c \neq j} = [f_{k_1}, \cdots, f_{k_{j-1}}, f_{k_{j+1}}, \cdots, f_{k_d}]^T \in \Re^{d-1}$, and $\sigma_{l_1 l_2}^{f_{k_c}}$ is a Pearson's correlation coefficient between $\langle f_j \rangle^{l_1}$ and $\langle f_{k_c} \rangle^{l_2}$. Note that we use both the training and testing samples in estimating inter-feature for robustness by taking advantage of the information from testing samples. However, the testing samples are not further involved in the following steps, i.e., feature selection and classifier learning.

Let $x_a(i) = [\langle f_j^i \rangle^1, \cdots, \langle f_j^i \rangle^L]$, $x_b(i) = [\langle f_{k_1}^i \rangle^1, \cdots, \langle f_{k_1}^i \rangle^L, \cdots, \langle f_{k_{d-1}}^i \rangle^1, \cdots, \langle f_{k_{d-1}}^i \rangle^L]$, and $\mathbf{w} = [\frac{1}{1!}, \frac{1}{2!}, \ldots, \frac{1}{L!}]$. We integrate the self-coupled interaction $\mathbf{R_a(f_j)}$ and the inter-feature coupling $\mathbf{R_b(f_j)}$ to obtain the coupled feature representation of the j-th feature for the i-th sample as follows:

$$\mathbf{u}_i(f_j) = x_a(i) \odot \mathbf{w} \otimes [\mathbf{R_a(f_j)}]^T + x_b(i) \odot [\mathbf{w}, \mathbf{w}, \cdots, \mathbf{w}] \otimes [\mathbf{R_b(f_j)}]^T$$

where $\mathbf{w} = [\frac{1}{1!}, \frac{1}{2!}, \ldots, \frac{1}{L!}]$ is a constant $1 \times L$ vector, $[\mathbf{w}, \mathbf{w}, \cdots, \mathbf{w}]$ is a $1 \times L \cdot (d-1)$ vector concatenated by $d-1$ constant vectors $\mathbf{w}$. $\odot$ and $\otimes$ denote, respectively, a Hadamard product and a matrix multiplication. Therefore, the final coupled feature representation for the i-th sample can be represented as follows:

$$\mathbf{u}_i = [u_i(f_1), u_i(f_2), \cdots, u_i(f_d)]^T \in \Re^{L \cdot d \times 1} \tag{1}$$

We then combine the high-level information inherent in the data with our hypergraph model for further feature selection.

3 Coupled Feature Hypergraph Analysis for Feature Selection

3.1 Hypergraph Construction via Multiple Feature Correlation

For our hypergraph construction, we regard each feature in the data set as a vertex on hypergraph $H = (V, E, \mathbf{W})$, where $V = \{f_1, f_2, \ldots, f_{L \cdot d}\}$ is the vertex set, E is set of non-empty subsets of V or hyperedges and $\mathbf{W}$ is a weight function which associated a real value with each hyperedge. Assume there are $L \cdot d$-dimensional features in the data set, and thus, the generated hypergraph contains $L \cdot d$ vertices. Following the recent developments on sparse representation and ℓ_1-regularized models [10], we generate hyperedges by linking features and their correlated features. Specifically, each feature vector can be regarded as a response vector, and can be estimated by a linear combination of the remaining $L \cdot d - 1$ feature vectors, i.e.,

$$f_i = F_i \beta_i + \varepsilon_i, i = 1, 2, \ldots, L \cdot d \tag{2}$$

where $F_i = [f_1, f_2, \ldots, f_{i-1}, 0, f_{i+1}, \ldots, f_{L \cdot d}]$ denote a feature sets including all the features excluding the i-th feature (we put 0 in its location), β_i essentially contains the combination coefficients for different features in approximating f_i, and $\varepsilon_i \in \Re^n$ is a noise term. Therefore, a natural method to determine sparse solution of β_i is to solve the following problem:

$$\min_{\beta_i} \|f_i - F_i \beta_i\|_2 + \lambda \|\beta_i\|_1 \tag{3}$$

where $\lambda > 0$ is a regularization parameter controlling the sparsity of β_i. Due to the nature of ℓ_1-norm penalty, some coefficients become zero if λ is large enough. In this case, we can generate a hyperedge containing the most correlated features (corresponding to the non-zero coefficients in β_i) with respect to f_i. Different λ values corresponds to different sparsity solutions and so instead of generating a single hyperedge for each feature f_i, we generate a group of hyperedges by varying value of λ in a specified range. Specifically, in our experiment, we vary the value of λ from 0.1 to 0.9 with an incremental step of 0.1.

3.2 Computing Hyperedge Weight by Higher-Order Features Correlation

Given a set of features $f_{i_1}, f_{i_2}, \ldots, f_{i_K}$, the interaction information among them can be measured by joint entropy [11]:

$$H(f_{i_1}, f_{i_2}, \ldots, f_{i_K}) = - \sum_{f_{i_1}, f_{i_2}, \ldots, f_{i_K}} P(f_{i_1}, f_{i_2}, \ldots, f_{i_K}) \log_2 P(f_{i_1}, f_{i_2}, \ldots, f_{i_K}) \tag{4}$$

where $P(f_{i_1}, f_{i_2}, \ldots, f_{i_K})$ is the probability of features $f_{i_1}, f_{i_2}, \ldots, f_{i_K}$ occurring together. According the above definition, the joint entropy is always positive which they measure the amount of information contained in the correlated features. Based on the joint entropy, a new measure called multivariate mutual information is defined to measure the high-order correlation among features, i.e.

$$I(f_{i_1}, f_{i_2}, \ldots, f_{i_K}) = \sum_{k=1}^{K} (-1)^{k-1} \sum_{F \subset f_{i_1}, f_{i_2}, \ldots, f_{i_K}, |F|=k} H(F) . \quad (5)$$

In Equation 5, F is a subset of features $\{f_{i_1}, f_{i_2}, \ldots, f_{i_K}\}$, and $H(F)$ represents the joint entropy of a discrete random variable. It is clear that the greater the value of $I(f_{i_1}, f_{i_2}, \ldots, f_{i_K})$ is, the more relevant the K features are. On the contrary, if $I(f_{i_1}, f_{i_2}, \ldots, f_{i_K}) = 0$, the features are unrelated.

For the constructed hypergraph $G = (V, E, \mathbf{W})$, we determine the weight of each hyperedge using an normalized multivariate mutual information which can measure the relevance degree contained in the features of each hyperedge with respect to class label C:

$$\mathbf{W}(f_{i_1}, f_{i_2}, \ldots, f_{i_K}; C) = K \frac{I(f_{i_1}, f_{i_2}, \ldots, f_{i_K}; C)}{H(f_{i_1}) + H(f_{i_2}) + \cdots + H(f_{i_K})} . \quad (6)$$

Therefore, a large value of $\mathbf{W}(f_{i_1}, f_{i_2}, \ldots, f_{i_K}; C)$ means $\{f_{i_1}, f_{i_2}, \ldots, f_{i_K}\}$ are strongly relevant with respect to the class label C.

3.3 Relevant Feature Subset Selection

For the constructed hypergraph, the vertex-edge incident matrix $H \in \Re^{|V| \times |E|}$ can be defined as:

$$H(v, e) = \begin{cases} 1 \text{ if } v \in e \\ 0 \text{ otherwise.} \end{cases} \quad (7)$$

Let $\mathbf{W}$ be the diagonal matrix containing the weights of the hyperedges, and the adjacency matrix $\mathbf{A}$ is

$$\mathbf{A} = HWH^T \quad (8)$$

where H^T is the transpose of H.

Given the hypergtaph adjacency matrix $\mathbf{A}$ and $L \cdot d$-dimensional feature indicator vector γ with γ_i representing the i-th element, we can locate the relevant feature subset (RFS) by finding the solutions of the following maximization problem:

$$\max f(\gamma) = \sum_{i=1}^{L \cdot d} \sum_{j=1}^{L \cdot d} \gamma_i \gamma_j A_{i,j} . \quad (9)$$

subject to $\gamma \in \triangle$, where the multidimensional solution vector γ fall on the simplex $\triangle = \{\gamma \in \Re^{L \cdot d} : \gamma \geq 0 \text{ and } \sum_{i=1}^{L \cdot d} \gamma_i = 1\}$ and $A_{ii} = 0$, i.e., all diagonal entries of $\mathbf{A}$ are set to zero. Our idea is motivated by graph-based clustering which groups the most dominant vertices into a cluster. On the other hand, in our work, the feature subset $\{f_i | 1 \leq i \leq L \cdot d, \gamma_i > 0\}$ is the most coherent subset of the initial feature set, with maximum internal homogeneity of the feature relevance (6). According to the value of γ, all features F fall into two disjoint subsets, $S_1(\gamma) = \{f_i | \gamma_i = 0\}$ and $S_2(\gamma) = \{f_i | \gamma_i > 0\}$. We refer to the set of nonzero variables $S_2(\gamma)$ as the relevant feature subset (RFS), because the objective function (9) selects RFS by maximizing features' average relevance.

The objective function (9) is typical quadratic program, and here we apply discrete-time first-order replicator equation to approximating the solution for the RFS.

$$\gamma_i^{new} = \frac{\gamma_i \sum_{j=1}^{L \cdot d} \gamma_j A_{i,j}}{\sum_{i=1}^{L \cdot d} \sum_{j=1}^{L \cdot d} \gamma_i \gamma_j A_{i,j}} \ . \tag{10}$$

where γ_i^{new} corresponded to the i-th feature vector after the update process. The complexity of finding the RFS is $O(t|E|)$, where $|E|$ is the number of edges of the feature graph constructed above and t is the average number of iteration needed to converge.

Algorithm 1. Coupled Feature Hypergraph Analysis for Feature Selection

Input: Coupled featured representation $\mathbf{U} \in \Re^{L \cdot d \times n}$ with all features $L \cdot d$
Output: The selected feature subset corresponding to the non-zero elements of
$\quad\quad\quad L \cdot d$-dimensional indicator vector γ
1) Hypergraph construction via sparse representation using Equation (3)
2) The weight matrix $\mathbf{W}$ for the hyperedges is computed using Equation (6) ;
3) The adjacent matrix of hypergraph $\mathbf{A}$ is comuted by Equation (8) ;
4) Compute γ based on $\mathbf{A}$ by Equation (9) and iteratively update γ by
Equation (10) ;
5) Select the relevant feature subset (RFS) from the non-zero elements of
$L \cdot d$-dimensional indicator vector γ.

---

4   Experiments and Comparisons

The data sets used to test the performance of our proposed algorithm are benchmark data sets from the UCI Machine Learning Repository. Table. 1 summarizes the extents and properties of the six data-sets.

4.1   Coupled Feature Representation Evaluation

To validate the advantage of the coupled feature representation, we employ 10-fold cross-validation strategy with C-SVM classifier on coupled feature representation and original feature representation respectively, and compare their performance. Figure 1 shows the performance of L (i.e. L=3,4,5) on six data sets. It is clear that the C-SVM classifier on coupled representation always outperforms that built on the original representation. That is, the coupled feature

Table 1. Summary of UCI benchmark data sets and their coupled feature representation with expansion parameter $L = 4$

Data-set	Sample	Original Features.no	Coupled features.no ($L = 4$)	Classes
Wine	178	13	52	3
Vowel	528	10	40	11
Planning	182	12	48	2
Ionosphere	351	34	136	2
Hepatitis	155	19	76	2
Blood	748	4	16	2

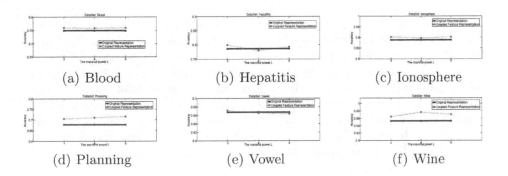

(a) Blood   (b) Hepatitis   (c) Ionosphere

(d) Planning   (e) Vowel   (f) Wine

Fig. 1. The effectiveness of coupled feature representation

representation is helpful to improve the classification accuracy. Regarding the expansion parameter L, a small L deteriorates the classification performance while a large L increases the unnecessary computation burden. In the following experiments, we fixed L to 4.

4.2   Relevant Feature Subset Evaluation

We compare the classification result from the features captured by the relevant feature subset (RFS) with those obtained using both alternative MI-based criterion method and sparsity regularization based methods. These methods are the MRMR algorithm, the Lasso algorithm and ElasticNet method. We use 10-fold cross-validation for the C-SVM classifier on the feature subsets obtained by the feature selection algorithms to verify their classification performance.

We summarize the classification accuracy rate of different methods in Table. 2. In the last row, the classification accuracy for the features from the relevant feature subset which referred as RFS and the automatically determined size of RFS are reported. Suppose that the determined size of RFS is k. To make a fair comparison, for each alternative method, we measure the classification accuracy for $k-1$, k and $k+1$ features, and take the best result as the baseline performance. As shown by the results in Table. 2, our extracted RFS consistently outperforms other feature subsets obtained by the alternative methods on all six data sets.

52   Z. Zhang et al.

Table 2. Performance comparison of accuracy rate around the size of features in RFS selected by different methods

Dataset	Blood	Hepatitis	Ionosphere	Planning	Vowel	Wine
MRMR	75.8%	78.3%	84.7%	71.1%	67.3%	97.1%
Lasso	73.2%	76%	83.7%	71.7%	64.4%	95.29%
ElasticNet	75.8%	78.7%	77.8%	70%	81%	94.1%
All couple features	75.8%	78%	**89.7%**	71.1%	**86.5%**	97.6%
RFS	**75.9%(3)**	**80%(10)**	87.1%(11)	**72.2%(8)**	83.3%(7)	**98.5%(9)**

The results verify that our proposed method is effective to locate the relevant feature subset. There are two reasons for this improvement in performance. First, the multivariate mutual information is applied to measure the features relevance, and this can capture the effects of higher-order feature dependencies between the features and the class. Second, the extraction of the relevant feature subset simultaneously considers the information-contribution for each feature together with the correlation between features. Thus structural information latent in the data can be effectively identified. As a result the optimal feature combinations can be located so as to group the greatest number of relevant features into homogenous subset.

5 Conclusion

In this paper, we devised a coupled feature representation with self-coupled and inter-feature coupling relationships by means of a matrix expansion. Based on the new representation of data, we have proposed a new nonuniform hypergraph construction with adaptive feature correlations, which can effective capture the higher-order feature relationship among features. This is completely different from the conventional uniform hypergraph construction methods, where all hyperedge have the same cardinality. Experimental results on six benchmark datasets indicate that our proposed feature selection method can not only improve the classification accuracy, buy also automatically determine the size of relevant feature subset.

Acknowledgments. This work is supported by National Natural Science Foundation of China (Grant No.61402389). Edwin R. Hancock is supported by a Royal Society Wolfson Research Merit Award.

References

1. Zhou, D., Huang, J., Scholkopf, B.: Learning with hypergraphs: Clustering, classification, and embedding. Advances in Neural Information Processing Systems 19, 1601–1608 (2006)

2. Sun, L., Ji, S., Ye, J.: Hypergraph spectral learning for multi-label classification. In: Proceedings of the 14th ACM SIGKDD International Conference on Knowledge Discovery and Data Mining, pp. 668–676 (2008)
3. Chung, F.: Spectral Graph Theory. American Mathematical Society (1992)
4. Huang, Y., Liu, Q., Lv, F., Gong, Y., Metaxas, D.N.: Unsupervised image categorization by hypergraph partition. IEEE Transactions on Pattern Analysis and Machine Intelligence 33(6), 1266–1273 (2011)
5. Zhou, D., Huang, J., Scholkopf, B.: Learning with hypergraphs: Clustering, classification, and embedding. Advances in Neural Information Processing Systems 19, 1601–1608 (2006)
6. Klimmek, R., Wagner, F.: A simple hypergraph min cut algorithm. Technical Report B 96-02, University Berlin, Germany (March 1996)
7. Bollegala, D., Matsuo, Y., Ishizuka, M.: Relation adaptation: learning to extract novel relations with minimum supervision. In: Proceedings of the Twenty-Second International Joint Conference on Artificial Intelligence, vol. 3, pp. 2205–2210 (2011)
8. Wang, C., Cao, L.B., Wang, M.C., Li, J.J., Wei, W., Ou, Y.M.: Coupled nominal similarity in unsupervised learning. In: Proceedings of the 20th ACM International Conference on Information and Knowledge Management, pp. 973–978 (2011)
9. Wang, C., She, Z., Cao, L.B.: Coupled attribute analysis on numerical data. In: Proceedings of the Twenty-Third International Joint Conference on Artificial Intelligence, pp. 1736–1742 (2013)
10. Wright, J., Yang, A.Y., Ganesh, A., Sastry, S.S., Ma, Y.: Robust face recognition via sparse representation. IEEE Transactions on Pattern Analysis and Machine Intelligence 31(2), 210–227 (2009)
11. MacKay, D.: Information theory, inference and learning algorithms. Cambridge Univ. Press, Cambridge (2003)
12. Baum, L.E., Eagon, J.A.: An inequality with applications to statistical estimation for probabilistic functions of Markov processes and to a model for ecology. Bull. Amer. Math. Soc. 73(3), 360–363 (1967)
13. Battiti, R.: Using Mutual Information for Selecting Features in Supervised Neural Net Learning. IEEE Transactions on Neural Networks, 537–550 (2002)
14. Peng, H., Long, F., Ding, C.: Feature Selection Based on Mutual Information: Criteria of Max-Dependency, Max-Relevance, and Min-Redundancy. IEEE Transactions on Pattern Analysis and Machine Intelligence 27(8), 1226–1238 (2005)
15. Tibshirani, R.: Regression shrinkage and selection via the Lasso. Journal of the Royal Statistical Society. Series B (Methodological) 58(1), 267–288 (1996)
16. Yuan, M., Lin, Y.: Model selection and estimation in regression with grouped variables. Journal of the Royal Statistical Society: Series B (Statistical Methodology) 68(1), 49–67 (2006)
17. Zou, H.: The adaptive lasso and its oracle properties. Journal of the American Statistical Association 101, 1418–1429 (2006)
18. Zou, H., Hastie, T.: Regularization and variable selection via the elastic net. Journal of the Royal Statistical Society: Series B (Statistical Methodology) 67(2), 301–320 (2005)
19. Zhang, Z., Hancock, E.R.: Hypergraph based Information-theoretic Feature Selection. Pattern Recognition Letters 33, 1991–1999 (2012)

Reeb Graphs Through Local Binary Patterns

Ines Janusch[✉]and Walter G. Kropatsch

Pattern Recognition and Image Processing Group
Institute of Computer Graphics and Algorithms
Vienna University of Technology, Vienna, Austria
{ines,krw}@prip.tuwien.ac.at

Abstract. This paper presents an approach to derive critical points of a shape, the basis of a Reeb graph, using a combination of a medial axis skeleton and features along this skeleton. A Reeb graph captures the topology of a shape. The nodes in the graph represent critical points (positions of change in the topology), while edges represent topological persistence. We present an approach to compute such critical points using Local Binary Patterns. For one pixel the Local Binary Pattern feature vector is derived comparing this pixel to its neighbouring pixels in an environment of a certain radius. We start with an initial segmentation and a medial axis representation. Along this axis critical points are computed using Local Binary Patterns with the radius, defining the neighbouring pixels, set a bit larger than the radius according to the medial axis transformation. Critical points obtained in this way form the node set in a Reeb graph, edges are given through the connectivity of the skeleton. This approach aims at improving the representation of flawed segmented data. In the same way segmentation artefacts, as for example single pixels representing noise, may be corrected based on this analysis.

Keywords: Reeb Graphs · Local Binary Patterns · Local Features · Critical Points · Shape Representation · Image Segmentation

1 Introduction

Reeb graphs capture a shape's topology: the connected components (the connectivity of the shape) are represented by the edges, while positions of change in topology are represented by nodes in the graph [1]. Reeb graphs are for example used as a tool of skeletonisation in [2], a tool of segmentation in [3] or a tool of shape analysis in [4]. Moreover, the compact shape description provided by a Reeb graph may be used for shape comparison and retrieval.

For a Reeb graph representation critical points are usually computed on the shape according to a Morse function [1]. The obtained Reeb graph is dependent on the applied Morse function and its properties. In contrast, we present a novel approach to derive the edges of a Reeb graph through the topology (connectivity) captured by a medial axis, the nodes are computed based on local features at certain positions in the shape.

In [4] we presented an analysis of roots for the purpose of plant phenotyping using Reeb graphs. Our results showed that Reeb graphs are suitable for such an analysis. However, the sensitivity of this representation to segmentation errors is likely to falsify results. To reduce the influence of segmentation errors, the computation of the Reeb graph should not be based solely on segmented data.

© Springer International Publishing Switzerland 2015
C.-L. Liu et al. (Eds.): GbRPR 2015, LNCS 9069, pp. 54–63, 2015.
DOI: 10.1007/978-3-319-18224-7_6

In order to analyse a shape using a Morse function, the input image needs to be binary segmented. Such a pre-processing may introduce artefacts e.g. spurious branches in the graph representation. Post-processing methods applied to the graph representation, for example graph pruning, may be used to discard spurious branches.

Instead of a post-processing procedure we propose an approach that is based on an initial shape representation according to a segmentation of the input data. For this initial representation the medial axis is used. It is obtained as the centers of maximally inscribed circles on a shape. The medial axis (in 2D and 3D) as well as the more sophisticated curve skeletons [5] are used as compact shape descriptor as they capture the topological characteristics of a shape. However, the medial axis is sensitive to noise due to segmentation artefacts. Thus, the skeleton is only used as a first representation to guide the computation of a Reeb graph representation.

Based on the skeleton, critical points that form the node set of a Reeb graph, are computed using Local Binary Patterns (LBPs) [6] centered along the skeleton. For the computation of critical points according to a Morse function, Morse conditions need to be kept. One such Morse condition requires the Morse function values of two different critical points to differ in order to derive a unique graph representation. However, when representing shapes in a discrete space (e.g. 2D pixel or 3D voxel space) this condition is likely to be violated as discussed in [7].

LBPs analyse and represent an image region, the neighbourhood at certain radius, around a pixel. A common application for LBPs is given by texture classification as originally presented in [8]. Moreover, LBPs have been used for face recognition in [9]. We use LBPs to computed nodes of a Reeb graph. When determined as the critical points of a Morse function, these nodes may be classified according to the different changes in topology they represent. Here, the different types of nodes correspond with the neighbourhood configurations that can be represented by an LBP[1]. LBPs allow to base the representation on the original unsegmented data. Starting from an initial segmentation and an initial skeleton representation LBPs may be computed on the unsegmented image. Segmentation artefacts may be detected and corrected in this way.

The rest of the paper is structured as follows: Section 2 provides an introduction to LBPs, Section 3 to Reeb graphs. The computation of a Reeb graph through LBPs is defined in Section 4. Results obtained are discussed in Section 5. Section 6 concludes the paper and gives an outlook to future work.

2   Principle of Local Binary Patterns

LBPs were introduced by Ojala, Pietikäinen and Harwood in 1994 [8] as a tool of texture classification. Due to its computational simplicity and its robustness to spurious color gradients e.g. due to lighting conditions, LBPs are popular texture operators.

A simple example how LBPs work is shown in Figure 1: The center pixel is compared to its subsampled neighbourhood. The relations of this comparison are stored as a bit pattern: In case the value of a neighbouring pixel is larger than or equal to the

---

[1] Although identical names are used for Reeb graph node types and LBP types an identical meaning is not guaranteed (e.g. an LBP of type saddle may not represent a saddle node in a Reeb graph). We indicate Reeb graph node types by [R], LBP types by [L] (e.g. saddle[R], saddle[L]).

$x_1 \geq c$	$x_2 < c$	$x_3 < c$	1	0	0	2^0	2^1	2^2	**LBP** = 10001111
$x_8 \geq c$	c	$x_4 < c$	1	c	0	2^7	c	2^3	$= 2^0 + 2^4 + 2^5 +$
$x_7 \geq c$	$x_6 \geq c$	$x_5 \geq c$	1	1	1	2^6	2^5	2^4	$2^6 + 2^7$
									= **241**

(a) center pixel and neighbourhood (b) comparison result (c) neighbourhood pattern (d) LBP operator for center pixel c

Fig. 1. Simple LBP computation

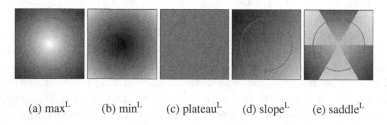

(a) max^L (b) min^L (c) $plateau^L$ (d) $slope^L$ (e) $saddle^L$

Fig. 2. Neighbourhood configuration detected by LBPs. The red circle indicates the neighbourhood used in the LBP computation for the pixel at its center.

value of the center pixel its bit is set to 1 otherwise to 0. The neighbourhood pattern is encoded as the position of each neighbourhood pixel in a binary data item [8].

Moreover, the configuration of the neighbourhood around a pixel encodes the local topology. The region may be a (local) maximum^L (the bit pattern contains only 0s), a (local) minimum^L (the bit pattern contains only 1s), a plateau^L (the bit pattern contains only 1s, but all pixels of the region have the same gray value), a slope^L (the bit pattern of the region contains one connected component of 1s and one connected component of 0s) or a saddle^L point otherwise [10]. Figure 2 shows examples for these region configurations that may be encoded by LBPs. The approach presented in this paper uses the different region configurations to derive critical points of a shape in order to represent it using a Reeb graph.

LBPs are not only defined for the eight immediate neighbours of a pixel. These eight direct neighbours of a center pixel are at radius 1 from the center pixel but the radius at which an LBP operator is computed may also be larger than 1.

3   Morse Theory and Reeb Graphs

Reeb graphs describe the topological structure of a shape (e.g. 2D or 3D content) as the connectivity of its level sets [11]. A shape is analysed according to a Morse function to derive a Reeb graph. Two common Morse functions are the height function and the geodesic distance. The nodes of a Reeb graph correspond to critical points computed on a shape according to a Morse function. At critical points the topology of the analysed shape changes, thus the number of connected components in the level-set changes.

Edges connecting critical points represent the connected components and thus describe topological persistence.

A point $p = (a, b)$ of a function $f(x, y)$ is called a *critical point* if both derivatives $f_x(a, b)$ and $f_y(a, b)$ are equal to 0 or if one of these partial derivatives does not exist. Such a critical point p is called degenerate if the determinant of the Hessian matrix at that point is zero, otherwise it is called non-degenerate (or Morse) critical point [12].

A Morse functions is defined in the continuous domain as follows:

A smooth, real-valued function $f : M \rightarrow \mathbb{R}$ is called a Morse function if it satisfies the following conditions for a manifold M with or without boundary:

- *M1*: all critical points of f are non-degenerate and lie inside M,
- *M2*: all critical points of f restricted to the boundary of M are non-degenerate,
- *M3*: for all pairs of distinct critical points p and q, $f(p) \neq f(q)$ must hold [13].

Although originally defined in the continuous domain, Reeb graphs have been extended to the discrete domain:

- Two point sets are connected if there exists a pair of points (one point of each point set) with a distance between these two points below a fixed threshold.
- If all non-empty subsets of a point set, as well as its complements, are connected, such a point set is called *connective*.
- A group of points that have the same Morse function value and that form a connective point set, is called a *level-set curve* [3].

The nodes in a discrete Reeb graph represent level-set curves, the edges connect two adjacent level-set curves, therefore the underlying point sets are connected [3]. In 2D three types of nodes in a Reeb graph correspond to critical points: minima^R, maxima^R or saddles^R [13]. Minimum^R and maximum^R nodes are of degree 1, due to the conditions for Morse functions (especially condition $M3$) saddle^R nodes are of degree 3. An example Reeb graph containing all possible types of nodes is shown in Figure 3.

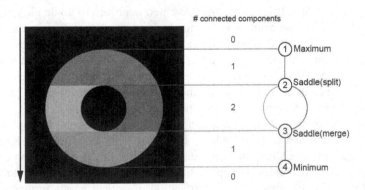

Fig. 3. Reeb graph of a ring according to the height function; the foreground region is colour labeled according to the connected components

4   Computation of Reeb Graphs Based on LBPs

Reeb graphs are derived on binary segmented 2D or 3D data using an analysis based on a Morse function. Different Morse functions applied to the same image may result in different Reeb graphs. Moreover, the computed Reeb graph depends on the properties of the Morse function used. A height function for example is not rotational invariant. Due to the rotational invariance property of the medial axis, Reeb graphs computed according to the approach presented in this paper are rotational invariant (apart from artefacts due to the discrete pixel space).

In order to analyse unsegmented data local descriptors, as for example LBPs, may identify the critical points directly on this data. For Reeb graphs based on a Morse function Morse conditions [13] apply. When computing the critical points (the nodes in a Reeb graph) according to LBPs, the following conditions apply: a critical point is determined by an LBP of type maximum^L, minimum^L, slope^L or saddle^L as described in Section 2. For LBPs of type saddle^L the neighbourhood configuration of the LBP may not be divided into more than six connected components (therefore three foreground regions, as saddle^R nodes are of degree 3). LBPs of type plateau^L do not represent critical points but regular points which do not represent any change in topology.

The approach presented here works on a segmented image as an input. The computation may be extended to unsegmented data. Nevertheless, a segmented image may still be needed as a first input to guide the computation of the critical points. In case the position of the critical points can be estimated (e.g. in video data based on the position in a previous frame) the segmentation is not necessary.

The computation of a Reeb graph according to LBPs in general works as follows:

1. initial binary segmentation and medial axis representation of foreground
2. computation of LBPs along skeleton pixels
3. determination of critical points
4. connection of critical points according to skeleton to obtain Reeb graph

Theorem 1. *This approach determines a Reeb graph representing the foreground shape.*

Proof. A conventional Reeb graph is defined as a topological graph which describes the evolution of level-sets of a real valued function on a manifold. The medial axis captures the topology of a shape. It is used in the presented approach to guide the graph computation and to define the connectivity. The obtained graph therefore is a topological graph. Our proposed method follows the pixels along a skeleton. The geodesic distance (along the skeleton, starting from an arbitrary skeleton pixel) serves as the function analysing the shape. Although, here one level-set only consists of single skeleton pixels at a certain distance to the starting point, the evolution of the level-sets is just as well described by the branching points and end points of the skeleton.

Changes of topology (e.g. an increase in the number of connected components due to a branch) are detected at the boundary of the shape when computing a Reeb graph according to a common Morse function (e.g. the height function or the geodesic distance). The medial axis guides our approach; critical points are positioned on the skeleton, therefore inside the shape. Our identification of critical points further considers LBPs along this skeleton. These LBPs are computed according to the medial axis radii, plus an increase of ϵ. Therefore, the boundary of the shape is taken into account.   □

4.1   Initial Skeleton

An initial binary segmentation of the input image is needed to compute the medial axis for the foreground region. This axis is formed by the centers of maximal circles that cover the shape completely. Therefore, the medial axis implicitly provides a measure of width, for each point p_i along the medial axis the radius r_i of the inscribed maximal circle (the distance to the boundary) is known [14]. Figure 4a shows a segmented example image (from the the mythological creatures database [15]) together with the medial axis representing it.

4.2   LBPs Along Skeleton

LBPs are computed for each skeleton pixel. According to the radius r_i stored with each skeleton pixel p_i, the LBP is computed for each p_i with a radius of $r_i * 1.5$. This enlargement by $\epsilon = 50\%$ was experimentally determined. ϵ may be adjusted according to the desired output, as it regulates the detection of spurious branches. It serves as Nyquist limit, as branches smaller than ϵ are discarded by this approach.

This radius is likely to be 15 pixel or more for the images in our dataset and in general. Therefore, it may not be possible to store the final LBP operator: For a radius of 15 pixel the LBP is (in our case) computed based on 64 neighbours along a circle of radius 15. As described in Section 2 the final LBP operator is obtained by converting the binary data item representing the neighbourhood of a pixel to the decimal system. This may result in numbers larger than 2^{64}, which cannot be represented in most programming languages. However, this LBP operator is not needed for the presented approach. Instead of computing LBP operators, we only compute the type of LBP neighbourhood configuration for each skeleton pixel. Figure 4b shows the skeleton pixels colour labeled according to the LBP neighbourhood configuration around these pixels.

In contrast to the LBP neighbourhood configurations shown in Figure 2 we only encounter the following three neighbourhood configurations (shown in Figure 5) when

(a) Binary segmented input image   (b) LBPs: red = slope^L, green = and its medial axis drawn in red   ridge^L, blue = branch^L.

Fig. 4. Computation of LBPs along the skeleton

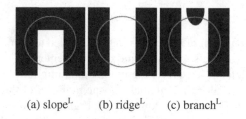

(a) slope^L (b) ridge^L (c) branch^L

Fig. 5. Possible LBP neighbourhood configuration along skeleton pixels of a binary segmented image (LBP radius according to the medial axis radius but enlarged)

computing LBPs along a skeleton of a binary segmented image: slope^L, ridge^L (special case of saddle^L) and branch^L (again special case of saddle^L). They are determined as follows: the LBP bit pattern according to the pixels along the LBP radius consists of one connected component of 0s and one connected component of 1s for a slope^L, two connected components each for a ridge^L and three each for a branch^L.

4.3 Critical Points According to LBPs

To determine the position of critical points, the skeleton pixels are analysed according to the type of LBP and the number of neighbouring skeleton pixels. A critical point of type saddle^R is detected as skeleton pixel for which the LBP shows a branch^L and which has three neighbouring skeleton pixels. Critical points of type minimum^R/maximum^R are given as skeleton pixels with a corresponding LBP of type slope^L and only one neighbouring skeleton pixel. Skeleton pixels with an LBP indicating a ridge^L and two neighbouring skeleton pixels correspond to regular points. At such positions nodes of degree two may be added along an edge in a Reeb graph. However, as nodes of degree two do not describe any changes in topology, they are typically disregarded in a topological representation.

(a) Nodes based on LBPs: red = saddle^R, green = min^R/max^R (b) Final Reeb graph based on LBPs along the skeleton (c) Reeb graph using geodesic distance as Morse function

Fig. 6. Computation of the Reeb graph according to the LBPs along the skeleton and according to the geodesic distance as Morse function

In case the LBP and the number of neighbouring skeleton pixels do not correspond according to the above given definition, artefacts may be detected: In case a skeleton pixel with three neighbouring skeleton pixels has an LBP of a type different from branch^L, small curvature changes along the boundary may have introduced a spurious branch in the skeleton. By computing the LBP using the enlarged medial axis radius, small spurious branches can be detected and discarded. Artefacts are further caused by circular curvature changes along the boundary that may assign an LBP of type branch^L to a skeleton pixel with only two neighbouring skeleton pixels. Such pixels are corrected to be not represented as a critical point but as a regular node in the Reeb graph (note that regular nodes are discarded in the final graph representation).

Figure 6a shows the critical points compute for the skeleton and the LBPs in Figure 4. Depending on the size of the LBP radius at a certain skeleton pixel, branches may be detected as spurious branches. In the example image this is visible at the tail of the mythological horse. Here a saddle^R node (red) was labeled as minimum^R/maximum^R node (green). Such branches are discarded in the final graph representation (Figure 6b).

4.4   Reeb Graph According to Skeleton and LBPs

The critical points are used as nodes in the Reeb graph. Edges are determined by the skeleton: The skeleton is traced from each branch point (saddle^R) to all neighbouring nodes. The resulting graph for the example image is shown in Figure 6b. For comparison Figure 6c shows a Reeb graph obtained using an alternative approach based on the geodesic distance as Morse function [4]. Apart from a difference in the positions of the nodes, the numbers of the represented connected components are equivalent.

5   Results

The approach presented in this paper shows work in progress. The results are therefore preliminary results.

Our approach was tested on the 99 images of the shape database presented in [16] and on the 15 images of the mythological creatures database [15]. Out of these 114 images the obtained Reeb graph correctly represents the topology for 60 images. In case the shape was not correctly represented, this can be ascribed to the graph pruning that is automatically included and dependent on the chosen LBP radius. Moreover, compact shapes with a circular boundary line may corrupt the representation. Table 1 shows Reeb graphs computed using the presented approach for 10 images of the evaluated database [16]. For simplicity the edges of the Reeb graph are drawn as straight lines although their precise geometry can be taken from the medial axis. For image 6 the head of the animal is not represented in the final graph . Due to the circular contour line the according skeleton segment was incorrectly detected as a spurious branch and therefore rejected. For image 10 no graph representation could be found, as well in this case the circular contour line rejected a branch in the skeleton as a true branch. Such representational artefacts may be avoided by computing more than one LBP operator according to different radii around one skeleton pixel, as well as computing the LBP operators on an unsegmented grayscale image instead of the segmented binary image.

Table 1. Reeb graph representation obtained for silhouette images

However, this Reeb graph representation, just as any other topological graph representation is best used for elongated, branched or articulated objects. Compact shapes with a smooth, rounded contour line (circular, no elongation and no dents along it) hold only little characteristic topological information (no significant branches) with therefore little discriminative power. Thus, also the representational power of a topological graph representing such a shape is limited. Compact shapes do not present suitable applications for Reeb graphs in general, but are well suited for shock graphs [16].

6   Conclusion and Future Work

The presented approach does not analyse the binary segmented input image using a Morse function to compute a Reeb graph. Instead the computation of the Reeb graph is based on LBPs along a skeleton. In this way critical points, which form the node set of the Reeb graph, and the connectivity of the nodes (the edges) representing the shape's topology can be derived. The results show that this approach derives Reeb graph representations for shapes that may be well represented using a graph (for example articulated objects).

In contrast to Morse functions, which need to be evaluated for every pixel of the shape, this new approach determines the Reeb graph by evaluating a much smaller number of locations: only the axis points. A future goal is to further limit these evaluation positions (e.g. to use the end points and branching points of the axis only).

The silhouette images that were used in the experiments only present a first test dataset. For future work the intended input data are grayscale images. An initial segmentation may still be needed in order to derive the skeleton which guides the Reeb graph computation. Nevertheless, the LBP computation may be performed on the unsegmented data. In this way segmentation artefacts may be detected in the graph and discarded. The segmentation may even be corrected based on the observed LBP operators.

Acknowledgements. We thank the anonymous reviewers for their constructive comments.

References

1. Biasotti, S., Giorgi, D., Spagnuolo, M., Falcidieno, B.: Reeb graphs for shape analysis and applications. Theoretical Computer Science 392(1-3), 5–22 (2008)
2. Ge, X., Safa, I.I., Belkin, M., Wang, Y.: Data skeletonization via Reeb graphs. In: Shawe-Taylor, J., Zemel, R.S., Bartlett, P., Pereira, F.C.N., Weinberger, K.Q. (eds.) Advances in Neural Information Processing Systems 24, pp. 837–845 (2011)
3. Werghi, N., Xiao, Y., Siebert, J.: A functional-based segmentation of human body scans in arbitrary postures. IEEE Transactions on Systems, Man, and Cybernetics, Part B: Cybernetics 36(1), 153–165 (2006)
4. Janusch, I., Kropatsch, W.G., Busch, W., Ristova, D.: Representing roots on the basis of reeb graphs in plant phenotyping. In: Agapito, L., Bronstein, M.M., Rother, C. (eds.) ECCV 2014 Worshops, Part IV. LNCS, vol. 8928, pp. 75–88. Springer, Heidelberg (2015)
5. Nyström, I., Sanniti di Baja, G., Svensson, S.: Curve skeletonization by junction detection in surface skeletons. In: Arcelli, C., Cordella, L.P., Sanniti di Baja, G. (eds.) IWVF 2001. LNCS, vol. 2059, pp. 229–238. Springer, Heidelberg (2001)
6. Pietikäinen, M.: Image analysis with local binary patterns. In: Kalviainen, H., Parkkinen, J., Kaarna, A. (eds.) SCIA 2005. LNCS, vol. 3540, pp. 115–118. Springer, Heidelberg (2005)
7. Janusch, I., Kropatsch, W.G., Busch, W.: Reeb graph based examination of root development. In: Proceedings of the 19th Computer Vision Winter Workshop, pp. 43–50 (February 2014)
8. Ojala, T., Pietikäinen, M., Harwood, D.: Performance evaluation of texture measures with classification based on kullback discrimination of distributions. In: Proceedings of the 12th IAPR International Conference on Computer Vision and Image Processing, vol. 1, pp. 582–585 (October 1994)
9. Ahonen, T., Hadid, A., Pietikäinen, M.: Face recognition with local binary patterns. In: Pajdla, T., Matas, J. (eds.) ECCV 2004. LNCS, vol. 3021, pp. 469–481. Springer, Heidelberg (2004)
10. Gonzalez-Diaz, R., Kropatsch, W., Cerman, M., Lamar, J.: Characterizing configurations of critical points through lbp. In: SYNASC 2014 Workshop on Computational Topology in Image Context (2014)
11. EL Khoury, R., Vandeborre, J.P., Daoudi, M.: 3D mesh Reeb graph computation using commute-time and diffusion distances. In: Proceedings SPIE: Three-Dimensional Image Processing (3DIP) and Applications II, vol. 8290, pp. 82900H–82900H–10 (2012)
12. Bott, R.: Lectures on Morse theory, old and new. Bulletin of the American Mathematical Society 7(2), 331–358 (1982)
13. Doraiswamy, H., Natarajan, V.: Efficient algorithms for computing Reeb graphs. Computational Geometry 42(6-7), 606–616 (2009)
14. Lee, D.T.: Medial axis transformation of a planar shape. IEEE Transactions on Pattern Analysis and Machine Intelligence PAMI-4(4), 363–369 (1982)
15. Bronstein, A.M., Bronstein, M.M., Kimmel, R.: Numerical geometry of non-rigid shapes. Monographs in Computer Science. Springer New York (2009)
16. Sebastian, T., Klein, P., Kimia, B.: Recognition of shapes by editing shock graphs. In: IEEE International Conference on Computer Vision, vol. 1, p. 755. IEEE Computer Society (2001)

Incremental Embedding Within a Dissimilarity-Based Framework

Rachid Hafiane[1(✉)], Luc Brun[2], and Salvatore Tabbone[1]

[1] Université de Lorraine, LORIA-UMR 7503
BP 239 LORIA - Campus scientifique,
54506, Vandœuvre-lès-Nancy Cedex, France
{rachid.hafiane,tabbone}@loria.fr
[2] ENSICAEN
GREYC UMR CNRS 6072
6 boulevard du Maréchal Juin
14050, Caen, Cedex, France
luc.brun@ensicaen.fr

Abstract. Structural pattern recognition methods based on strings or graphs provide a natural encoding of objects' relationships but can usually be combined only with a few set of machine learning methods. This last decade has seen majors advancements aiming to link these two fields. The two majors research fields in this direction concern the design of new graph and string kernels and different explicit embedding schemes of structural data. Explicit embedding of structural data can be combined with any machine learning methods. Dissimilarity representation methods are important because they allow an explicit embedding and the connection with the kernel framework. However these methods require the whole universe to be known during the learning phase and to obtain a Euclidean embedding, the matrix of dissimilarity encoding dissimilarities between any pair of objects should be regularized. This last point somehow violates the usual separation between training and test sets since both sets should be jointly processed and is an important limitation in many practical applications where the test set is unbounded and unknown during the learning phase. Moreover, requiring the whole universe represents a bottleneck for the processing of massive dataset. In this paper, we propose to overcome these limitations following an incremental embedding based on dissimilarity representations. We study in this paper, the pros and cons of two methods, which allow computing implicitly, and separately the embedding of points in the test set and in the learning set. Conclusions are set following experiments performed on different datasets.

1 Introduction

Structural (Graph or String) kernels [1, 2] and dissimilarity based representations [3–5] aim to define either implicitly or explicitly a vectorial representation from structural objects. Both approaches are dual one from the other since a kernel may be understood as a similarity function.

The basic idea of dissimilarity based representations consists in encoding each object of a set by a set of dissimilarities, hence avoiding to base explicitly pattern recognition

© Springer International Publishing Switzerland 2015
C.-L. Liu et al. (Eds.): GbRPR 2015, LNCS 9069, pp. 64–73, 2015.
DOI: 10.1007/978-3-319-18224-7_7

methods on particular features of objects. Two main strategies are usually followed within this framework:

The first strategy [6] consists in defining a set of prototypes and in describing each object by a vector of distances to object's prototypes. This strategy encodes explicitly each object as a vector and any machine learning method may be combined with this representation of objects. However, the optimization of a set of object's prototypes according to a classification or regression task remains an open question mainly solved through different heuristics.

The second strategy [7, 8] encodes each object of a set by its dissimilarities to all other objects of the set. Then the whole dataset may be described by a matrix of dissimilarities between objects. Graph diffusion methods [8] encode then, each object as a node of a graph, edges between pairs of vertices being weighted by a decreasing function of the distance. Note that some edges may be not encoded if their associated distances are greater than a threshold or according to some other heuristics. Classification or regression problems, may then be solved using graph diffusion techniques by mapping iteratively to each node a value deduced from a weighted mean of the values of its neighbors.

One alternative family of methods [7] also based on a matrix of dissimilarities is closely related to multidimensional scaling [9] and consists in interpreting the matrix of dissimilarities as a matrix of squared distances. The problem turn then to attach a vector to each object so that the squared distance between each pair of vector is equal to the dissimilarity between the associated objects. Let us consider a $N \times N$ matrix of dissimilarities D and the space $\mathbb{S}_h^N$ of squared symmetric matrices with a 0 diagonal. We say that D is an Euclidean distance matrix [10] if $D \in \mathbb{S}_h^N$ and :

$$\forall c \in \mathbb{R}^N \ s.t \ e^t c = 0 \text{ and } \|c\| = 1 \text{ we have: } c^t D c \leq 0$$

where $e^t = (1, \ldots, 1)$.

Using usual dissimilarity functions, one can usually assume that $D \in \mathbb{S}_h^N$. However, if D is not an Euclidean distance matrix, it exists no Euclidean space so that we can attach a vector in this space to each object, the squared norm of the difference between two vectors being equal to the dissimilarity. Indeed, in this case the only vectorial space whose topology is compatible with matrix D is a Krein space where the norm of each vector is defined as a difference between the norms of two projections of this vector onto two orthogonal spaces. Such a space may thus allow negative norms and hence negative squared distances. Some methods [3] work nevertheless into this space which is strongly related with non positive definite kernels and several machine learning methods have been adapted to such spaces. Note however that theoretical conditions satisfied by solutions provided by machine learning method in Krein spaces are usually weaker than the ones defined in Euclidean spaces. For example, the SVM algorithm may still be applied on a Krein space but the problem solved by SVM becomes in this case non convex and the interpretation of its result should be modified. One alternative strategy consists in modifying the dissimilarity matrix D in order to insure that it corresponds to an Euclidean distance matrix.

Both graph diffusion and multidimensional scaling approaches are based on a set of dissimilarities and both approaches suppose that both training sets and test sets are available during the learning phase. However, this latter property may be an important drawback in many practical applications where one does not know in advance the composition of the test set which may also be unbounded. Online classification methods are a typical example of such a type of applications. Pekalska [5] proposes to embed incoming data in the Krein space defined on the learning set. This solution involves to work on a non Euclidean space with weaker mathematical properties. Using [11] a regularization of a matrix defined from the learning set and the incoming data may be performed for each new data in order to force our embedding to remain on an Euclidean space. This last solution involves important processing times when the size of the learning set is large. We propose in this paper a solution which allows to perform a regularization only once on the learning set. This solution is compared with a method without regularization but using a learning algorithm devoted to Krein spaces.

Our paper is structured as follows, in section 2 we recall the main definitions and results of dissimilarity representations based on multidimensional scaling. The processing of incoming data is then described in section 3. In section 4 we provide a way to avoid the regularization step and how to learn incrementally the data. Finally, we compare our method to the classical dissimilarity based approach in section 5 and conclusions are drawn in section 6.

2   Explicit Embedding

Given a $n \times n$ symmetric dissimilarity matrix D with a 0 diagonal, dissimilarity representation based on multidimensional scaling and an Euclidean embedding aim at defining a matrix $\tilde{D}$ close to D such that there exists a set of vectors $(x_i)_{i \in \{1, \ldots, n\}}$ with $\tilde{D}_{ij} = \|x_i - x_j\|^2$. The solution to this problem is of course not unique since any translation of vectors x_i would provide a same distance matrix. In order to overcome this problem we perform a centralization of matrix D by considering $S^c = -\frac{1}{2} D^c$, where $D^c = QDQ$ is the definition of the centralization and $Q = I_n - \frac{1}{n} e_n e_n^\intercal$ is the projection matrix on the orthogonal complement of $e_n = (1, \ldots, 1)^T$. Equation $S^c = -\frac{1}{2} D^c$ may also be written explicitly as follows:

$$S_{i,j}^c = -\frac{1}{2} \left[D_{ij} - \frac{1}{n} \sum_{k=1}^{n} D_{jk} - \frac{1}{n} \sum_{l=1}^{n} D_{jl} + \frac{1}{n^2} \sum_{k=1}^{n} \sum_{l=1}^{n} D_{lk} \right] \qquad (1)$$

Such a matrix S^c satisfies:

$$D_{ij}^c = S_{ii}^c + S_{jj}^c - 2S_{ij}^c \qquad (2)$$

If S^c is semi definite positive, its singular value decomposition is equal to $S^c = V \Lambda V^t$ where columns of V encode the eigenvectors of S^c and Λ is a diagonal matrix encoding its positive eigenvalues. Setting $X = V(\Lambda)^{\frac{1}{2}}$, we obtain $S^c = XX^t$. Hence, each element S_{ij}^c of S^c is equal to a scalar product $< x_i, x_j >$ between the lines i and j of X. Equation 2 may thus be interpreted as a classical result on Euclidean norms stating that the squared distance between two vectors is equal to the sum of the squared norms of these vectors minus twice their scalar product. Matrix S^c is thus in this case a matrix of scalar product and may be interpreted as a Gram matrix. The lines of matrix X, $(x_i)_{i \in \{1,...,n\}}$ provide thus a natural embedding of matrix D when S^c is definite positive.

However, if S^c is not definite positive some of its eigenvalues are negative and we can not consider anymore the squared root of Λ. In such a case, the space associated to our initial matrix D is a Krein space (Section 1). This space is defined as a direct sum of a positive eigenspace span by eigenvectors associated to positive eigenvalues and a negative eigenspace span by eigenvectors associated to negative eigenvalues. In order to force the embedding of our data within an Euclidean space, one usual solution [12], called constant shift, consists to shift all eigenvalues of matrix S^c by its lowest (negative) eigenvalue:

$$\tilde{S} = S^c - \lambda_n(S^c)I_n \tag{3}$$

where $\lambda_n(S^c)$ is the minimal eigenvalue of the matrix S^c.

The diagonal shift of the matrix S^c transforms the dissimilarity matrix D in a matrix representing squared Euclidean distances. The resulting embedding of D is defined by (minimal shift theorem):

$$\tilde{D}_{ij} = \tilde{S}_{ii} + \tilde{S}_{jj} - 2\tilde{S}_{ij} \iff \tilde{D} = D - 2\lambda_n(S^c)(e_n e_n^\mathsf{T} - I_n) \tag{4}$$

The difference between distances encoded by D and the ones encoded by $\tilde{D}$ is thus proportional to $2\lambda_n(S^c)$. The constant shift embedding should thus be applied only for small absolute values of $\lambda_n(S^c)$.

3   Incoming Test Data

If the gram matrix, defined in section 2 is not definite positive, the lowest eigenvalue of $\tilde{S}$ using equation 3, is equal to 0. This means that the norm associated to our embedding space is a semi norm where some non identical objects may have a 0 dissimilarity. Let us suppose that we define $\lambda_n(S^c)$ to be slightly lower (up to a value ϵ) than the lowest eigenvalue of S^c. Then all eigenvalues of $\tilde{S}$ become strictly positive and $\tilde{S}$ is a strictly definite positive matrix associated to a norm.

Under the above assumption, let us consider $\tilde{X}$ such that $\tilde{S} = \tilde{X}^\mathsf{T}\tilde{X}$. Each line of vector $\tilde{X}$, as the one of vector X remains associated to the embedding of a particular object. More precisely, each line i of $\tilde{X}$ corresponds to a centered vector $x_i - \bar{x}$ where $\bar{x}$ denotes the mean of $x_i, i \in \{1, \ldots, n\}$. Given a vector $Y = (y_i)_{i \in \{1,...,n\}}$ and the sequence of vectors $(x_i - \bar{x})_{i \in \{1,...,n\}}$, one may wonder if it exists a vector x such that the scalar product between x and $x_i - \bar{x}$ is equal to y_i for any i in $\{1, \ldots, n\}$. In other terms, one may wonder if we may found a vector whose scalar product with each object of our dataset is equal to a predefined value. In matrix form, this problem may be formulated as the existence of vector x solving $Y = \tilde{X}x^t$. This initial problem may be reformulated as follows:

$$Y = \tilde{X} x^t$$
$$X^t Y = \tilde{X}^t \tilde{X} x^t$$
$$X^t Y = \tilde{S} x^t$$

since $\tilde{S}$ is supposed to be strictly definite positive, this problem as a unique solution:

$$x^t = \tilde{S}^{-1} \tilde{X}^t Y. \tag{5}$$

Let us consider a new incoming object o whose dissimilarity with all objects o_i of our training set is encoded by a vector $(d(o, o_i))_{i \in \{1,\ldots,n\}}$. Let us first suppose that this object may be associated to a vector x so that:

$$\forall i \in \{1, \ldots, n\} \; d(o, o_i) = \|x - x_i\|^2 \tag{6}$$

where x_i corresponds to the embedding of o_i.

Using the vectorial embedding x of object o we obtain:

$$d(o, o_i) = \|x - x_i\|^2 = \|x\|^2 + \|x_i\|^2 - 2 < x, x_i >$$

Hence:

$$< x, x_i > = \frac{1}{2}(\|x\|^2 + \|x_i\|^2 - d(o, o_i)) \tag{7}$$

Let $\bar{x}$ be the mean of vectors $x_i, i \in \{1, \ldots, n\}$ and let us consider the scalar product: $< x - \bar{x}, x_i - \bar{x} >$. We have by basic calculus:

$$\begin{aligned}
< x - \bar{x}, x_i - \bar{x} > &= < x, x_i > - < x, \bar{x} > \\
&\quad - < \bar{x}, x_i > + < \bar{x}, \bar{x} > \\
&= < x, x_i > - \frac{1}{n} \sum_{k=1}^{n} < x, x_k > \\
&\quad - \frac{1}{n} \sum_{k=1}^{n} < x_i, x_k > \\
&\quad + \frac{1}{n^2} \sum_{k=1, l=1}^{n} < x_k, x_l >
\end{aligned}$$

Using equation 7 and again some basic calculus, scalar products in the above equation may be replaced by dissimilarity values:

$$< x - \bar{x}, x_i - \bar{x} > = -\frac{1}{2}\left(d(o, o_i) - \frac{1}{n}\sum_{k=1}^{n} d(o, o_k) \right.$$

$$-\frac{1}{n}\sum_{k=1}^{n} d(o_i, o_k)$$

$$\left. +\frac{1}{n^2}\sum_{k=1,l=1}^{n} d(o_k, o_l) \right)$$

(8)

where $d(o_i, o_k)$ refers to the element D_{ik} of matrix D if S^c is definite positive and to $\tilde{D}_{ik}$ otherwise.

This last equation corresponds to equation 1 but is defined between an incoming data o encoded by x and all the objects of our dataset. The important point to underline, is that the only hypothesis made to obtain equation 8 is that object o may be associated to an embedding x satisfying equation 6. A straightforward extension of equation 8 consists to compute the squared norm of $x - \bar{x}$. In such a case we have to replace $d(o_i, o_k)$ by $d(o, o_k)$ in the second sum of equation 8.

Let us show that an embedding of o satisfying eq. 6 exists. According to equation 5, since $\tilde{S}$ is strictly definite positive, it exists an unique vector $x - \bar{x}$ satisfying (eq. 8) for any $i \in \{1, \ldots, n\}$. If such a vector exists we have:

$$\|x - x_i\|^2 = < x - \bar{x}, x - \bar{x} > + < x_i - \bar{x}, x_i - \bar{x} >$$
$$-2 < x - \bar{x}, x_i - \bar{x} >$$

Basic calculus using equation 8 show that this last equation is equal to $d(o, o_i)$. Hence, considering the right part of equation 8 we can insure the existence of a vector x whose squared distance to any vector x_i is equal to $d(o, o_i)$.

The matrix of scalar products $\tilde{S}$ may be interpreted as a Gram matrix used within the kernel framework. Moreover, the values of the scalar products computed using equation 8 may be interpreted as kernel values between our input data o and the objects $(o_i)_{i\in\{1,\ldots,n\}}$ of our train set. Given the matrix $\tilde{S}$ we can thus apply all kernel based machine learning methods such as the SVM, the kernel PCA or the kernel ridge regression. Using this approach, for each input data, we only have to evaluate its similarities to the data of our training set using equation 8. The explicit embedding of an incoming data, is never computed but its existence is insured by equation 5.

4   Indefinite Kernels

The strategy proposed in Section 3 consists in applying a regularization step to the Gram matrix encoding the similarity of the whole learning set. Each data of the test set is then compared to this regularized learning set using any kernel method such as

Table 1. Main statistics of our datasets

Database	mean d_{ij}	$max\lambda_i(S^c)$	$min\lambda_i(S^c)$	$\sum_{\lambda_i<0}\lambda_i(S^c)$	$\sum_{\lambda_i>0}\lambda_i(S^c)$
AIDS (250)	50.5	3267.8	-5.23	-99.29	6385.1
LETTER LOW(750)	3.8	223.8	-35.4	-955.88	2365.27
LETTER MED(750)	4.3	179.8	-22.9	-1602.88	3217.43
LETTER HIGH(750)	4.6	188.85	-18.5	-1609.5	3332.9

SVM. This strategy allows to work on an Hilbert space whose mathematical properties are well stated and to apply a large set of machine learning methods.

This strategy has however two limitations. Firstly, as mentioned in Section 2 a regularization step should be applied only if the absolute value of the lowest eigenvalue of the Gram matrix S^c is low. Using this strategy with high values of the lowest eigenvalue lead to perform a uniform translation of all distances between not identical objects (eq. 4). Such a translation may thus alter the initial metric defined by the dissimilarity function. Secondly, using very large datasets, the initial learning set may be large. This last point may prevent us to encode all pairwise dissimilarities in a matrix which should be regularized.

In order to solve these problem and to test the limits of the proposed approach, we propose as in [5] a second strategy which consists in avoiding the regularization step. We thus work in a Krein space and combine our non definite Gram matrix S^c with a machine learning method designed for this kind of space such as the PSVM [13].

However, unlike [5], we propose to extend this framework to online learning. Given a dataset of n datum encoded by a similarity matrix S^c and a new labeled data associated to a vector x our aim consists to obtain a new similarity matrix S'^c encoding similarities between $n + 1$ datum. In order to avoid to reconsider the whole matrix S^c we neglect the modification of the mean induced by the addition of a new data. Hence, the part of matrix S'^c encoding similarities between the n previous data remains unchanged. Moreover, the similarities $< x - \overline{x}, x_i - \overline{x} >$ between the new data and the n initial data is provided by equation 8. Finally, the value of $< x - \overline{x}, x - \overline{x} >$ is provided by the modification of equation 8 described in Section 3. The new matrix S'^c is thus build from matrix S^c by adding one line and one column defined as:

$$\forall i \in \{1, \ldots, n\}\ (S'^c)_{n+1,i} = (S'^c)_{i,n+1} \qquad = < x - \overline{x}, x_i - \overline{x} > \text{ and}$$
$$(S'^c)_{n+1,n+1} = < x - \overline{x}, x - \overline{x} > \tag{9}$$

5 Experimental Section

In order to evaluate the two proposed strategies we evaluated each strategy both on a bi-classification problem and on a multi-class problem using a one against all method. The two datasets used for these experiments[1] are both composed of labeled graphs:

- The AIDS dataset describes in [14] is composed of 2000 chemical compounds. These chemical compounds have been screened as active or inactive against HIV

---

[1] All the datasets in this section are available on the IAPR-TC 15 webpage:
http://iapr-tc15.greyc.fr/links.html

and are split into three different sets. A train set is composed of 250 compounds used to train the classifiers, 250 compounds to validate the classifier's parameters and 1500 compounds for the test.

– The LETTER dataset set is divided in three subsets according to the level of added noise (LOW, MED and HIGH) each subset is composed of 2250 graphs representing 15 different geometric letters. These subsets are decomposed into a training set, a validation set and a test set each set being composed of 750 elements.

For each dataset, the dissimilarity matrices, are obtained using an approximate graph edit distance [15]. The main statistics of each dataset are displayed in Table 1. More precisely, Table 1 provides the mean distance computed on the learning set (column 2), the maximal and minimal eigenvalues of matrices S^c (columns 3 and 4) and the sum of negative (respectively positive) eigenvalues of matrices S^c in column 5 (respectively 6). These last two columns allow to measure the amount of information of the dataset associated to positive and negative eigenvalues.

As shown in Table 1, the lowest eigenvalue of matrix S^c on the learning set of AIDS dataset is approximately equal to 10 % of the mean distance. This last point indicates that the regularization of matrix S^c should not modify drastically the initial distances. This point is confirmed by the two last columns on this lines which show that most of the information on the AIDS dataset is contained in the positive eigenspace.

Table 2 shows the classification accuracy obtained on the test set of AIDS dataset using a regularization of matrix S^c (Section 3) combined with the SVM method [16] and the results obtained without any regularization (Section 4) using PSVM algorithm [13]. As expected the regularization does not induce a loss of information and both results with and without regularization are equivalent. Note however, that the use of a definite positive matrix S^c allows to use a larger set of machine learning methods. The following lines of Table 2 illustrate the results obtained by our online learning method (Section 4). For each line, from line 2 to line 5, we select randomly a subset of the learning set whose size is indicated on the first column. The Gram matrix S^c corresponding to this initial set is then completed using (see eq. 9) up to the size 250×250 which corresponds to the whole learning set. Classification accuracy show that our completion of the Gram matrix does not induces a significant loss of accuracy.

As shown in Table 1, the ratio between the minimal eigenvalue and the mean distance varies on the LETTER dataset between 4 and 9. In such a case, regularizing matrix S^c would lead to shift our matrix of dissimilarities by a value greater than most of its dissimilarity values (see eq. 4) hence inducing an important loss of information.

Table 2. Classification results on AIDS dataset and kernels computation time. Each learning set is completed incrementally up to the size 250X250. 1500 compounds are used for the test set.

Size of the learning set	Regularized S^c	Non regularized S^c	Computation time (ms)
250	99.67%	99.67%	1264
200		99.67%	1202
150		99.60%	817
100		99.53%	317
50		99.67%	118

This last point is confirmed by the last column of Table 1 which show that about 30% of the information of these datasets is contained into the negative eigenspace.

As discussed in Section 3 in such a case we can not perform a regularization without an important loss of information. Therefore, we only present in Table 3 the results obtained by our method without regularization (Section 4). For each dataset, the first line represent the classification accuracy on the test set computed with a matrix S^c defined on the whole learning set. We can notice an expected decrease of the classification accuracy according to the complexity of the datasets. The remaining lines show the same classification accuracy for a matrix S^c defined on a subset of the learning set chosen randomly and completed using eq. 9 in order to obtain a 750×750 matrix defined on the whole learning set. As already noticed on AIDS dataset, this incremental completion of matrix S^c does not induce a significant decrease of the classification accuracy.

Table 3. Classification results on LETTER datasets and kernels computation time. Each learning set is completed incrementally up to the size 750X750. 750 letters are used in each subset for the test set.

Databases	Size of the initial learning sets	Non regularized S^c	Computation time (ms)
LETTER	750	93.87%	6803
LOW	600	93.87%	4567
	450	93.87%	2187
	300	93.60%	1852
	150	93.20%	928
LETTER	750	86.67%	6996
MED	600	86.80%	3781
	450	86.27%	2448
	300	86.80%	2167
	150	86.67%	769
LETTER	750	76.27%	6390
HIGH	600	76.93%	3759
	450	75.73%	2293
	300	76.13%	1632
	150	76.13%	1011

6   Conclusion and Perspectives

We have proposed in this paper two methods within the dissimilarity based pattern recognition framework which allow to differentiate the learning set from the test set. Our first method performs a SVD on the learning set in order to obtain a positive definite Gram matrix which may be combined with many machine learning methods in order to test incoming data without further processing. This method may however not work if a large part of the information contained in the dataset lies in the negative eigen space. In this case our second method working in Krein space avoids any regularization and may be used for online learning. Both methods avoid to update a singular value decomposition (SVD) for each incoming data. Such a SVD is a bottleneck for the processing of massive dataset.

References

1. Mahé, P., Vert, J.P.: Graph kernels based on tree patterns for molecules. Machine Learning 75, 3–35 (2008)
2. Gaüzère, B., Brun, L., Villemin, D.: Two New Graphs Kernels in Chemoinformatics. Pattern Recognition Letters 33, 2038–2047 (2012)
3. Calana, Y.P., Orozco-Alzate, M., Reyes, E.B.G., Duin, R.P.W.: Selecting feature lines in generalized dissimilarity representations for pattern recognition. Digital Signal Processing 23, 902–911 (2013)
4. Pekalska, E., Duin, R.: Classifiers for dissimilarity-based pattern recognition. In: Proceedings of the 15th International Conference on Pattern Recognition, vol. 2, pp. 12–16 (2000)
5. Pekalska, E., Paclik, P., Duin, R.P.W.: A generalized kernel approach to dissimilarity-based classification. J. Mach. Learn. Res. 2, 175–211 (2002)
6. Riesen, K., Neuhaus, M., Bunke, H.: Graph embedding in vector spaces by means of prototype selection. In: Escolano, F., Vento, M. (eds.) GbRPR 2007. LNCS, vol. 4538, pp. 383–393. Springer, Heidelberg (2007)
7. Jouili, S., Tabbone, S.: Graph Embedding Using Constant Shift Embedding. In: Ünay, D., Çataltepe, Z., Aksoy, S. (eds.) ICPR 2010. LNCS, vol. 6388, pp. 83–92. Springer, Heidelberg (2010), The original publication is available at http://www.springerlink.com
8. Lezoray, O., Elmoataz, A., Bougleux, S.: Graph regularization for color image processing. Computer Vision and Image Understanding (CVIU) 107, 38–55 (2007)
9. Cox, T.F., Cox, M.: Multidimensional Scaling, 2nd edn. Chapman and Hall/CRC (2000)
10. Dattorro, J.: Convex Optimization & Euclidean Distance Geometry. Meboo Publishing USA (2011)
11. Chin, T.J., Suter, D.: Incremental kernel principal component analysis. IEEE Transactions on Image Processing 16, 1662–1674 (2007)
12. Roth, V., Laub, J., Kawanabe, M., Buhmann, J.M.: Optimal cluster preserving embedding of nonmetric proximity data. IEEE Trans. Pattern Anal. Mach. Intell. 25, 1540–1551 (2003)
13. Hochreiter, J., Obermayer, K.: Support vector machines for dyadic data. Neural Comput. 18, 1472–1510 (2006)
14. Riesen, K., Bunke, H.: IAM graph database repository for graph based pattern recognition and machine learning. In: da Vitoria Lobo, N., Kasparis, T., Roli, F., Kwok, J.T., Georgiopoulos, M., Anagnostopoulos, G.C., Loog, M. (eds.) SSPR&SPR 2008. LNCS, vol. 5342, pp. 287–297. Springer, Heidelberg (2008)
15. Raveaux, R., Burie, J.C., Ogier, J.M.: A graph matching method and a graph matching distance based on subgraph assignments. Pattern Recognition Letters 31, 394–406 (2010)
16. Chang, C.C., Lin, C.J.: LIBSVM: A library for support vector machines. ACM Transactions on Intelligent Systems and Technology 2, 27:1–27:27 (2011)

Graph Matching

A First Step Towards Exact Graph Edit Distance Using Bipartite Graph Matching

Miquel Ferrer[1]([⊠]), Francesc Serratosa[2], and Kaspar Riesen[1]

[1] Institute for Information Systems, University of Applied Sciences and Arts,
Riggenbachstrasse 16, Olten, 4600, Switzerland
miquel.ferrer@fhnw.ch
[2] Departament d'Enginyeria Informàtica i Matemàtiques, Universitat Rovira i Virgili,
Avda. Països Catalans 26, 43007 Tarragona, Spain

Abstract. In recent years, a powerful approximation framework for graph edit distance computation has been introduced. This particular approximation is based on an optimal assignment of local graph structures which can be established in polynomial time. However, as this approach considers the local structural properties of the graphs only, it yields sub-optimal solutions that overestimate the true edit distance in general. Recently, several attempts for reducing this overestimation have been made. The present paper is a starting point towards the study of sophisticated heuristics that can be integrated in these reduction strategies. These heuristics aim at further improving the overall distance quality while keeping the low computation time of the approximation framework. We propose an iterative version of one of the existing improvement strategies. An experimental evaluation clearly shows that there is large space for further substantial reductions of the overestimation in the existing approximation framework.

1 Introduction

Graph edit distance [1, 2] is one of the most flexible and versatile approaches to error-tolerant graph matching. In particular, graph edit distance is able to cope with directed and undirected, as well as with labeled and unlabeled graphs. In addition, no constraints have to be considered on the alphabets for node and/or edge labels. Moreover, through the concept of cost functions graph edit distance can be adapted and tailored to diverse applications [3, 4]. An extensive survey about graph edit distance can be found in [5].

The major drawback of graph edit distance is its high computational complexity that restrict its applicability to graphs of rather small size. In fact, graph edit distance belongs to the family of *quadratic assignment problems* (QAPs), which belong to the class of $\mathcal{NP}$-*complete* problems. Therefore, exact computation of graph edit distance can be solved in exponential time complexity only.

In recent years, a number of methods addressing the high computational complexity of graph edit distance have been proposed (e.g. [6–9]). Beyond these works, an algorithmic framework based on bipartite graph matching has been

© Springer International Publishing Switzerland 2015
C.-L. Liu et al. (Eds.): GbRPR 2015, LNCS 9069, pp. 77–86, 2015.
DOI: 10.1007/978-3-319-18224-7_8

introduced recently [10, 11]. The main idea behind this approach is to convert
the difficult problem of graph edit distance to a *linear sum assignment problem*
(LSAP). LSAPs basically constitute the problem of finding an optimal assignment between two independent sets of entities, for which a collection of polynomial algorithms exists [12]. In [10, 11] the LSAP is formulated on the sets of
nodes including local edge information. The main advantage of this approach is
that it allows the approximate computation of graph edit distance in a substantially faster way than traditional methods. However, during the node assignment
only local instead of global structural information is taken into account. Hence,
this might lead to incorrect node assignments compared with an exact matching
and thus, the derived edit distance is equal to, or larger than, the exact graph
edit distance.

In order to overcome this problem and reduce the overestimation of the true
graph edit distance, a variation of the original framework [10] has been proposed in [13]. Given the initial assignment found by the bipartite framework,
the main idea is to introduce a post-processing step such that the number of
incorrect assignments is decreased (which in turn reduces the overestimation).
The proposed post-processing varies the original node assignment by systematically swapping the target nodes of two node assignments. In order to search the
space of assignment variations a *beam search* (i.e. a tree search with pruning) is
used. One of the most important observations derived from [13] is that given an
initial node assignment, one can substantially reduce the overestimation using
this local search method. Yet, beam search is sub-optimal in the sense of possibly
pruning the optimal solution in an early stage of the search process.

Now the crucial question arises, how the space of assignment variations could
be explored such that promising parts of the search tree are not (or at least not
too early) pruned. In [13] the initial assignment is systematically varied without
using any kind of heuristic or additional information to keep the best potential assignment unpruned in the tree. In particular it is not taken into account
that certain nodes and/or local assignments have greater impact than other on
the graph edit distance approximation and should thus be considered first in
the beam search process. Clearly, considering more important node assignments
and/or nodes in an early stage of the beam search process might reduce the risk
of pruning the optimal assignment. In this sense we argue that the introduction
of procedures and heuristics that guide the order in which the assignments are
varied during beam search is a rewarding line of research.

The main objective of the present paper is to start the investigation towards
new heuristics that improve the overall quality of the node assignment. In particular, we propose an iterative version of [13] to derive randomized permutations
of the original mapping which serve as starting point for beam search improvements. Hence, the present paper introduces a heuristic in order to answer the
general question to what extent the ordering of the nodes affects the mapping
quality.

Next, in Section 2, the original bipartite framework for graph edit distance approximation [10] as well as its recent extension [13], named *BP-Beam*, are summarized. In Section 3 our novel iterative version of *BP-Beam* is described. An experimental evaluation on diverse data sets is carried out in Section 4. Finally, in Section 5 we draw conclusions and outline some possible tasks and extensions for future work.

2 Approximate Graph Edit Distance Computation

Given two graphs, g_1 and g_2, the basic idea of graph edit distance is to transform g_1 into g_2 using edit operations, namely, *insertions*, *deletions*, and *substitutions* of both nodes and edges. The substitution of two nodes u and v is denoted by $(u \rightarrow v)$, the deletion of node u by $(u \rightarrow \epsilon)$, and the insertion of node v by $(\epsilon \rightarrow v)^1$. A sequence of edit operations $e_1, \ldots, e_k$ that transform g_1 completely into g_2 is called an edit path between g_1 and g_2.

To find the most suitable edit path out of all possible edit paths between two graphs, a cost measuring the strength of the corresponding operation is introduced. The edit distance between two graphs g_1 and g_2 is then defined by the minimum cost edit path between them. Exact computation of graph edit distance is usually carried out by means of a tree search algorithm (e.g. A*) which explores the space of all possible mappings of the nodes and edges of the first graph to the nodes and edges of the second graph.

2.1 Bipartite Graph Edit Distance Approximation

The computational complexity of exact graph edit distance is exponential in the number of nodes of the involved graphs. That is considering n nodes in g_1 and m nodes in g_2, the set of all possible edit paths contains $O(n^m)$ solutions to be explored. This means that for large graphs the computation of edit distance is intractable. In order to reduce its computational complexity, in [10], the graph edit distance problem is transformed into a linear sum assignment problem (LSAP). To this end, based on the node sets $V_1 = \{u_1, \ldots, u_n\}$ and $V_2 = \{v_1, \ldots, v_m\}$ of g_1 and g_2 respectively, a cost matrix C is first established as follows:

$$
C = \begin{bmatrix}
c_{11} & c_{12} & \cdots & c_{1m} & c_{1\epsilon} & \infty & \cdots & \infty \\
c_{21} & c_{22} & \cdots & c_{2m} & \infty & c_{2\epsilon} & \cdots & \infty \\
\vdots & \vdots & \ddots & \vdots & \vdots & \vdots & \ddots & \vdots \\
c_{n1} & c_{n2} & \cdots & c_{nm} & \infty & \infty & \cdots & c_{n\epsilon} \\
c_{\epsilon 1} & \infty & \cdots & \infty & 0 & 0 & \cdots & 0 \\
\infty & c_{\epsilon 2} & \cdots & \infty & 0 & 0 & \cdots & 0 \\
\vdots & \vdots & \ddots & \infty & \vdots & \vdots & \ddots & \vdots \\
\infty & \infty & \cdots & c_{\epsilon m} & 0 & 0 & \cdots & 0
\end{bmatrix}
$$

---
[1] Similar notation is used for edges.

Entry c_{ij} denotes the cost of a node substitution $(u_i \rightarrow v_j)$, $c_{i\epsilon}$ denotes the cost of a node deletion $(u_i \rightarrow \epsilon)$, and $c_{\epsilon j}$ denotes the cost of a node insertion $(\epsilon \rightarrow v_j)$. The left upper corner of the cost matrix represents the costs of all possible node substitutions, the diagonal of the right upper corner the costs of all possible node deletions, and the diagonal of the bottom left corner the costs of all possible node insertions. In each entry c_{ij}, not only the cost of node operation is taken into account but also the minimum sum of edge edit operation costs, implied by the corresponding node operation. That is, the matching cost of the local edge structure is encoded in the individual entries of C.

In a second step of [10], an assignment algorithm is applied to the square cost matrix $C = (c_{ij})$ in order to find the minimum cost assignment of the nodes (and their local edge structure) of g_1 to the nodes (and their local edge structure) of g_2. Note that this task exactly corresponds to an instance of an LSAP and can thus be optimally solved in polynomial time by several algorithms [12].

Any of the LSAP algorithms will return a permutation $(\varphi_1, \ldots, \varphi_{n+m})$ of the integers $(1, 2, \ldots, (n+m))$, which minimizes the overall mapping cost $\sum_{i=1}^{(n+m)} c_{i\varphi_i}$. This permutation corresponds to the mapping

$$\psi = \{u_1 \rightarrow v_{\varphi 1}, u_2 \rightarrow v_{\varphi 2}, \ldots, u_{m+n} \rightarrow v_{\varphi_{m+n}}\}$$

of the nodes of g_1 to the nodes of g_2. Note that ψ does not only include node substitutions $(u_i \rightarrow v_j)$, but also deletions and insertions $(u_i \rightarrow \epsilon)$, $(\epsilon \rightarrow v_j)$ and thus perfectly reflects the definition of graph edit distance (mappings of the form $(\epsilon \rightarrow \epsilon)$ can be dismissed, of course). Hence, mapping ψ can be interpreted as partial edit path between g_1 and g_2, which considers operations on nodes only.

In a third step, the partial edit path ψ between g_1 and g_2 is completed with respect to the edges. This can be accomplished since edge edit operations are implied by edit operations on their adjacent nodes. That is, whether an edge is substituted, deleted, or inserted, depends on the edit operations performed on its adjacent nodes. The total cost $d_{\langle \psi \rangle}(g_1, g_2)$ of the completed edit path between graphs g_1 and g_1 is finally returned as approximate graph edit distance. We refer to this graph edit distance approximation algorithm as $BP(g_1, g_2)$.

2.2   Improving the Approximation Using Beam Search

Several experimental evaluations indicate that the suboptimality of BP, i.e. the overestimation of the true edit distance, is very often due to a few incorrectly assigned nodes in ψ with respect to the optimal edit path. The extension presented in [13] ties in at this observation. In particular, the node assignment ψ is used as a starting point for a subsequent search in order to improve the quality of the distance approximation.

In [13], the original node assignment ψ is systematically varied by swapping the target nodes v_{φ_i} and v_{φ_j} of two node assignments $(u_i \rightarrow v_{\varphi_i}) \in \psi$ and $(u_j \rightarrow v_{\varphi_j}) \in \psi$. For each swap it is verified whether (and to what extent) the derived distance approximation stagnates, increases or decreases. For a systematic variation of mapping ψ a tree search is used.

Algorithm 1. *BP-Beam*(g_1, g_2, ψ, b)

---

1. $d_{best} = d_{\langle\psi\rangle}(g_1, g_2)$
2. Initialize $open = \{(\psi, 0, d_{\langle\psi\rangle}(g_1, g_2))\}$
3. **while** $open$ is not empty **do**
4.   Remove first tree node in $open$: $(\psi, q, d_{\langle\psi\rangle}(g_1, g_2))$
5.   **for** $j = (q+1), \ldots, (m+n)$ **do**
6. $\psi' = \psi \setminus \{u_{q+1} \to v_{\varphi_{q+1}}, u_j \to v_{\varphi_j}\} \cup \{u_{q+1} \to v_{\varphi_j}, u_j \to v_{\varphi_{q+1}}\}$
7.   Derive approximate edit distance $d_{\langle\psi'\rangle}(g_1, g_2)$
8. $open = open \cup \{(\psi', q+1, d_{\langle\psi'\rangle}(g_1, g_2))\}$
9.   **if** $d_{\langle\psi'\rangle}(g_1, g_2) < d_{best}$ **then**
10. $d_{best} = d_{\langle\psi'\rangle}(g_1, g_2)$
11.   **end if**
12.   **end for**
13.   **while** size of $open > b$ **do**
14.   Remove tree node with highest approximation value $d_{\langle\psi\rangle}$ from $open$
15.   **end while**
16. **end while**
17. **return** d_{best}

The tree nodes in the search procedure correspond to triples $(\psi, q, d_{\langle\psi\rangle})$, where ψ is a certain node assignment, q denotes the depth of the tree node in the search tree and $d_{\langle\psi\rangle}$ is the approximate distance value corresponding to ψ. The root node of the search tree refers to the optimal node assignment ψ found by *BP*. Hence, the root node (with depth $= 0$) is given by the triple $(\psi, 0, d_{\langle\psi\rangle})$. Subsequent tree nodes $(\psi', q, d_{\langle\psi'\rangle})$ with depth $q = 1, \ldots, (m+n)$ contain node assignments ψ' with swapped element $(u_q \to v_{\varphi_q})$.

As usual in tree search based methods, a set *open* is employed that holds all of the unprocessed tree nodes. The tree nodes in *open* are kept sorted in ascending order according to their depth in the search tree (known as *breadth-first search*). As a second order criterion the approximate edit distance $d_{\langle\psi\rangle}$ is used.

The extended framework with the tree search based improvement is given in Alg. 1. As long as *open* is not empty, we retrieve (and remove) the triple $(\psi, q, d_{\langle\psi\rangle})$ at the first position in *open*, generate the successors of this specific tree node and add them to *open*. To this end all pairs of node assignments $(u_{q+1} \to v_{\varphi_{q+1}})$ and $(u_j \to v_{\varphi_j})$ with $j = (q+1), \ldots, (n+m)$ are individually swapped resulting in two new assignments $(u_{q+1} \to v_{\varphi_j})$ and $(u_j \to v_{\varphi_{q+1}})$. In order to derive node mapping ψ' from ψ, the original node assignment pair is removed from ψ and the swapped node assignment is added to ψ'. Since index j starts at $(q+1)$ we also allow that a certain assignment $u_{q+1} \to v_{\varphi_{q+1}}$ remains unaltered at depth $(q+1)$ in the search tree.

Since every tree node in our search procedure corresponds to a complete solution and the cost of these solutions neither monotonically decrease nor increase with growing depth in the search tree, we need to buffer the best possible distance approximation found during the tree search in d_{best} (which is returned as soon as *open* is empty)

As stated before, given a mapping ψ from *BP*, the derived edit distance over-estimates the true edit distance in general. This overestimation is due to some incorrect node mappings in ψ. Hence, the objective of any post-processing should be to find a variation ψ' of the original mapping ψ such that $d_{\langle\psi'\rangle} < d_{\langle\psi\rangle}$. However, the search space of all possible permutations of ψ contains $(n + m)!$ possibilities, making an exhaustive search (starting with ψ) both unreasonable and intractable. Therefore, only the b assignments with the lowest approximate distance values are kept in *open* at all time (known as *beam search*). Note that parameter b can be used as trade-off parameter between run time and approximation quality. That is, it can be expected that larger values of b lead to both better approximations and increased run time (and vice versa). From now on we refer to this variant of the approximation framework as *BP-Beam(g_1, g_2, ψ, b)*.

3   Iterative BP-Beam

Note that the the successors of tree node $(\psi, q, d_{\langle\psi\rangle})$ are generated in fixed order in *BP-Beam*. In particular, the assignments of the original node matching ψ are processed according to the depth q of the current search tree node. That is, at depth q the assignment $(u_q \to v_{\varphi_q})$ is processed and swapped with other assignments. Note that beam search prunes quite large parts of the tree during the search process. Hence, processing correct node assignments at the top of the search tree is somewhat useless and moreover runs the risk of potentially pruning crucial parts of the tree at an early stage of the search. That is, the fixed order processing, which does not take any information about the individual node assignments into account, is a clear drawback of the procedure described in [13].

Clearly, it would be highly favorable to process important node assignments as early as possible in the tree search. Our hypothesis is that there should exist some heuristics that indicate which node assignments of ψ are the most important or critical ones and should thus be processed first. Finding such heuristics that indicate the impact of a single node assignment on the approximation quality turns out to be a highly non-trivial task. Moreover, it is not yet proven whether the order of the assignment processing actually has a great impact on the resulting distance quality.

As a starting point towards the question of whether or not the ordering of the assignment processing in *BP-Beam* has great influence on the quality of the distance approximation, we propose a procedure with random re-orderings of the individual assignments. Thus, given a mapping ψ obtained from *BP*, we propose to perform several iterations over *BP-Beam*. In every iteration the original mapping ψ is randomly reordered and fed into *BP-Beam*. This leads to an iterative version of *BP-Beam* called *IBP-Beam* from now on. *IBP-Beam* takes two parameters, namely the number of iterations k and the beam size b.

The algorithm *IBP-Beam* is given in Alg. 2. The first three lines correspond to the three major steps of the original approximation. Then, the main loop is carried out k times. In each iteration a newly ordered assignment ψ' is generated by randomly permuting the original assignment ψ derived from *BP*. Then, the assignment and the corresponding distance are possibly improved using *BP-Beam* taking ψ' as starting point for the tree search. Whenever the *BP-Beam*

Algorithm 2. IBP-Beam(g_1,g_2, k, b)

---

1. Build cost matrix $C = (c_{ij})$ according to the input graphs g_1 and g_2
2. Compute optimal node assignment $\psi = \{u_1 \rightarrow v_{\varphi 1}, \ldots, u_{m+n} \rightarrow v_{\varphi_{m+n}}\}$ on C
3. $d_{best} = d_{\langle \psi \rangle}(g_1, g_2)$
4. $i = 1$
5. **while** $i \leq k$ **do**
6. $\psi' = RandomPermutation(\psi)$
7. $d_{beam} = BP\text{-}Beam(g_1, g_2, \psi', b)$
8. $d_{best} = \min\{d_{best}, d_{beam}\}$
9. $i++$
10. **end while**
11. **return** d_{best}

---

is able to further decrease the distance approximation, the best solution d_{best} is replaced by the novel approximation.

Clearly, further reductions of the overestimation with this random iterative procedure, would highly encourage the hypothesis that there might be heuristics that would solve the same task of reordering in a deterministic manner (in particular when the number of iterations k needed remains small).

4 Experimental Evaluation

The goal of the experimental evaluation is to verify whether the proposed extension is able to reduce the overestimation of graph edit distance approximation returned by *BP* and in particular *BP-Beam*, and how the iterative process affects the computation time. Three data sets from the IAM graph database repository involving molecular compounds (AIDS), fingerprint images (Fingerprint), and symbols from architectural and electronic drawings (GREC) are used to carry out this experimental part. For further details about these data sets we refer to [14]. For all data sets, subsets of 100 graphs are randomly selected on which 10,000 pairwise graph edit distance computations are performed.

4.1 Impact of Meta Parameter b and k

In this first experiment we aim at researching the impact of the two parameters b and k on *IBP-Beam*. To this end we perform several executions of the *IBP-Beam* with $b, k \in \{1, 5, 10, 15, 20\}$, leading to 25 different distance approximations. We compute the sum of the difference between the exact distance (A^*) and the distance obtained by *IBP-Beam*. Figure 1 shows such sum of differences as a function of the number of iterations k (x-axis) and the beam size b (y-axis).

First, we observe that the sum of differences is monotonically reduced as long as k and b are increased. We can also observe that we are able to obtain distance values very close to the exact distance on all data sets. Finally (and probably most importantly), we note that the major part of the reduction in the sum of differences is already obtained with values of k and b of 5. Further increases of both parameters lead to relatively small further reductions of the overestimation.

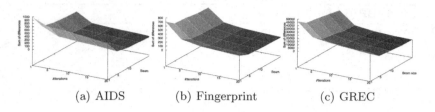

(a) AIDS   (b) Fingerprint   (c) GREC

Fig. 1. Sum of difference between A^* and *IBP-Beam(k,b)* as a function of the number of iterations k (x-axis) and the beam size b (y-axis)

4.2   Relative Overestimation and Computation Time

Next we measure the mean relative overestimation ϕo [%] and the mean computation time ϕt [ms] for all algorithms (see Table 1). The mean relative overestimation ϕo of a certain approximation is computed as the relative difference to the sum of distances returned by A^*. The relative overestimation of A^* is thus zero and the value of ϕo for *BP* is taken as reference value and corresponds to 100%. The computation times ϕt measures the average matching time for a pair of graphs. Note that for both *BP-Beam* and *IBP-Beam* the beam size b is fixed to 5 (thus this parameter is not shown in the Table 1, but only k).

Table 1. The mean relative overestimation of the exact distance (ϕo) in %, and the mean run time for one matcing (ϕt) in ms. for each data set and for a given algorithm. The beam size for *BP-Beam* and *IBP-Beam* algorithm is set to 5, and parameter k is varied from 5 to 20 for *IBP-Beam*.

Algorithm	AIDS		Fingerprint		GREC	
	ϕo	ϕt	ϕo	ϕt	ϕo	ϕt
A^*	0.00	25750.22	0.00	31645.08	0.00	7770.81
BP	100.00	0.28	100.00	0.35	100.00	0.27
BP-Beam	15.09	1.82	24.57	1.45	16.98	2.62
IBP-Beam(5)	9.27	7.81	8.63	5.70	8.53	12.01
IBP-Beam(10)	6.11	15.27	6.11	10.98	5.72	23.57
IBP-Beam(15)	4.85	22.79	5.21	16.14	4.57	35.81
IBP-Beam(20)	4.26	30.41	4.51	20.94	3.73	47.19

Regarding the overestimation ϕo we observe a substantial improvement of the distance quality using *BP-Beam* rather than *BP*. For instance, on the AIDS data the overestimation is reduced by 85% (similar results are obtained on the other data sets). By using *IBP-Beam* further substantial reductions of the overestimation are possible on all data sets. For instance, on the AIDS data set using *IBP-Beam* with $k = 5$ rather than *BP-Beam* enables a reduction from 15.09% to 9.27% (similar or even better results are obtained on the other data sets). Increasing the number of iterations further decreases the overestimation such that a distance accuracy very near to the exact edit distance is possible with our novel approach. These substantial reductions of the overestimation from *BP* to *BP-Beam* and from *BP-Beam* to *IBP-Beam* can also be seen in Figure 2 where for each pair of graphs the exact distance (x-axis) is plotted vs. the distance obtained by an approximation algorithm (y-axis). In fact, for *IBP-Beam* the line-like scatter plot along the diagonal suggests that the approximation is very near to the optimal distance.

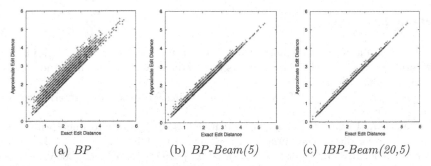

      (a) *BP*   (b) *BP-Beam(5)*   (c) *IBP-Beam(20,5)*

Fig. 2. Exact (x-axis) vs. approximate (y-axis) edit distance on the AIDS dataset computed with (a) *BP*, (b) *BP-Beam(5)*, (c) *IBP-Beam(20,5)*

Regarding the computation time ϕt we can report that *BP* provides the lowest computation time on all data sets (approximately 0.3ms per matching on all data sets). Yet, remember that this fast matching time is at the expense of the highest overestimation. *BP-Beam* increases the computation time to approximately 2ms per matching and *IBP-Beam* further increases the average run time to several milliseconds per matching. As expected, the run time of *IBP-Beam* linearly grows with parameter k. However, it is important to remark that in all cases the computation time is much lower than those of A^*. Overall *IBP-Beam(5)* seems to be a good trade-off between computation time and reduction of the overestimation.

5   Conclusions and Future Work

In recent years a framework based on bipartite graph matching to derive approximate solutions of the graph edit distance has been presented. In its original version it suffers from a high overestimation of the computed distance with respect to the true edit distance. In this paper, we propose an iterative extension of one of the existing bipartite-based graph edit distance approximation algorithm. The aim of the paper is to empirically investigate the influence of the order in which the assignments are explored in a post processing search process on the distance quality. The experimental evaluation on three different databases verifies that this order is actually one of the critical factors to improve the overall distance quality. Though the run times are increased when compared to our former framework (as expected), they are still far below the run times of the exact algorithm. The presented approach can be seen as a first step towards finding determinant heuristics to guide the search through the space of possible assignment variants (starting with ψ).

Acknowledgements. This work has been supported by the Swiss National Science Foundation (SNSF) project Nr. 200021_153249, the Hasler Foundation Switzerland, and by the Spanish CICYT project DPI2013–42458–P and TIN2013–47245–C2–2–R.

References

1. Sanfeliu, A., Fu, K.-S.: A distance measure between attributed relational graphs for pattern recognition. IEEE Transactions on Systems, Man and Cybernetics SMC-13(3), 353–362 (1983)
2. Bunke, H., Allermann, G.: Inexact graph matching for structural pattern recognition. Pattern Recognition Letters 1(4), 245–253 (1983)
3. Neuhaus, M., Bunke, H.: A graph matching based approach to fingerprint classification using directional variance. In: Kanade, T., Jain, A., Ratha, N.K. (eds.) AVBPA 2005. LNCS, vol. 3546, pp. 191–200. Springer, Heidelberg (2005)
4. Robles-Kelly, A., Hancock, E.R.: Graph edit distance from spectral seriation. IEEE Trans. Pattern Anal. Mach. Intell. 27(3), 365–378 (2005)
5. Gao, X., Xiao, B., Tao, D., Li, X.: A survey of graph edit distance. Pattern Anal. Appl. 13(1), 113–129 (2010)
6. Boeres, M.C., Ribeiro, C.C., Bloch, I.: A randomized heuristic for scene recognition by graph matching. In: Ribeiro, C.C., Martins, S.L. (eds.) WEA 2004. LNCS, vol. 3059, pp. 100–113. Springer, Heidelberg (2004)
7. Sorlin, S., Solnon, C.: Reactive tabu search for measuring graph similarity. In: Brun, L., Vento, M. (eds.) GbRPR 2005. LNCS, vol. 3434, pp. 172–182. Springer, Heidelberg (2005)
8. Justice, D., Hero, A.O.: A binary linear programming formulation of the graph edit distance. IEEE Trans. PAMI 28(8), 1200–1214 (2006)
9. Neuhaus, M., Riesen, K., Bunke, H.: Fast suboptimal algorithms for the computation of graph edit distance. In: Yeung, D.-Y., Kwok, J.T., Fred, A., Roli, F., de Ridder, D. (eds.) SSPR&SPR 2006. LNCS, vol. 4109, pp. 163–172. Springer, Heidelberg (2006)
10. Riesen, K., Bunke, H.: Approximate graph edit distance computation by means of bipartite graph matching. Image Vision Comput. 27(7), 950–959 (2009)
11. Serratosa, F.: Fast computation of bipartite graph matching. Pattern Recognition Letters 45, 244–250 (2014)
12. Burkard, R.E., Dell'Amico, M., Martello, S.: Assignment Problems. SIAM (2009)
13. Riesen, K., Fischer, A., Bunke, H.: Combining bipartite graph matching and beam search for graph edit distance approximation. In: El Gayar, N., Schwenker, F., Suen, C. (eds.) ANNPR 2014. LNCS, vol. 8774, pp. 117–128. Springer, Heidelberg (2014)
14. Riesen, K., Bunke, H.: IAM graph database repository for graph based pattern recognition and machine learning. In: da Vitoria Lobo, et al. (eds.) [15], pp. 287–297
15. da Vitoria Lobo, N., Kasparis, T., Roli, F., Kwok, J.T., Georgiopoulos, M., Anagnostopoulos, G.C., Loog, M. (eds.): SSPR&SPR 2008. LNCS, vol. 5342. Springer, Heidelberg (2008)

Consensus of Two Graph Correspondences Through a Generalisation of the Bipartite Graph Matching

Carlos Francisco Moreno-García$^{(\boxtimes)}$, Francesc Serratosa, and Xavier Cortés

Universitat Rovira i Virgili, Tarragona, Spain
carlosfrancisco.moreno@estudiants.urv.cat,
{xavier.cortes,francesc.serratosa}@urv.cat

Abstract. One of the most important processes related to structural pattern recognition is to compare the involved objects through representing them as attributed graphs and using error-tolerant graph matching methods. To do so, it is needed a first step to extract the graphs given the original objects and deduct the whole attribute values of nodes and edges. Depending on the application, there are several methods to obtain these graphs and so, the object at hand can be represented by several graphs, not only with different nodes and edges but also with different attribute domains. In the case that we have several graphs to represent the same object, we can deduct several correspondences between graphs. In this work, we want to solve the problem of having these correspondences by exploding this diversity to announce a final correspondence in which the incongruences introduced in the graph extraction and also the graph matching could be reduced. We present a consensus method which, given two correspondences between two pairs of attributed graphs generated by separate entities and with different attribute domains, enounces a final correspondence consensus considering the existence of outliers. Our method is based on a generalisation of the Bipartite graph matching algorithm that minimises the Edit cost of the consensus correspondence while forcing (to the most) to be the mean correspondence of the two original correspondences.

Keywords: Bipartite graph matching · Graph correspondence · Consensus correspondence

1 Introduction

When two parties decide to solve the assignment problem between two images, differences in the mappings may occur. In [1], it is explained how consensus methodologies are used to combine two different mappings between images to obtain a final consensus mapping. That work was inspired on [2], where a weighted mean consensus of a pair of clustering was obtained. In this paper, we generalise the paper presented in [1] since the input of our method is composed of two correspondences between attributed graphs instead of two correspondences between two sets of points

This research is supported by the Spanish CICYT project DPI2013-42458-P, by project TIN2013-47245-C2-2-R and by Consejo Nacional de Ciencia y Tecnologías (CONACyT Mexico)

© Springer International Publishing Switzerland 2015
C.-L. Liu et al. (Eds.): GbRPR 2015, LNCS 9069, pp. 87–97, 2015.
DOI: 10.1007/978-3-319-18224-7_9

that represent images. In this way, attributed graphs can represent any type of object, being images one of them. This is a useful property, since in the image registration domain, some techniques have appeared that represent images as an attributed graph and then a correspondence between graphs is deducted, such as in [3] and [4].

Suppose we have two images and we have used two different salient point extractors (for instance, SIFT and SURF) on these two images to represent them as a set of points. Figure 1.a shows the two images and the extracted salient points. Red circles and blue squares represent the two different extractors. We first realise there are some points that have been selected from both methods but other ones have been selected by only one of the methods. Then, from each image and from each set of points we generate a graph using any structural method such as Delaunay triangulation or K-Nearest neighbours. Therefore, we have graphs G^a and G^b that represent the left image and graphs G'^a and G'^b that represent the right image (graph edges are not drawn). Nodes in G^a and G'^a are drawn in red circles and nodes in G^b and G'^b are drawn in blue squares. If we want to compare these images, we need to apply an error-tolerant graph matching method to deduct the node correspondences and also a distance. There are some options, for instance, Graduated Assignment [5] or Bipartite Graph Matching [6], [7], [8], [9], [10]. Figure 1.b shows the obtained correspondences between both pairs of graphs. Note there are some discrepancies not only on the selected points, but also in the node mappings. Moreover, some new nodes (we call them null nodes) have been introduced to assure correspondences f^a (red lines) and f^b (blue lines) be bijective. Figure 1.c shows the final obtained consensus $f^{a,b}$. Notice $G^{a,b}$ has one null node and $G'^{a,b}$ has two null nodes.

In summary, the input of our method consists of two bijective correspondences $f^a: G^a \rightarrow G'^a$ and $f^b: G^b \rightarrow G'^b$, as well as two node mappings that mark which points in $[G^a$ and $G^b]$ and $[G'^a$ and $G'^b]$ have to be considered the same ones, forcing the node intersection between them. Contrarily of the method presented in [1] that has to be used on images, our method is independent of the domain of these graphs.

(a): Two feature extractors on two images   (b): f^a and f^b correspondences   (c): $f^{a,b}$ correspondence

Fig. 1. The process of obtaining a consensus correspondence from the original extracted salient points

One of the most well-known and practical options to reduce the complexity of a combinatorial calculation of all the possible consensus options is combinatorial optimisation. The concept of optimisation is related to the selection of the "best" configuration or set of parameters to achieve a certain goal [9]. Functions involved in an optimisation problem can be either conformed by continuous values or discrete values, often called as combinatorial scenarios. These scenarios have been largely studied and applied for matching problems, particularly the case of the Hungarian algorithm [10] and the Jonker-Volgenant solver [11]. This method converts a combinatorial problem into a correspondence problem, which will eventually derive in an optimal configuration for a cost-based correspondence. Recently, some collaborative methods have been proposed that given a set of classifiers; return the most promising class [12]. These methods learn some weights that gauge the importance of each classifier and also of the sample through several techniques, such as voting [13] or hierarchical methods [14]. Nevertheless, these methods cannot be directly adapted to our problem, since their output is a class index and our output is a whole correspondence between two sets of salient points.

To generate a correspondence between graphs, several proposals can be found on literature. One of the most relevant in recent years is Bipartite Graph Matching (BP) [8], which has demonstrated to be the efficient and error-tolerant. Also, in [6] and [7] a version to calculate BP in a fast and efficient form has been presented, called Fast bipartite and Square Fast Bipartite (FBP and SFBP). In the Bipartite algorithms, it is important to consider which local sub-structure has been used. This situation, acknowledged in [15] and [16], is also a matter of discussion on this paper.

The paper is structured as follows. In section 2 we briefly define and explain attributed graphs and Bipartite graph matching. In section 3 and 4 we explain how we generalise the BP algorithm to obtain the consensus correspondence and we demonstrate our method. In section 5 we show the experimental evaluation. There is a first explanation of the used database, which we defined and made public [17]. Its main feature is that it is composed of pairs of graphs and some ground truth correspondences between them. Moreover, there is also some information of which nodes of different graphs have to be considered the same ones in the consensus process. Finally, section 6 concludes the paper and presents our future work.

2   Attributed Graphs and Bipartite Graph Matching

Let $G^a = (\Sigma_v^a, \Sigma_e^a, \gamma_v^a, \gamma_e^a)$ and $G'^a = (\Sigma_v'^a, \Sigma_e'^a, \gamma_v'^a, \gamma_e'^a)$ be two attributed graphs. To allow maximum flexibility in the matching process, these graphs have been extended with null nodes to be of order n^a. $\Sigma_v^a = \{v_i^a \mid i = 1, ..., n^a\}$ is the set of vertices and $\Sigma_e^a = \{e_{i,j}^a \mid i, j \in 1, ..., n^a\}$ is the set of edges. Functions $\gamma_v^a: \Sigma_v^a \to \Delta_v^a$ and $\gamma_e^a: \Sigma_e^a \to \Delta_e^a$ assign attribute values in any domain to vertices and edges. Coherent definitions hold for $G'^a = (\Sigma_v'^a, \Sigma_e'^a, \gamma_v'^a, \gamma_e'^a)$. One of the most widely used methods to evaluate an error-correcting graph isomorphism is the Graph edit distance [18], [19], [20]. The dissimilarity is defined as the minimum amount of required distortion to transform one graph into the other. To this end, a number of distortion or edit operations, consisting of insertion, deletion and substitution of nodes and edges are defined. Edit cost functions are introduced to quantitatively evaluate the edit operations. The basic idea is to assign a penalty cost to each edit operation according to the amount of distortion that it introduces in the transformation. Deletion and insertion operations are transformed to assignations of a non-null node of the first or second graph to a null node of the second or first graph. Substitutions simply indicate node-to-node assignments. Using this transformation, given two graphs G^a and G'^a, and a bijection between their nodes f^a, the graph edit cost $EditCost(G^a, G'^a, f^a)$ is computed [20]. It is based on the following constants and functions: C_{vs} is a function that represents the cost of substituting node v_i^a of G^a by node $f^a(v_i^a)$ of G'^a. C_{es} is a function that represents the cost of substituting edge $e_{i,k}^a$ of G^a by edge $f^a(e_{i,k}^a)$ of G'^a. Constant K_v is the cost of deleting node v_i^a of G^a (mapping it to a null node) or inserting node $v_j'^a$ of G'^a (or being mapped from a null node). Likewise for the edges, K_e is the cost of assigning edge $e_{i,k}^a$ of G^a to a null edge of G'^a or assigning edge $e_{j,p}'^a$ of G'^a to a null edge of G^a. Note that we have not considered the cases in which two null nodes or null edges are mapped; this is because this cost is zero by definition.

The Graph edit distance $EditDist$ is defined as the minimum cost under any bijection in T:

$$EditDist(G^a, G'^a) = \min_{f^a \in T}\{EditCost(G^a, G'^a, f^a)\}$$

(1)

We say the optimal bijection, f^{a*}, is the one that obtains the minimum cost.

BP algorithm [8] is composed of three main steps. The first step defines a cost matrix, the second step applies a linear solver such as the Hungarian method or the Jonker-Volgenant method to this matrix and obtains the correspondence f^{a*}. The third step computes the Edit distance cost given this correspondence between both graphs, $EditDist(G^a, G'^a) = EditCost(G^a, G'^a, f^{a*})$. Figure 2 shows the cost matrix of the BP algorithm.

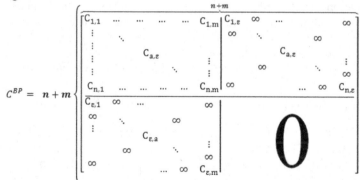

Fig. 2. Cost matrix of the BP algorithm

Quadrant Q1 denotes the combination of substituting costs $C_{i,j}$ and their local sub-structures. The diagonal of quadrant Q2 denotes the whole costs $C_{i,\varepsilon}$ of deleting nodes v_i^a and its local sub-structures. Similarly, the diagonal of quadrant Q3 denotes the whole costs $C_{\varepsilon,j}$ of inserting nodes $v_j'^a$ and its local sub-structures. Q4 quadrant is filled with zero values since the substitution between null nodes has a zero cost. In this paper, we propose a method to perform a consensus of two initial correspondences using the most used local sub-structures, viz. node, degree and clique. In the node case, edges are not considered. The degree is composed of the set of neighbouring edges and the clique is composed of the set of neighbouring edges and also the neighbouring nodes. Other structures have been presented in [15], [16].

3   Consensus of a Pair of Correspondences Between Graphs

Assume $f^a \colon \Sigma_v^a \to \Sigma_v'^a$ and $f^b \colon \Sigma_v^b \to \Sigma_v'^b$ are two correspondence functions between nodes of two attributed graphs $G^a = (\Sigma_v^a, \Sigma_e^a, \gamma_v^a, \gamma_e^a)$ and $G^b = (\Sigma_v^b, \Sigma_e^b, \gamma_v^b, \gamma_e^b)$ and two attributed graphs $G'^a = (\Sigma_v'^a, \Sigma_e'^a, \gamma_v'^a, \gamma_e'^a)$ and $G'^b = (\Sigma_v'^b, \Sigma_e'^b, \gamma_v'^b, \gamma_e'^b)$. The order of G^a and G'^a is n^a and the order of G^b and G'^b is n^b since the correspondences f^a and f^b are defined to be bijective (some null nodes in these graphs may have been added to consider insertion and deletion operations). We can only assure $\Delta_v^a = \Delta_v'^a$ and $\Delta_v^b = \Delta_v'^b$ yet Δ_v^a may be deferent of Δ_v^b and $\Delta_v'^a$ may be deferent of $\Delta_v'^b$ (and similarly for the edges). Moreover, we assume there is some level of intersection between both input node sets and also both output node sets, although it is not strictly necessary, and also it may happen that $n^a \neq n^b$. Note this intersection is imposed through mappings $\zeta \colon \Sigma_v^a \times \Sigma_v^b \to \{0,1\}$ and $\zeta' \colon \Sigma_v'^a \times \Sigma_v'^b \to \{0,1\}$. Mapping $\zeta(v_i^a, v_j^a) = 1$ means v_i^a has to be considered the same node as v_j^b and 0 means they are not the same node. In the same way, mapping $\zeta'(v_i'^a, v_j'^a) = 1$ means $v_i'^a$ has to be considered the same node as $v_j'^b$ and 0 means they are not the same node. If this function is expressed in a matrix form, there is only one cell with a value of 1 in each row and column. Moreover, the number of 1's in the matrix is the number of nodes that are considered the same in both graphs.

Note these two mappings relate nodes but not edges since they may cause some edge inconsistencies. The problem at hand is to define a consensus correspondence $f^{a,b} \colon \Sigma_v^{a,b} \to \Sigma_v'^{a,b}$ given the four graphs G^a, G^b, G'^a and G'^b, bijections f^a and f^b and mappings ζ and ζ'. The set $\Sigma_v^{a,b}$ is composed of the union of sets Σ_v^a and Σ_v^b but the ones mapped by ζ are considered only once. In this case, some null nodes are added to have the possibility of deleting the whole graph. $n^{a,b}$ is the cardinality of $\Sigma_v^{a,b}$ and $\Sigma_v'^{a,b}$. The set $\Sigma_v'^{a,b}$ is composed of the union of sets $\Sigma_v'^a$ and $\Sigma_v'^b$ where the elements mapped by ζ' are considered only once. For this set, some null nodes are added to have the possibility of inserting the whole graph. In this work, we do not want to find the attributed graphs $G^{a,b}$ and $G'^{a,b}$ that could represent the union of graphs G^a and G^b and also of graphs G'^a and G'^b. There are some methods, such as median graphs [21], dedicated to it. This is because we suppose the nature of graphs G^a and G^b and also of graphs G'^a and G'^b is different and, as commented, they have different attribute domains. The only operation that we do on these four graphs is to extend them with null nodes to have the same order $n^{a,b}$. The extended graphs are called $\hat{G}^a$, $\hat{G}^b$, $\hat{G}'^a$ and $\hat{G}'^b$. Accordingly to these graph extensions, the correspondences f^a and f^b are also extended to $\hat{f}^a$ and $\hat{f}^b$ such that the new null nodes are mapped to each other. Thus, $EditCost(G^a, G'^a, f^a) = EditCost(\hat{G}^a, \hat{G}'^a, \hat{f}^a)$ and $EditCost(G^b, G'^b, f^b) = EditCost(\hat{G}^b, \hat{G}'^b, \hat{f}^b)$.

We seek for the correspondence $f^{a,b}$ such that $EditCost(\hat{G}^a, \hat{G}'^a, f^{a,b})$ and $EditCost(\hat{G}^b, \hat{G}'^b, f^{a,b})$ are minimised, and also that it is restricted to be a mean of bijections f^a and f^b. The degree of restriction depends on weight λ, which is a real positive number. Moreover d_H is the Hamming distance.

$$f^{a,b^*}_{\lambda} = \operatorname{argmin}_{\forall f^{a,b} \colon \Sigma_v^{a,b} \to \Sigma_v'^{a,b}} \left\{ \begin{matrix} EditCost(\hat{G}^a, \hat{G}'^a, f^{a,b}) + EditCost(\hat{G}^b, \hat{G}'^b, f^{a,b}) + \\ \lambda \cdot [\, d_H(\hat{f}^a, f^{a,b}) + d_H(\hat{f}^b, f^{a,b})\,] \end{matrix} \right\} \quad (2)$$

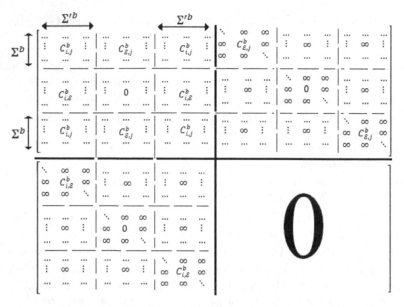

Fig. 3. Generalised cost matrix C^b given graphs G^b and G'^b

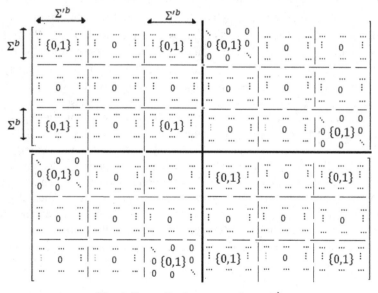

Fig. 4. Generalised correspondence F^b

In the next section, we demonstrate that minimising the above functional is the same than minimising the following one expressed through matrices instead of functions.

$$f^{a,b^*}_{\lambda} = \mathrm{argmin}_{\forall f^{a,b}: \Sigma^{a,b}_v \to \Sigma'^{a,b}_v} \left\{ C^a_{f^{a,b}} + C^b_{f^{a,b}} + \lambda \cdot \left[\left(\mathbf{1}_{f^{a,b}} - F^a_{f^{a,b}} \right) + \left(\mathbf{1}_{f^{a,b}} - F^b_{f^{a,b}} \right) \right] \right\} \quad (3)$$

where C^a and C^b are extended cost matrices and F^a and F^b are extended correspondence matrices. **1** is a matrix with all ones. The four matrices have been extended to have $n^{a,b} \times n^{a,b}$ cells. That is, to assure the whole combinations of substituting, deleting and inserting nodes is possible for the whole nodes. Nevertheless, these matrix extensions have to consider mappings ζ and ζ' between nodes. Figure 3 shows the extended cost matrix C^b. Rows have been split depending on nodes belong to Σ_v^a, Σ_v^b, both or none of them. And similarly for columns and nodes in $\Sigma_v'^a$ and $\Sigma_v'^b$. Correspondingly, figure 4 shows the extended matrix F^b.

The consensus method we propose is based on applying a linear solver such as the Hungarian method or the Jonker-Volgenant method to obtain $f_\lambda^{a,b}{}^*$ as it is defined in equation 3, but using the following matrix and the BP algorithm [8].

$$H = C^a + C^b + \lambda \cdot [2 - (F^a + F^b)] \tag{4}$$

where **2** is a matrix with all two.

4   Minimising the Functional Through Matrices

We have to demonstrate that $EditCost(\hat{G}^a, \hat{G}'^a, f^{a,b}) + EditCost(\hat{G}^b, \hat{G}'^b, f^{a,b}) + \lambda \cdot [d_H(\hat{f}^a, f^{a,b}) + d_H(\hat{f}^b, f^{a,b})]$ equals to $C^a{}_{f_{a,b}} + C^b{}_{f_{a,b}} + \lambda \cdot [(1_{f_{a,b}} - F^a{}_{f_{a,b}}) + (1_{f_{a,b}} - F^b{}_{f_{a,b}})]$.   By   construction, $EditCost(\hat{G}^a, \hat{G}'^a, f^{a,b}) + EditCost(\hat{G}^b, \hat{G}'^b, f^{a,b}) = C_{f_{a,b}}^a + C_{f_{a,b}}^b$ and also $1_{f_{a,b}} = n^{a,b}$, for this reason we only need that,

$$d_H(\hat{f}^a, f^{a,b}) + d_H(\hat{f}^b, f^{a,b}) = 2 \cdot n^{a,b} - F^a{}_{f_{a,b}} - F^b{}_{f_{a,b}} \tag{5}$$

Table 1. Four cases of nodes on $\hat{G}^a$ respectcorrespondences $\hat{f}^a$ and $\hat{f}^b$

Respect $\hat{f}^a$	Respect $\hat{f}^b$	Nodes in $\Sigma_v^{a,b}$	$d_H(\hat{f}^a, f^{a,b})$	$d_H(\hat{f}^b, f^{a,b})$	$F^a{}_{f_{a,b}}$	$F^b{}_{f_{a,b}}$
$\hat{f}^a(v_i^a)$ $\neq f^{a,b}(v_i^a)$	$\hat{f}^b(v_i^b)$ $\neq f^{a,b}(v_i^b)$	A	A	A	0	0
$\hat{f}^a(v_i^a)$ $\neq f^{a,b}(v_i^a)$	$\hat{f}^b(v_i^b)$ $= f^{a,b}(v_i^b)$	B	B	0	0	B
$\hat{f}^a(v_i^a)$ $= f^{a,b}(v_i^a)$	$\hat{f}^b(v_i^b)$ $\neq f^{a,b}(v_i^b)$	C	0	C	C	0
$\hat{f}^a(v_i^a)$ $= f^{a,b}(v_i^a)$	$\hat{f}^b(v_i^b)$ $= f^{a,b}(v_i^b)$	D	0	0	D	D
TOTAL:		$n^{a,b}$	A+B	A+C	C+D	B+D

holds. Then, we can confirm that it is valid to use equation 4 to solve our problem. Considering the relation between correspondences $f^{a,b}$, $\hat{f}^a$ and $\hat{f}^b$, we have split the node set $\Sigma_v^{a,b}$ in four cases. The first two rows of table 1 show these four combinations. The third row shows the supposed number of nodes that hold this case (we use A, B, C and D to represent these number of nodes). Clearly, the addition of these four values is the number of nodes in $\Sigma_v^{a,b}$. In the next two columns, we show the Hamming distance between the subsets of nodes that hold each specific case. And in the last two columns, we show the obtained values of the correspondence matrices applied only to the specific nodes.

Considering the third and fourth columns, we deduct $d_H(\hat{f}^a, f^{a,b}) + d_H(\hat{f}^b, f^{a,b}) = 2A + B + C$. Besides, considering the last two columns, we obtain $F^a_{fa,b} + F^b_{fa,b} = 2D + B + C$. Therefore, $2 \cdot n^{a,b} - F^a_{fa,b} - F^b_{fa,b} = 2A + 2B + 2C + 2D - 2D - B - C = 2A + B + C$ which is exactly the same number than the one obtained for $d_H(\hat{f}^a, f^{a,b}) + d_H(\hat{f}^b, f^{a,b})$ ∎

5   Experimental Validation

5.1   Database Used

To validate our method, we decided to use the "Tarragona Exteriors" dataset [17], defined through five public image databases called "BOAT", "EAST_PARK", "EAST_SOUTH", "RESIDENCE" and "ENSIMAG [22]. These databases are composed of a sequence of images taken from the same object, but from different positions and using a different zoom. Together with the images, the homography estimations h^i that convert the first image (img00) of the set into the other ones (img01 through img10) are provided. From each of the images, the 50 most reliable salient points were extracted using 5 methodologies: FAST, HARRIS, MINEIGEN, SURF (native Matlab 2013b libraries) and SIFT (own library). From these five sets of salient points, we were able to build five representative graph of each image, where the nodes represented the position of the salient points, and the edges were conformed using the Delaunay triangulation method. Notice that the key difference between the salient point data and the graph data is solely the addition of the edges, since the location and features of both the salient points and the graph's nodes is the same.

Between the first image (img00) of the sequence and the other ten images (img01 through img10), we computed correspondences using the five different extracted structures and four different matching functions which are: a) the Matlab's *MatchFeatures* function for salient points' matching (native Matlab 2013b libraries) and the *FastBipartite* function for graph matching using b) node, c) degree and d) clique local sub-structures (own library). Note that for the *MatchFeatures* function, the *MaxRatio* parameter was set to 1 to find as many mappings as possible, although we removed the non-bijective labellings, since this function often maps a salient point more than twice. Thus, the database has a total of 5 sequences × 10 pairs of images × 5 extractor methods × 4 matching options = 1000 quartets Q_i composed of two structures S^1_i and S^2_i (each representing the salient point's location and features plus the graph's edges) for two given images, one correspondence f^i and one homography h^i, resulting in $Q_i = \{S^1_i, S^2_i, f^i, h^i\}$, where i ∈ [1 ... 1000].

5.2   Results and Interpretation

The purpose of this section is to show that performing a consensus using FBP ([6] and [7]) and considering the local sub-structure (node, degree or clique) is better than performing a consensus on the set of points as done in [1]. Notice that for the consensus of the graph matching, we only show the configuration that delivered the best results. C_{vs} is the normalised Euclidean distance and $C_{es} = 0$ due to edges do not have attributes. Node: $K_v = 50$, Degree: $K_v = K_e = 50$ and Clique: $K_v = K_e = 250$.

Table 2. Average number of correct inlier mappings obtained by our consensus strategy with sub-structures Node, Degree and Clique combining two feature extractors ($E_{n,m}$). We have added the number of inliers published in [1] in which the consensus method is based on points.

		$E_{1,2}$	$E_{1,3}$	$E_{1,4}$	$E_{1,5}$	$E_{2,3}$	$E_{2,4}$	$E_{2,5}$	$E_{3,4}$	$E_{3,5}$	$E_{4,5}$	Average
	CLIQUE	84	78	104	70	96	116	89	128	57	160	88.2
	DEGREE	74	72	82	60	68	109	61	111	52	42	73.1
	NODE	70	65	84	63	56	89	70	62	42	37	63,8
BOAT	POINTS	10	10	48	9	9	50	9	46	9	66	26,6
	CLIQUE	35	27	84	36	22	94	43	83	41	85	55
	DEGREE	29	27	31	27	33	42	42	35	40	32	33.8
EAST	NODE	11	23	43	35	18	38	32	46	37	46	32.9
PARK	POINTS	2	1	38	4	2	47	1	44	2	67	20.8
	CLIQUE	8	19	20	12	16	27	11	27	10	30	18
	DEGREE	11	7	34	18	3	29	14	25	12	30	18.3
EAST	NODE	9	2	18	10	4	17	10	19	12	18	11.9
SOUTH	POINTS	1	1	22	1	1	20	1	22	1	32	10.2
	CLIQUE	32	24	124	27	25	129	26	122	23	117	64.9
	DEGREE	16	19	36	26	12	26	20	29	21	21	22.5
	NODE	10	15	24	15	12	24	16	24	19	17	17.6
RESID	POINTS	1	1	59	1	2	51	1	39	1	106	26.2
	CLIQUE	4	3	47	4	5	44	6	43	5	52	21.3
	DEGREE	4	5	34	3	1	37	5	35	3	35	16.2
ENSI	NODE	2	3	30	5	3	37	6	31	5	36	15.8
MAG	POINTS	1	1	42	1	1	38	1	36	1	53	17.5
	CLIQUE	32.6	30.2	75.8	29.8	32.8	82	35	80.6	27.2	68.8	
	DEGREE	25.8	23.8	147.6	27.6	23.8	123.6	28.4	120	25	170.8	
Av	NODE	20.4	21.6	39.8	25.6	18.6	41	26.8	36.4	23	30.8	
	POINTS	3	2.8	41.8	3.2	3	41.2	2.6	37.4	2.8	64.8	

Table 2 shows the number of correct mappings found after applying the consensus for each possible combination of the five feature extractors. Consider 1=FAST, 2=HARRIS, 3=MINEIGEN, 4=SURF, 5=SIFT.

In this experimental validation, the results show that when performing a consensus between two graph matching correspondences and using the new model, we obtain a bigger improvement than when performing salient point's correspondence consensus using the previous method. For example, in the case of the "BOAT" dataset using $E_{3,4}$, the clique consensus obtained 128 correct inliers (31 more than the sum of the individual correspondences) whereas the previous consensus obtained 46 correct inliers, even though the sum of the individual methods is of 73 correct inliers.

6   Conclusions and Further Work

We have presented a consensus method that obtains a correspondence between two attributed graphs given two correspondences between two pairs of attributed graphs generated by separate entities. The method is based on a generalisation of the BP algorithm. We have shown in the experimental section the validity of our method. In this paper, we are able to show that graph representation and matching helps not only to increase the number of correct mappings in the initial proposals, but also to increase the improvement of a consensus correspondence respect representing the objects as a set of points. As a future work, we propose to extend this method such that the consensus can be applied to several correspondences and not only on a pair of them as done in [23] for the salient point case and in [24] for the clustering case. To do so, we are investigating on weighting and voting consensus methods. Nevertheless, the method we present is the first step for the several-correspondences method. Since we have defined the basic mechanism of the method, a several-correspondences method could be applied simply by using the 2-correspondence method iteratively. As we have seen, our method achieves a good accuracy when there are discrepancies between both labellings. Due to in the several-correspondences case the number of discrepancies would increase, our first intuition is that our method would obtain a good consensus correspondence.

References

1. Moreno-García, C.F., Serratosa, F.: Weighted Mean Assignment of a Pair of Correspondences Using Optimisation Functions. In: Fränti, P., Brown, G., Loog, M., Escolano, F., Pelillo, M. (eds.) S+SSPR 2014. LNCS, vol. 8621, pp. 301–311. Springer, Heidelberg (2014)
2. Franek, F., Jiang, X., He, C.: Weighted Mean of a Pair of Clusterings. Pattern Analysis Applications (2012)
3. Sanromà, G., Alquézar, R., Serratosa, F., Herrera, B.: Smooth Point-set Registration using Neighbouring Constraints. Pattern Recognition Letters 33, 2029–2037 (2012)
4. Sanromà, G., Alquézar, R., Serratosa, F.: A New Graph Matching Method for Point-Set Correspondence using the EM Algorithm and Softassign. Computer Vision and Image Understanding 116(2), 292–304 (2012)
5. Gold, S., Rangarajan, A.: A Graduated Assignment Algorithm for Graph Matching. Pattern Analysis and Machine Intelligence 18(4), 377–388 (1996)
6. Serratosa, F.: Speeding up Fast Bipartite Graph Matching through a New Cost Matrix. International Journal of Pattern Recognition & Artificial Intelligence 29(2) (2015)
7. Serratosa, F.: Fast Computation of Bipartite Graph Matching. Pattern Recognition Letters 45, 244–250 (2014)
8. Riesen, K., Bunke, H.: Approximate Graph Edit Distance Computation by Means of Bipartite Graph Matching. Image Vision Comput. 27(7), 950–959 (2009)

9. Serratosa, F., Cortés, X.: Interactive Graph-Matching using Active Query Strategies. Pattern Recognition 48, 1360–1369 (2015)
10. Cortés, X., Serratosa, F.: An Interactive Method for the Image Alignment problem based on Partially Supervised Correspondence. Expert Systems With Applications 42(1), 179–192 (2015)
11. Papadimitriou, C., Steiglitz, K.: Combinatorial Optimization: Algorithms and Complexity. Dover Publications (1998)
12. Kuhn, H.W.: The Hungarian Method for the Assignment Problem Export. Naval Research Logistics Quarterly 2(1-2), 83–97 (1955)
13. Jonker, R., Volgenant, T.: Improving the Hungarian Assignment Algorithm. Operations Research Letters 5(4), 171–175 (1986)
14. Timofte, R., Van Gool, L.: Adaptive and Weighted Collaborative Representations for Image Classification. Pattern Recognition Letters (Available Online 2014)
15. Saha, S., Ekbal, A.: Combining Multiple Classifiers using Vote Based Classifier Ensemble Technique for Named Entity Recognition. Data & Knowledge Engineering 85, 15–39 (2013)
16. Serratosa, F., Cortés, X., Solé, A.: Component Retrieval Based on a Database of Graphs for Hand-Written Electronic-Scheme Digitalisation. Expert Systems with Applications 40, 2493–2502 (2013)
17. Riesen, K., Bunke, H., Fischer, A.: Improving Graph Edit Distance Approximation by Centrality Measures. In: International Conference on Pattern Recognition (2014)
18. Cortés, X., Serratosa, F.: Learning Graph-Matching Edit-Costs based on the Optimality of the Oracle's Node Correspondences. Pattern Recognition Letters (2015)
19. http://deim.urv.cat/~francesc.serratosa/databases/
20. Sanfeliu, A., Fu, K.S.: A Distance Measure between Attributed Relational Graphs for Pattern Recognition. IEEE Transactions on Systems, Man, and Cybernetics 13(3), 353–362 (1983)
21. Gao, X., Xiao, B., Tao, D., Li, X.: A Survey of Graph Edit Distance. Pattern Analysis and Applications 13(1), 113–129 (2010)
22. Solé, A., Serratosa, F., Sanfeliu, A.: On the Graph Edit Distance Cost: Properties and Applications. Intern. Journal of Pattern Recognition and Artificial Intell. 26(5) (2012)
23. Ferrer, M., Valveny, E., Serratosa, F.: Median Graph: A New Exact Algorithm using a Distance Based on the Maximum Common Subgraph. Pattern Recognition Letters 30(5), 579–588 (2009)
24. http://www.featurespace.org
25. Moreno-García, C.F., Serratosa, F., Cortés, X.: Iterative Versus Voting Method to Reach Consensus Given Multiple Correspondences of Two Sets. In: IbPRIA 2015 (2015)
26. Franek, L., Jiang, X.: Ensemble Clustering by means of Clustering Embedding in Vector Space. Pattern Recognition 47(2), 833–842 (2014)

Revisiting Volgenant-Jonker for Approximating Graph Edit Distance

William Jones[(✉)], Aziem Chawdhary, and Andy King

University of Kent, Canterbury, CT2 7NF, UK
{wrj2,a.m.king}@kent.ac.uk, aziem@chawdhary.co.uk

Abstract. Although it is agreed that the Volgenant-Jonker (VJ) algorithm provides a fast way to approximate graph edit distance (GED), until now nobody has reported how the VJ algorithm can be tuned for this task. To this end, we revisit VJ and propose a series of refinements that improve both the speed and memory footprint without sacrificing accuracy in the GED approximation. We quantify the effectiveness of these optimisations by measuring distortion between control-flow graphs: a problem that arises in malware matching. We also document an unexpected behavioural property of VJ in which the time required to find shortest paths to unassigned nodes decreases as graph size increases, and explain how this phenomenon relates to the birthday paradox.

1 Introduction

Graph edit distance (GED) [5] measures the similarity of two graphs as the minimum number of edit operations needed to convert one graph to another. More precisely, suppose $G = \langle V, E, \ell \rangle$ is a labelled directed graph where $E \subseteq V \times V$ and $\ell : V \rightarrow \Sigma$ assigns each vertex to a label drawn from an alphabet Σ. (In the general case, edges can also be similarly attributed.) An edit operation on a graph G_1 inserts or deletes an isolated vertex, inserts or deletes an edge, or relabels a vertex, to obtain a new graph G_2. Applying a sequence of $n - 1$ edit operations gives a sequence of n graphs $G_1, G_2, \ldots, G_n$. Since the cost of edit operations is not necessarily uniform, in the more general form, each edit operation has an associated edit cost as defined by a cost function. The GED between two graphs G and G' is the minimum sum of edit operation costs. GED has proven to be useful [2,4,7,8] because it is an error tolerant measure of similarity.

However, computing GED is equivalent to finding an optimal permutation matrix [11], which is NP-hard. Fast but suboptimal approaches have thus risen to prominence [8], in which GED is approximated by solving a linear sum assignment problem. Of those algorithms proposed for solving this problem, the Volgenant-Jonker (VJ) algorithm [6] is the most efficient. This paper takes VJ as the starting point, and explores how it can be improved for the specific task of GED computation. Several similar works have been attempted before [9,10]. Our work differs from these previous attempts because our changes attack the highly regular structure of the cost matrix and the redundancy that this implies for the

© Springer International Publishing Switzerland 2015
C.-L. Liu et al. (Eds.): GbRPR 2015, LNCS 9069, pp. 98–107, 2015.
DOI: 10.1007/978-3-319-18224-7_10

VJ algorithm instead of approaching the problem by using a refined concept of edit distance. Our paper makes the following contributions:

- It shows how the VJ algorithm can be tuned to GED computation;
- It quantifies the ensuing speedup and decrease in memory requirements;
- It reports an emergent behaviour in which the time taken on the shortest path calculations decreases as the problem size increases;
- It gives an explanation to the above phenomenon based on the well known birthday problem.

The paper is structured as follows: To keep the paper self-contained, section 2, explains how GED is related to the linear sum assignment problem, and section 3 describes the classical VJ algorithm. Section 4 introduces the proposed optimisations, and Section 5 presents the experimental results, comparing the improved algorithm with the original. Finally, Section 6 conclusions.

2   The Linear Assignment Problem and GED

Given an $n \times n$ cost matrix C, the linear assignment problem [3] is that of finding a bijection $f : \{1, \ldots, n\} \rightarrow \{1, \ldots, n\}$ which minimises $\sum_{i=1}^{n} C_{i,f(i)}$. When $|V_1| = n = |V_2|$ the GED between $G_1 = \langle V_1, E_1, \ell_1 \rangle$ and $G_2 = \langle V_2, E_2, \ell_2 \rangle$ can be approximated by solving a linear assignment problem using an $n \times n$ matrix C where $C_{i,j}$ denotes the cost of substituting vertex i for vertex j. Approximating GED with this minimum requires $|V_1| = |V_2|$ and is imprecise since it only considers vertex substitutions. A more general approach [8] addresses both these problems by working on an extended cost matrix defined as follows:

$$
C = \left[
\begin{array}{cccc|cccc}
c_{1,1} & c_{1,2} & \cdots & c_{1,m} & c_{1,\epsilon} & \infty & \cdots & \infty \\
c_{2,1} & c_{2,2} & \cdots & c_{2,m} & \infty & c_{2,\epsilon} & \ddots & \vdots \\
\vdots & \vdots & \ddots & \vdots & \vdots & \ddots & \ddots & \infty \\
c_{n,1} & c_{n,2} & \cdots & c_{n,m} & \infty & \cdots & \infty & c_{n,\epsilon} \\
\hline
c_{\epsilon,1} & \infty & \cdots & \infty & 0 & 0 & \cdots & 0 \\
\infty & c_{\epsilon,2} & \cdots & \infty & 0 & 0 & \ddots & 0 \\
\vdots & \ddots & \ddots & \infty & \vdots & \ddots & \ddots & 0 \\
\infty & \cdots & \infty & c_{\epsilon,m} & 0 & \cdots & 0 & 0
\end{array}
\right]
$$

where $n = |V_1|$ and $m = |V_2|$. The top left hand corner of $c_{i,j}$ describes the cost of vertex substitution. The top right hand corner $c_{i,\varepsilon}$ the cost of vertex deletion u_i. The bottom left hand corner $c_{\varepsilon,j}$ denotes the cost of vertex insertion v_j. The bottom right hand corner is uniformly zero (henceforth called the null quadrant).

A further extension [2] uses E_k and ℓ_k to compute a lower bound on GED, by finding labels on the incoming neighbours of a given vertex j in G_k using $In_k(j) = \{\ell_k(i) \mid \langle i, j \rangle \in E_k\}$. The matrix is defined $c_{i,j} = d_{i,j} + e_{i,j}$ where $d_{i,j} = 1$ if $\ell_1(i) \neq \ell_2(j)$ otherwise 0, and $e_{i,j} = \max(|In_1(i) - In_2(j)|, |In_2(i) - In_1(j)|)$. Then $c_{i,j}$ accounts for any difference in labelling between vertex i and vertex j and also removing edges and relabelling their incoming neighbours. The diagonals $c_{i,\epsilon}$ and $c_{\epsilon,i}$ are degenerative and defined as above with $In_k(\epsilon) = \emptyset$.

$$
\begin{bmatrix} 1 & 7 & 6 \\ 5 & 2 & 3 \\ 8 & 9 & 4 \end{bmatrix} \qquad \begin{bmatrix} 0 & 5 & 3 \\ 4 & 0 & 0 \\ 7 & 7 & 1 \end{bmatrix} \qquad \begin{bmatrix} 3 & 5 & 3 \\ 7 & 0 & 0 \\ 10 & 7 & 1 \end{bmatrix} \qquad \begin{bmatrix} 0 & 2 & 0 \\ 7 & 0 & 0 \\ 10 & 7 & 1 \end{bmatrix}
$$
$$
\quad (a) \qquad\qquad\quad (b) \qquad\qquad\quad (c) \qquad\qquad\quad (d)
$$

Fig. 1. (a) Example cost matrix; (b) After column reduction; (c) After anti-column reduction; (d) After row reduction (reduction transfer)

3   The Classical VJ Algorithm

The VJ algorithm [6] is a shortest path algorithm solved by a dual method. We describe the essence of the algorithm (though not the detail). The algorithm consists of two main steps, which are outlined in the sub-sections that follow:

1. Initialisation in three stages: (a) column reduction; (b) reduction transfer; and (c) augmenting row reduction.
2. Augmentation until complete, in which alternating paths are found where each path is from an unassigned row to an unassigned column.

3.1   Initialisation

Column reduction The first step of initialisation is a column reduction, in which a positive value is subtracted from each element of a column. Starting at the last column, the VJ algorithm reduces each column by its minimum element, so that each column contains a zero. Figure 1(b) illustrates the result of column reduction. As the matrix is scanned right-to-left, each column is assigned, whenever possible, to a unique row that contains a zero in that column. Column 3 is assigned to row 2 (and vice versa), and column 1 is assigned to row 1 (and vice versa), but column 2 will remain unassigned (as does row 3).

Reduction Transfer. The second step of initialisation is reduction transfer, which is applied to enable row reduction, in which a positive value is subtracted from each element of a row. Row reduction cannot be applied to row 1 of Figure 1(b), without introducing a negative entry. Thus an inverse of column reduction is applied to row 1, to give the matrix depicted in Figure 1(c). Row reduction is then applied, the result of which is illustrated in Figure 1(d), albeit at the expense of column reduction. This exchange in reduction value between a column and a row, in this case by 3, is called reduction transfer.

Augmenting Row Reduction. In the third phase of initialisation, an attempt is made to find a set of (alternating) paths where each path starts in an unassigned row and ends in an unassigned column. For a given unassigned row i, VJ finds a column j_1 that contains the minimum entry e_1 and another column j_2 that contains the least entry e_2 such that $e_2 \geq e_1$. Row i is then reduced by e_2. If $e_2 > e_1$ then this incurs a negative value in column j_1, in which case, anti-column reduction is applied to column j_1 to eliminate the negative entry. Row

$$\begin{bmatrix} 3 & 1 & 4 & 1 & \infty & \infty \\ 5 & 9 & 2 & \infty & 6 & \infty \\ 5 & 3 & 5 & \infty & \infty & 8 \\ \hline 9 & \infty & \infty & 0 & 0 & 0 \\ \infty & 7 & \infty & 0 & 0 & 0 \\ \infty & \infty & 9 & 0 & 0 & 0 \end{bmatrix} \qquad \begin{bmatrix} 3 & 1 & 4 & 1 \\ 5 & 9 & 2 & 6 \\ 5 & 3 & 5 & 8 \\ \hline 9 & 0 & 0 & 0 \\ 7 & 0 & 0 & 0 \\ 9 & 0 & 0 & 0 \end{bmatrix} \qquad \begin{bmatrix} 3 & 1 & 4 & 1 \\ 5 & 9 & 2 & 6 \\ 5 & 3 & 5 & 8 \\ 9 & 7 & 9 & \end{bmatrix}$$

$$(a) \qquad\qquad\qquad (b) \qquad\qquad\qquad (c)$$

Fig. 2. (a) Original cost matrix. (b) Row-by-row representation with zeroes. (c) Row-by-row representation without zeroes.

i is then assigned to column j_1 regardless of whether this column is already assigned or not. If j_1 was previously assigned to a row k, then row k becomes unassigned and the procedure continues from row k. This repeats until either row k is matched to an unassigned column, or it becomes impossible to transfer reduction to the selected row k. Observe how reduction transfer provides a vehicle for constructing a path that alternates between rows and columns, hence the name.

3.2   Augmentation

For each unassigned row, the augmentation phase finds a shortest alternating path (of the type previously described) to an unassigned column. VJ modifies Dijkstra's algorithm to search for these shortest paths, where the notion of distance between a row and a column is the entry in the cost matrix. Search starts at an unassigned row, say row i, and a shortest edge is found from row i to a column j. If column j was previously assigned to row k, then row k becomes unassigned (though no changes are made until a complete path to an unassigned column is found) and search resumes from row k. Unlike classical Dijkstra, search continues in this fashion until such a column is found. After augmentation, the assignments to the cost matrix are updated so that all assignments in the current solution correspond to minimum entries in each row of the cost matrix.

4   The Improved VJ Algorithm

This section explores several improvements to the classical VJ algorithm, most of which follow from the regular structure of the cost matrix.

4.1   Representation

A brief foreword to this section. It should be noted that while we change the representation in memory, we do not just naïvely iterate over it. Our change is simply to simplify many operations; we are still essentially calculating assignments and solutions in their "real" positions.

$$C = \begin{bmatrix} c_{1.1} & c_{1.2} & \cdots & c_{1.m} & c_{1,\epsilon} \\ c_{2.1} & c_{2.2} & \cdots & c_{2.m} & \vdots \\ \vdots & \vdots & \ddots & \vdots & \vdots \\ c_{n,1} & c_{n,2} & \cdots & c_{n,m} & c_{n,\epsilon} \\ c_{\epsilon,1} & c_{\epsilon,2} & \cdots & c_{\epsilon,m} & \end{bmatrix} \qquad C = \begin{bmatrix} c_{1.1} & c_{1.2} & \cdots & c_{1.m} & c_{1,\epsilon} \\ c_{2.1} & c_{2.2} & \cdots & c_{2.m} & \vdots \\ \vdots & \vdots & \ddots & \vdots & \vdots \\ c_{n,1} & c_{n,2} & \cdots & c_{n,m} & c_{n,\epsilon} \\ c_{\epsilon,1} & c_{\epsilon,2} & \cdots & c_{\epsilon,m} & \end{bmatrix}$$

(a)   (b)

Fig. 3. (a) The split for the left hand and right hand block (the left hand block is highlighted). (b) The split for the top and bottom blocks (the top is highlighted).

Given two reasonably-sized graphs G_1 and G_2, the largest data structure by far is the cost matrix requiring $(n + m)^2$ entries for $n = |V_1|$ and $m = |V_2|$. However, most of these entries are either zero or infinity, as is illustrated in Figure 2(a). Given the operations that must be applied to the cost matrix, there are two natural compressed representations: a row-by-row representation in which each row of the matrix is represented by only storing non-infinite values, as depicted in Figure 2(b); and an analogous column-by-column representation.

There is also the question of whether to explicitly store the zeros in the bottom right of Figure 2(b). We choose to discard them as we found no algorithmic benefit in retaining them. With this change, the row-by-row representations reduces to Figure 2(c). The net effect is that the cost matrix is represented in space $(n + 1) \times (m + 1)$. Although row equality is no longer preserved, operations on the effected rows can still be performed in constant time since there is only one variable entry per row. Moreover, this representation homogenises the column-by-column and row-by-row representations, which means that it simultaneously benefits both row- and column-based operations.

4.2 Column Reduction

This representation simplifies column reduction in two ways: First, almost half of the costs in the matrix are infinite and so will never be chosen as a minimum in a column. Second, nearly a quarter of the costs will be 0, and these zeros dictate that the column minimum will be zero. Hence only the position of the minimum need be computed in column reduction (rather than its position and value). To take advantage of this, the matrix is considered as two separate blocks with different operations provided for each. see Figure 3a. The leftmost entries are handled as before thanks to the new data-structure. The rightmost entries that are stored in a single column, see Figure 2(c), correspond to the top-right diagonal above the null quadrant. These entries are only compared to zeros and hence the reduction value will always be zero and thus only the position of the zero need be found. This can be further simplified in the case that deletion costs are non-zero because this removes the need for position computation too.

4.3   Reduction Transfer

Reduction transfer actually becomes slightly more complicated to accommodate the new data-structure. However, this is a worthwhile trade as reduction transfer typically takes up a minute fraction of the total run time.

4.4   Augmenting Row Reduction

Augmenting row reduction can be optimised in a very similar fashion to column reduction. Once again, for each row being considered there is no point considering infinite costs and similarly the presence of the null quadrant simplifies many of the calculations. The only difference is that since augmenting row reduction considers rows instead of columns, the most effective way to operate is a top-to-bottom split (see Figure 3b), where the top block and the bottom block are handled separately. We can consider similar simplifications as before if we can assume that vertex addition will always have a non-zero cost.

4.5   Augmentation

Augmentation is slightly quicker with these improvements, particularly over very large graphs. There are a number of operations that can be simplified by clever use of the new data-structure, but these refinements have little to no benefit. The only significant changes are those that simply replace variable lookups when the outcome is known, for example, looking up an entry in the null quadrant.

5   Experimental Results

To empirically assess the proposed improvements to VJ, two versions were implemented: the original version (VJ-ORG) and an improved version (VJ-IMP). Both were compared against a version developed by Jonker himself (VJ-CTRL), which was used as a control. All three versions were implemented in C++.

5.1   Evaluation on Random Data

Initially random square ($n = m$) cost matrices were used to provide a large corpus of data for comparing all versions of VJ. The improvements have least effect on square matrices and thus, if anything, the setup is biased against VJ-IMP. Costs and matrix sizes were chosen to approximate what might typically be encountered in malware matching. In this context a control-flow graph (CFG) is extracted from a binary for comparison against a database of CFGs derived from malware. BinSlayer [2] was used to derive CFGs from several medium-sized binaries. These possessed between 465 to 6984 vertices (basic blocks), and produced matrices where the costs rarely exceeded 2000 and never exceeded 3000. To cover a range of scenarios, matrices were populated with random values from cost ranges varying over 1-500 to 1-3000. Matrix sizes were also varied between

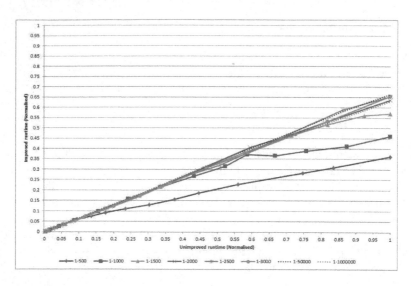

Fig. 4. VJ-IMP against of the fastest of VJ-ORG and VJ-CTRL for various cost ranges and matrix sizes

1000 and 14000, again to simulate CFGs. We also tested the resilience of the improvements over some much larger cost ranges (1-50000 and 1-10^9) for the full range of matrix sizes.

Figure 4 gives a series of plot lines, one for each cost range. Each data point on a line corresponds to the runtimes for a different matrix size, though the horizontal axis is normalised by the runtime for solving a 14000 × 14000 matrix. The vertical axis is normalised relative to the unimproved times, so that the instantaneous gradient quantifies the improvement over a range of costs and matrix sizes. Note that VJ-IMP is uniformly faster than both VJ-ORG, and the third party implementation by Jonker, VJ-CTRL. While we find the improved version of algorithm to be about twice as fast; the difference being more striking

CFG 1	CFG 2	Size Ratio	VJ-ORG	VJ-CTRL	VJ-IMP	Speedup
bash	BinSlayer	0.27	201	205	125	62%
BinSlayer	bash	3.75	2944	3164	1820	68%
comaker	BinSlayer	0.11	623	627	264	137%
BinSlayer	comaker	8.83	13221	13729	8905	51%
comaker	bash	0.42	3763	3901	3080	24%
bash	comaker	2.35	54308	58333	43576	29%
GB	BinSlayer	0.07	1862	1656	588	199%
BinSlayer	GB	15.02	34020	35468	22127	57%
GB	bash	0.25	4555	5087	3155	53%
bash	GB	4.00	151457	160074	114091	37%
GB	comaker	0.59	17987	18495	15291	19%
comaker	GB	1.70	209562	228467	164442	33%

Fig. 5. Runtimes in milliseconds, where the size ratio is $|V_1|/|V_2|$

at lower costs and higher sizes. Although not represented by this data, there is a small performance advantage to VJ-ORG over VJ-CTRL, which suggests that our implementations and experiments are robust.

5.2 Evaluation on CFGs

Figure 5 summaries some CFG comparisons for four different binaries, where the CFGs were derived using BinSlayer. Comparing $CFG_1 = \langle V_1, E_1, \ell_1 \rangle$ against $CFG_2 = \langle V_2, E_2, \ell_2 \rangle$ does not necessarily take the same time as comparing CFG_2 against CFG_1. This is because if $|V_1| < |V_2|$ then cost matrix will have a large deletion block (in its top right) and a small addition block (in its bottom left). It is notable that while all versions of VJ are faster when $|V_1| < |V_2|$, the benefits to VJ-IMP are more significant. Excluding augmentation, the runtime of each component of each algorithm is almost constant no matter whether $|V_1| < |V_2|$. However, the runtime of augmentation is smaller when $|V_1| < |V_2|$. Since column reduction is faster in VJ-IMP extra benefits follow from $|V_1| < |V_2|$ because column reduction is faster and the cost of augmentation is less dominant. We have found that the number of iterations in augmentation does not depend on $|V_1| < |V_2|$, and so the decreased time in augmentation is entirely a byproduct of an decrease in the runtime of each iteration.

5.3 Component Analysis

Figure 6 shows the time proportion spent in each component of VJ-ORG and VJ-IMP. Column reduction and augmenting row reduction benefit most from the improvements; augmentation is faster with VJ-IMP (in absolute terms).

We also see an interesting behaviour in the runtimes of augmentation and column reduction, especially at low cost ranges. Column reduction quickly increases as a percentage of total runtime as cost matrix size increases eventually

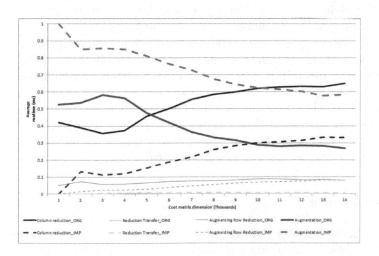

Fig. 6. Normalised runtime of components, for VJ-ORG and VJ-IMP, over range 1-500

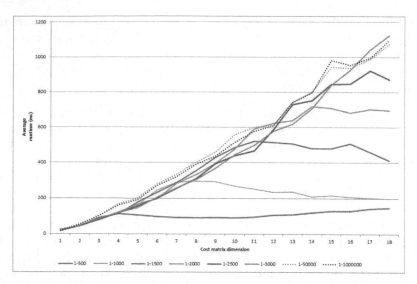

Fig. 7. Runtime of each iteration of augmentation (averaged over all three algorithms)

overtaking augmentation. On closer examination, it is apparent that this is the result of a reduction in the growth of the runtime of augmentation instead of a sharp increase in the runtime of column reduction. Furthermore this happens across all cost ranges (but is more visible for smaller ranges). This is surprising as augmentation is merely an implementation of Dijkstra's algorithm. Figure 7 suggests that this stems from an effect in which the growth in runtime of each iteration of augmentation actually tails off as the size of cost matrix increases.

We conjecture that this is because of a statistical property related to the birthday paradox. During initialisation the column reduction step works backward (from the highest index to the lowest), so low index columns have a higher chance of involving a collision and having a lowest element in the same position as another column. For a randomly generated matrix of total size $t \times t$, column index i will have a $\left(\frac{t}{t+1}\right)^{t-i}$ probability of not having a minimum element in the same row as another column, and thus low i are very likely to be unassigned. Thus a column with a given index is more likely to be unassigned as cost matrix size increases. Consequently not only are low indexed columns more likely to be unassigned, but across all columns this effect will increase disproportionally as matrix size increases. Since Dijkstra's algorithm scans from low indexed columns to high indexed ones, it will find assignments for most of its rows more quickly as cost matrix size increases, even though worse-case complexity remains in $O(n^3)$.

6   Conclusions

We have examined the VJ algorithm and studied improvements for approximating GED. We have shown that our improved algorithm is uniformly faster than its unimproved counterparts, both across randomly generated matrices and data sets that arise in call-graph comparison. The speedups, which are almost 200% in one case, suggest the that refinements are truly worthwhile. Moreover, the improved version has a smaller memory footprint, and incurs no loss of accuracy whatsoever. Finally, we have also documented and explained an anomaly in the runtime of the Dijkstra's shortest path search component of VJ. Future work will, among other things, empirically investigate how the relative sizes of the two graphs under comparison effect the overall runtime, and also explore the prospects for parallelisation [1].

Acknowledgements. We thank Earl Barr and David Clark for discussions on malware matching that motivated this work, and the reviewers of this paper who had several important insights into our work, and Laetitia Pelacchi for her support.

References

1. Balasn, E., Miller, D., Pekny, J., Toth, P.: A parallel shortest augmenting path algorithm for the assignment problem. JACM 38(4), 985–1004 (1991)
2. Bourquin, M., King, A., Robbins, E.: BinSlayer: Accurate Comparison of Binary Executables. In: Proceedings of Program Protection and Reverse Engineering Workshop. ACM (2013)
3. Burkard, R.E., Cela, E.: Linear Assignment Problems and Extensions. Springer (1999)
4. Conte, D., Foggia, P., Sansone, C., Vento, M.: Thirty Years of Graph Matching in Pattern Recognition. International Journal of Pattern Recognition and Artificial Intelligence 18(3), 265–298 (2004)
5. Gao, X., Xiao, B., Tao, D., Li, X.: A Survey of Graph Edit Distance. Pattern Analysis and Applications 13(1), 113–129 (2010)
6. Jonker, R., Volgenant, A.: A Shortest Augmenting Path Algorithm for Dense and Sparse Linear Assignment Problems. Computing 38(4), 325–340 (1987)
7. Myers, R., Wison, R.C., Hancock, E.R.: Bayesian graph edit distance. IEEE Transactions on Pattern Analysis and Machine Intelligence 22(6), 628–635 (2000)
8. Riesen, K., Bunke, H.: Approximate Graph Edit Distance computation by means of Bipartite Graph Matching. Image and Vision Computing 27(7), 950–959 (2009)
9. Serratosa, F.: Fast computation of bipartite graph matching. Pattern Recognition Letters 45, 244–250 (2014)
10. Serratosa, F., Cortés, X.: Edit Distance Computed by Fast Bipartite Graph Matching, pp. 253–262 (2014)
11. Zeng, Z., Tung, A.K.H., Wang, J., Feng, J., Zhou, L.: Comparing Stars: On Approximating Graph Edit Distance. VLDB Endowment 2(1), 25–36 (2009)

A Hypergraph Matching Framework for Refining Multi-source Feature Correspondences

He Zhang[(✉)], Bin Du, Yanjiang Wang, and Peng Ren[(✉)]

College of Information and Control Engineering, China University of Petroleum,
Qingdao 266580, China
zhangheupc@126.com, pengren@upc.edu.cn

Abstract. In this paper, we develop a hypergraph matching framework which enables feature correspondence refinement for multi-source images. For images obtained from different sources (e.g., RGB images and infrared images), we first extract feature points by using one feature extraction scheme. We then establish feature point correspondences in terms of feature similarities. In this scenario, mismatches tend to occur because the feature extraction scheme may exhibit certain ambiguity in characterizing feature similarities for multi-source images. To eliminate this ineffectiveness, we establish an association hypergraph based on the feature point correspondences, where one vertex represents a feature point pair resulted from the feature matching and one hyperedge reflects the higher-order structural similarity among feature point tuples. We then reject the mismatches by identifying outlier vertices of the hypergraph through higher order clustering. Our method is invariant to scale variation of objects because of its capability for characterizing higher order structure. Furthermore, our method is computationally more efficient than existing hypergraph matching methods because the feature matching heavily reduces the enumeration of possible point tuples for establishing hypergraph models. Experimental results show the effectiveness of our method for refining feature matching.

Keywords: Hypergraph matching · Feature matching · Multi-source image processing

1 Introduction

Establishing consistent correspondences between two sets of feature points is one of the most extensively studied topics in computer vision. Feature point correspondences are essential in applications such as 2D/3D registration [1], shape matching [2] and object recognition [3]. The main task of feature point matching is to establish a mapping between the feature points of one image and those of another. In this scenario, feature point matching methods can be categorized in two classes: feature-based matching methods and structure-based matching methods.

The recently developed feature-based matching methods, e.g., the Scale Invariant Feature Transform (SIFT) based matching method [4], the Speeded

© Springer International Publishing Switzerland 2015
C.-L. Liu et al. (Eds.): GbRPR 2015, LNCS 9069, pp. 108–117, 2015.
DOI: 10.1007/978-3-319-18224-7_11

Up Robust Features (SURF) based matching method [5], and the state-of-the-art Oriented FAST and Rotated BRIEF (ORB) based matching method [6] have achieved comparatively desirable matching performance in computer vision tasks. However, most of these feature-based matching methods mainly consider the feature similarities in images, regardless of the spatial layout of feature points. Moreover, their matching results tend to yield certain mismatches when it comes to matching features between multi-source images. One reason for this ineffectiveness is that common features, extracted by using one method, from multi-source images may appear quite different in the feature space. On the other hand, different features may appear similar to one another and form incorrect correspondences in the feature space spanned by the same feature extraction method.

Structure-based matching methods, generally formulated in terms of graph matching [11] or hypergraph matching [10], take global spatial layout of feature points into consideration and aim at establishing feature point correspondences. Thus, the graphs or hypergraphs established in these methods represent spatial correlations of feature points in each image. Furthermore, hypergraph matching is usually invariant to scale variation of objects or scenes because of its capability of characterizing higher order structures rather than pairwise relationships. Unfortunately, hypergraph matching methods may manifest high computational complexity when the hypergraphs are large.

In this paper, we seek to present a hypergraph matching framework for refining feature correspondences between multi-source images. Unlike the feature-based matching methods which neglect the spatial correlations of feature points, our method refines matched feature pairs subject to higher order spatial constraints in terms of a hypergraph matching formulation. Furthermore, compared with the hypergraph matching methods which are computationally expensive, our method reduces the computational complexity by taking advantages of the matching results obtained from feature-based matching and achieves high efficiency in operation.

We use the proposed framework for matching feature points from multi-source images. We first conduct feature-based matching for images from different sources. This results in a coarse feature point mapping with certain mismatches because the multi-source feature ambiguity may arise in one feature space. In this scenario, incorrect correspondences tend to violate the spatial consistence of feature points. We thus develop a hypergraph matching scheme, which encodes spatial correlations into hyperedges, for identifying the mismatches in the coarse matching results. The spatially inconsistent matches are then rejected from the refined matching results. Our matching framework is robust to scale variation because the hypergraphs characterize higher order spatial correlations which are scale invariant. In our work, we take RGB images and near infrared (NIR) images as multi-source images and aim at matching feature points for them. We compare the performance of state-of-the-art methods and our method and validate the effectiveness of our hypergraph framework.

2   Hypergraph Matching for Refining Feature Correspondences between Multi-source Images

For combining the merits of feature-based matching methods and structural-based matching methods, we first extract the feature points for multi-source images by using one shared feature extraction algorithm. Then, we apply feature-based matching methods to get the initial matching results and building hypergraphs based on the feature-based matching results. Then, we use the high order structural refinement to refine the initial matching results. Fig. 1 shows the framework of our hypergraph-based feature correspondence refining method for multi-source images.

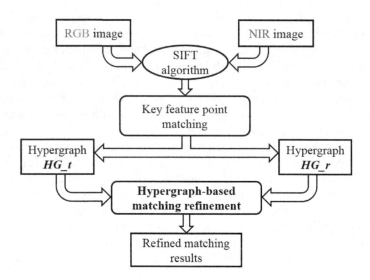

Fig. 1. The hypergraph matching framework for multi-source images feature correspondences

2.1   Feature-Based Matching

We employ SIFT [8] as the feature extraction scheme for multi-source images. Other feature extraction strategies can also be used in our framework. In order to measure the similarity between a pair of features extracted from the target image (e.g., the RGB image) and the reference image (e.g., the NIR image) separately, we compute the arc cosine of the dot product of the two features. The feature similarities for all pairs of features from multi-source images are calculated in the same way. Multi-source feature pairs with the smallest arc cosine values are considered as desirable correspondences for feature matching. We thus obtain the initial feature-based matching results between multi-source images.

2.2 Hypergraph Matching Based on Coarse Feature Correspondences

First, we get M SIFT feature points denoted as $p_1, \ldots, p_M$ in the target image matched with M feature points denoted as $q_1, \ldots, q_M$ in the reference image, respectively. We then exploit hypergraph matching strategies for refining the feature matching results obtained in Section 2.1. We use the set of M feature points in the target image to form one uniform hypergraph HG_t. Its vertices represent the feature points in the target image and the weight on its one hyperedge measures the spatial relationship among K vertices. In our work, we use three vertices (i.e., $K=3$) to form one hyperedge for the representation simplicity. Let $\mathcal{A}$ be the adjacency tensor for the hypergraph HG_t. Thus, its (i, j, k)th entry $a_{i,j,k}$ representing the weight on the hyperedge is determined by

$$a_{i,j,k} = U(p_i, p_j, p_k) \cdot W(p_i, p_j, p_k) \tag{1}$$

where i, j, k of p_i, p_j, p_k represent the indices of vertices in the hypergraph HG_t. We define contents $U(i, j, k)$ and $W(i, j, k)$ of (1) respectively as

$$U(p_i, p_j, p_k) = \det([v_i - v_k, v_j - v_k]) \tag{2}$$

and

$$W(p_i, p_j, p_k) = \sum_{i,j,k} \frac{1}{\sqrt{||v_i - v_k|| \cdot ||v_j - v_k||}} \tag{3}$$

where the two dimensional column vector v_i, v_j, v_k represent the coordinates of the vertices p_i, p_j, p_k separately. The value of $a_{i,j,k}$ equals to zero if there is no hyperedge encompassing p_i, p_j, p_k. According to (1), the value of $a_{i,j,k}$ is great when the three vertices are geometrically similar, and is small if these vertices are geometrically different [13].

Similarly, we use the set of M feature points in the reference image to form one uniform hypergraph HG_r, with each feature point representing a vertex. The adjacency tensor for HG_r is denoted as $\mathcal{B}$. The (i, j, k)th entry $b_{i,j,k}$ representing the weight on the hyperedge and the correlation between q_i, q_j, q_k is calculated as follows

$$b_{i,j,k} = U(q_i, q_j, q_k) \cdot W(q_i, q_j, q_k) \tag{4}$$

where i, j, k of q_i, q_j, q_k represent the indices of vertices in HG_r. The value of $b_{i,j,k}$ is zero when the corresponding hyperedge not exists. Fig. 2 shows the hypergraphs and the matching example for multi-source images.

Based on HG_t and HG_r, we establish the association hypergraph HG, whose M vertices represent the possible matching pairs between multi-source images and the weight on its hyperedge measures the similarity of the potential correspondences. We build the adjacency tensor $\mathcal{S}$ for HG with (i, j, k)th entry $S_{i,j,k}$ representing the hyperedge weight, which is defined as follows

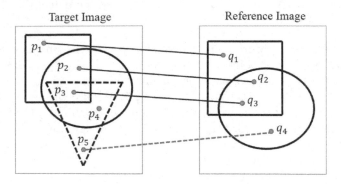

Fig. 2. Matching example for multi-source images. Different shape of borders represent hyperedges of each hypergraph. Additionally, the solid line and the dash line between feature points represent correct match and incorrect match, respectively.

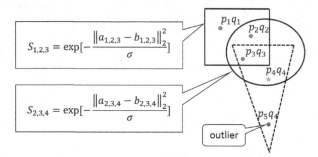

Fig. 3. The association hypergraph for the hypergraphs in Fig. 2. Four round form in blue represent the initial matching pairs and the one starlike form in gray represents the correct matching pair which is not established.

$$S_{i,j,k} = \exp[-\frac{\|a_{i,j,k} - b_{i,j,k}\|_2^2}{\sigma}] \tag{5}$$

where σ is a scaling parameter. Fig. 3 shows the association hypergraph HG for hypergraphs structured in Fig. 2.

According to (5), the $S_{i,j,k}$ characterizes the similarity between one form (i.e., a hyperedge compassing feature points $\{p_i, p_j, p_k\}$) in HG_t and the matched form (i.e., a hyperedge compassing the matched feature points $\{q_i, q_j, q_k\}$) in HG_r. Additionally, it characterizes the structural consistency between the hypergraphs established based on the target image and reference image separately.

2.3   Structural Refinement for Feature-Based Matching Results

The task of structurally refining the feature-based matching results can be transformed into removing outliers, i.e., the incorrect matching results, from a tight cluster in the subspace spanned by the adjacency tensor $\mathcal{S}$ of the association hypergraph HG. According to Ren et $al.$ [9], we apply the variation of dominant cluster analysis (DCA), the High Order Dominant Cluster Analysis (HO-DCA), to removing the outliers. Hence, we denote the column vector $\mathbf{x}$ to record the matching score, whose nth entry x_n ($n=1, \ldots, M$) indicating the degree of the structural consistency for the potential matching pairs $\{p_n, q_n\}$. Let $\mathbf{T}$ denote the subset of vertices in HG with vertices represent the correct matching pairs for HG_t and HG_r. As a result, the nth entry x_n of $\mathbf{x}$ also represent the probability for the nth vertex in HG belong to $\mathbf{T}$. For our three-dimensional tensor by using these ingredients, the optimal model can be formulated as

$$\widehat{\mathbf{x}} = \arg\max_{\mathbf{x}} \sum_{i=1}^{M}\sum_{j=1}^{M}\sum_{k=1}^{M} S_{i,j,k} \prod_{n=i,j,k} x_n \tag{6}$$

subject to the constraints $\forall n$, $x_n \geq 0$ and $\sum_{i=1}^{M} x_i = 1$.

According to (6), if the initial matching pair $\{p_n, q_n\}$ is an incorrect matching pair of $\mathbf{T}$, then the nth entry x_n of $\mathbf{x}$ will much less than 1. We refer to nonzero value x_n of $\mathbf{x}$ satisfying the optimality condition in (6) as the association degree for the matching pair $\{p_n, q_n\}$. Therefore, the problem of moving outliers can be solved as a constraint optimization problem.

According to [9], we adopt the following iterative formula to reach the convergence of x_i

$$x_i(t+1) = x_i(t) \frac{\sum_{j=1}^{M}\sum_{k=1}^{M} S_{i,j,k}\prod_{n=j,k} x_n(t)}{\sum_{i=1}^{M}\sum_{j=1}^{M}\sum_{k=1}^{M} S_{i,j,k}\prod_{n=i,j,k} x_n(t)} \tag{7}$$

where t indicate the tth iteration. The rest M-1 entries of $\mathbf{x}$ can be computed in the same iterative formula (7).

We use (7) to update the score vector $\mathbf{x}$ until we reach the convergence. At convergence the score vector $\mathbf{x}$ is the optimal solution to (6) and the nonzero elements in $\mathbf{x}$ corresponds to correct matches.

2.4   Computational Complexity

We consider the problem of matching two K-uniform hypergraphs with each having N vertices. The existing hypergraph matching strategies establish an association hypergraph with N^2 vertices by enumerating all potential matching pairs and $C_{N^2}^{K}$ hyperedges. These strategies are not feasible for practical applications, because they normally require the computational complexity $O((N^K)^2)$, especially when the value of N and K are large. In contrast, our hypergraph matching framework establish an association hypergraph only with N vertices and C_N^K hyperedges. The computational complexity of our method is $O(N^K)$. Key to this efficiency is that the feature-based matching establishes coarse correspondences between the hypergraph HG_t and HG_r, which heavily reduces the enumeration of possible matching pairs. Thus, our hypergraph matching framework improve the matching accuracy by getting rid of outliers from the coarse matching results with low computational complexity.

3   Experimental Validation

In this section, we compare state-of-the-art feature point matching methods and our hypergraph matching framework, for the purpose of matching feature points from multi-source images. All experiments are performed in a PC machine with Intel Core i3 CPU 330 (2.13Ghz) and 4GB memory.

Multi-source Images. We use RGB images and near infrared (NIR) images as multi-source images in our experiments. The database of the RGB/NIR images captured in different fields were released by The Chinese University of Hong Kong [7].

Parameter Setting. To empirically evaluate alternative methods for matching objects or scenes with different scales, we keep the target image as its original size and resize the reference image into 80% of its original size. Our hypergraph-based refining matching method has two parameters σ and θ, scaling the similarity between a pair of potential triple-point matches and the confidence of the matching results, respectively. To effectively measure the similarity, the value of σ is empirically set to less than 0.001. Small θ values allow more accuracy of the selection of potential matching points.

Results. For the evaluation, different matching methods for the target image and the reference image of different objects and scenes are tested in MATLAB. To make a quantitative experimental evaluation, we count the accuracy of matches for the SIFT-based matching, the mutual information (MI) [12] based cross-field matching and our hypergraph-based matching. The accuracy of different matching methods are given in Table 1, which shows that our method outperforms the other two matching methods and achieves high accuracy in refining multi-source feature correspondences. Fig. 4, 5 and 6 show the matching results of different matching methods. Each figure consists of the original multi-source images, the SIFT-based matching performances, the MI-based matching performances and our hypergraph-based matching performances.

Table 1. Comparison of matching results

	Books	Bowls	Cave	Day	Haze	Teapot
SIFT	0.90	0.90	0.83	0.36	0.92	0.96
MI	0.98	0.99	0.91	0.75	1.00	1.00
Ours	1.00	1.00	1.00	0.75	1.00	1.00

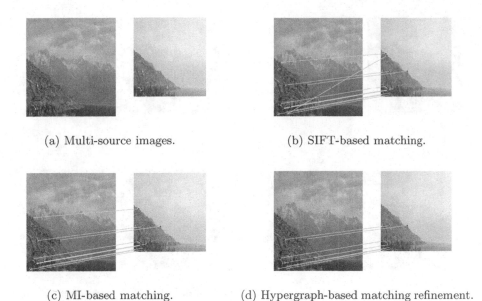

(a) Multi-source images.   (b) SIFT-based matching.

(c) MI-based matching.   (d) Hypergraph-based matching refinement.

Fig. 4. Matching results for haze images

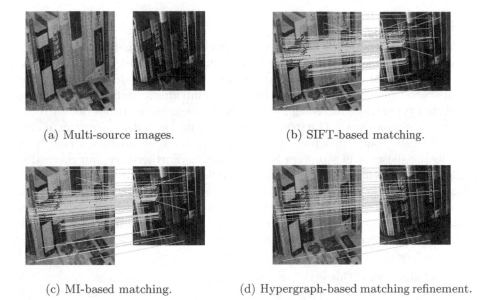

(a) Multi-source images.   (b) SIFT-based matching.

(c) MI-based matching.   (d) Hypergraph-based matching refinement.

Fig. 5. Matching results for books images

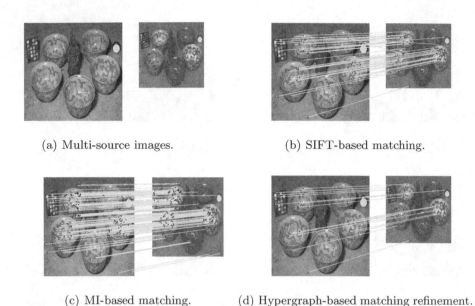

(a) Multi-source images.   (b) SIFT-based matching.

(c) MI-based matching.   (d) Hypergraph-based matching refinement.

Fig. 6. Matching results for bowls images

4   Conclusion

In this paper, a hypergraph matching framework for refining multi-source feature correspondences has been proposed. The framework we presented jointly considers both feature similarities and spatial feature layouts. Our framework takes feature-based matching results as the coarse matching results. Then we have conducted hypergraph matching for rejecting mismatches and thus obtained the refined correspondences. Our framework has exhibited effectiveness in multi-source feature points matching, because our method shares the virtues from both feature-based and structure-based strategies. Furthermore, our method is computationally efficient because the association hypergraph established based on coarse correspondences have considerable smaller number of vertices than those hypergrap-based matching methods without the coarse matching. Quantitative evaluation in our experiments has shown that our method achieves high accuracy than alternative matching methods.

Acknowledgement. This work was supported by grants from National Natural Science Foundation of China (Project No. 61305012), Shandong Outstanding Young Scientist Foundation (Project No. BS2013DX006) and Qingdao Fundamental Research Project (Project No. 13-1-4-256-jch).

References

1. Besl, P.J., McKay, N.D.: A method for registration of 3-D shapes. IEEE Trans. Pattern Anal. Mach. Intell. 14, 239–256 (1992)
2. Belongie, S., Malik, J., Puzicha, J.: Shape matching and object recognition using shape contexts. IEEE Trans. Pattern Anal. Mach. Intell. 24, 509–522 (2002)
3. Brown, M., Lowe, D.G.: Recognising panoramas. In: Internat. Conf. on Computer Vision (ICCV), vol. 2, pp. 1218–1225 (2003)
4. Lowe, D.G.: Distinctive image features from scale-invariant keypoints. Int. J. Comput. Vision 2, 91–110 (2004)
5. Bay, H., Ess, A., Tuytelaars, T., Van Gool, L.: SURF: speeded up robust features. Computer Vision and Image Understanding 110, 346–359 (2008)
6. Rublee, E., Rabaud, V., Konolige, K., Bradski, G.: ORB: an efficient alternative to SIFT or SURF. In: Internat. Conf. on Computer Vision (ICCV), pp. 2564–2571 (2011)
7. Yan, Q., Shen, X., Xu, L., Zhuo, S., Zhang, X., Shen, L., Jia, J.: Cross-field joint image restoration via scale map. In: Internat. Conf. on Computer Vision (ICCV), pp. 1537–1544 (2013)
8. Lowe, D.G.: Object Recognition from Local Scale-Invariant Features. In: Internat. Conf. on Computer Vision (ICCV), vol. 2, pp. 1150–1157 (1999)
9. Ren, P., Wilson, R.C., Hancock, E.R.: High Order Structural Matching Using Dominant Cluster Analysis. In: Maino, G., Foresti, G.L. (eds.) ICIAP 2011, Part I. LNCS, vol. 6978, pp. 1–8. Springer, Heidelberg (2011)
10. Duchenne, O., Bach, F., Kweon, I., Ponce, J.: A Tensor-Based Algorithm for High-Order Graph Matching. IEEE Trans. Pattern Anal. Mach. Intell. 33, 2383–2395 (2011)
11. Conte, D., Foggia, P., Sansone, C., Vento, M.: Thirty Years of Graph Matching in Pattern Recognition. Int. J. Pattern Recognition and Artificial Intelligence 18, 265–298 (2004)
12. Gong, M., Zhao, S., Jiao, L., Tian, D., Wang, S.: A novel coarse-to-fine scheme for automatic image registration based on SIFT and mutual information. IEEE Trans. Geoscience and Remote Sensing 52, 4328–4338 (2014)
13. Lerman, G., Whitehouse, J.T.: On d-dimensional d-semimetrics and simplex-type inequalities for high-dimensional sine functions. J. Approximation Theory 156, 52–81 (2009)

Kite Recognition by Means of Graph Matching

Kamel Madi[1], Hamida Seba[1(✉)], Hamamache Kheddouci[1],
Charles-Edmont Bichot[2], Olivier Barge[3], Christine Chataigner[3],
Remy Crassard[3], Emmanuelle Reganon[3], and Emmanuelle Vila[3]

[1] LIRIS, CNRS, UMR5205, Université de Lyon,
Université Lyon 1, 69622, Lyon, France
[2] LIRIS, UMR5205, Ecole Centrale de Lyon, 69134, Lyon, France
[3] CNRS, UMR 5133 Archéorient, Lyon, France
hamida.seba@univ-lyon1.fr

Abstract. Kites are remnants of long stone walls that outline the shape of a child's kite. But the kites are huge, their big size makes them often clearly visible on high-resolution satellite images. Identified at first in the Near East, their area of distribution is getting larger and larger. This wide distribution gives new dimensions in the interpretation of these structures. Consequently, a large scale recognition of kites will help archeologists to understand the functionality of these enigmatic constructions. In this paper, we investigate how the satellite imagery can be exploited in this purpose using a graph representation of the kites. We propose a similarity measure and a kite identification process that can highlights the preservation state of the kites. We also construct from real images a benchmark of kite graphs that can be used by other researchers.

Keywords: Kite recognition · graph matching · edit distance · satellite image

1 Background and Motivation

Kites are archaeological structures made of long walls converging on an enclosure that has small cells in its periphery (see Figure 1(a)). Kites have been discovered in the Middle East since the 1920s. Despite several studies, the issues related to their age and function remain without satisfactory answers [4, 8]. The kites are thus an underestimated phenomenon. Establishing the duration of their utilization, outlining their use and functioning, and trying to identify the population who are behind these constructions are the challenges that would highlight the significance of this unknown phenomenon. However, these issues cannot be seriously addressed without an almost exhaustive inventory of these structures.

In this paper, we investigate how the satellite imagery can be exploited in this purpose using a graph representation of the kites. Kites are naturally representable with graphs. A graph $G = (V, E)$ is a representation tool composed of a set of vertices V and a set of edges E with the cardinalities $|V(G)| = n$ and $|E(G)| = m$ where n is called the order of the graph and m its size. The set of edges E is a subset of $V \times V$ such that $(u, v) \in E$ means that vertices u and v

© Springer International Publishing Switzerland 2015
C.-L. Liu et al. (Eds.): GbRPR 2015, LNCS 9069, pp. 118–127, 2015.
DOI: 10.1007/978-3-319-18224-7_12

are connected. Usually, a finite number of labels are associated with vertices and edges. Graphs are commonly used as representation tools in many applications of structural pattern recognition and classification. For these kind of applications, graph comparison is a fundamental issue. Graph matching and more generally graph comparison is a powerful tool for structural pattern recognition [1, 5, 19]. Its has been used in the recognition process of architectonic/urbanistic elements in aerial/satellite images such as building or Gable-Roof detection [2, 20]. Graph comparison solutions are classified into two wide categories: exact approaches and fault-tolerant approaches. Exact approaches, such as those that test for graph isomorphism or sub-graph isomorphism [13, 18], refer to the methods that look for an exact mapping between the vertices and edges of a query graph and the vertices and edges of a target graph. Fault-tolerant graph comparison computes a distance between the compared graphs. This distance measures how similar are the graphs and allows to deal with the errors that are introduced by the processes needed to model objects by graphs. Several similarity measures are proposed in the literature using different approaches: graph kernels, graph embedding, maximum common subgraph, graph invariants, etc. We refer the reader to [1, 19] for more exhaustive surveys. We focus here on two main approaches that we use in the rest of the paper: graph edit distance (GED) and graph invariants. GED is one of the most powerful fault-tolerant graph matching tool. GED is based on a kind of graph transformations called edit operations. An edit operation is either an insertion, a suppression or a re-labeling of a vertex or an edge in the graph. A cost function associates a cost to each edit operation. The edit distance between two graphs is defined by the minimum costing sequence of edit operations that are necessary to transform one graph into the other [17]. This sequence is called an optimal edit path. However, computing the exact value of the edit distance between two graphs is NP-Hard for general graphs and induces a high computational complexity. This motivated several heuristics that approaches the exact value of GED in polynomial time using different methods such as dynamic programming and probability (see [6] for a detailed survey). In this context, [15] proposes a polynomial-time framework based on a fast bipartite assignment procedure mapping nodes and their local structures of one graph to nodes and their local structures of another graph. Similar approaches are also presented in [14]. Graph invariants have been efficiently used to solve the graph comparison problem in general and the graph isomorphism problem in particular. They are used for example in Nauty [13] which is one of the most efficient algorithm for graph and subgraph isomorphism testing. A vertex invariant, for example, is a number $i(v)$ assigned to a vertex v such that if there is an isomorphism that maps v to v' then $i(v) = i(v')$. Examples of invariants are the degree of a vertex, the number of cliques of size k that contain the vertex, the number of vertices at a given distance from the vertex, etc. Graph invariants are also the basis of graph probing [12] where a distance between two graphs is defined as the norm of their probes. Each graph probe is a vector of graph invariants. A generalization of this concept is also used in in [21] to compare biological data.

In this paper, we propose a graph comparison approach that deals with the following issues related to kite recognition:

1. to distinguish between two isomorphic graphs that have different geometry by using angles between two adjacent edges as well as distances between connected vertices in the similarity measure. In fact, almost all existing graph similarity measures compare the structures of graphs in terms of vertices, edges and and their labels but they do not consider the geometric forms of these graphs. This must be very important when the graphs represent objects with specific forms as for kite graphs presented later in the paper,
2. to deal with an important number of graphs: the challenge is to be able to process an important number of images in almost real-time.
3. to deal with disconnected graphs : depending on the state of preservation of the kites, the kite graphs may have several connected components.

The proposed approach combines the speed of graph invariant computation with the resilience of GED, in a multilevel framework consisting of: (1) a fast computable global similarity measure based on graph invariants computed on the compared graphs. This similarity aims to rapidly discard the graphs that can not be kites and avoid unnecessary and more costly comparisons; (2) a more accurate local similarity measure based on the geometric form and the structured features extracted from the graphs. This similarity uses GED to deal with the state of preservation of the kites; and (3) a reconstruction process that verifies if the different connected components of the graph constitute a kite.

We also provide a benchmark of kite graphs that we constructed from real images. This benchmark can be used to evaluate other graph matching methods.

The rest of the paper is organized as follows: in Section 2 we describe how the kites are represented by graphs and the proposed similarity measure. Section 3 gives the results of our experiments. Section 4 concludes the paper and describe our future work.

2   A Graph Based Approach for Kite Recognition

In this section, we present our contribution which is twofold:

1. The construction of a valuable dataset of kite graphs extracted from real images. This dataset comes enriching the existing graph databases used by researchers to evaluate graph matching and similarity measures.
2. A graph-based similarity measure for kite recognition.

We will use the notations summarized in Table 1 in the remaining of the paper. We consider simple undirected graphs with labels on both vertices and edges. Furthermore, we use the angles between two consecutive edges to record the geometric form of the graphs. We first describe the process of extracting kite graphs from images. Then, we give the detailed algorithms of our similarity measure.

Table 1. Notation

Symbol	Description
$G(V, E)$	undirected labeled graph with V its vertex set and E its edge set. Both vertices and edges are labeled.
$V(G)$	vertex set of graph G.
$E(G)$	edge set of graph G.
$deg(v)$	degree of vertex v.
$\Delta(G)$	the greatest vertex degree in graph G.
$\ell(e)$	the label of edge e.
$\mathcal{A}(G)$	the greatest angle in graph G.
$\mathcal{L}(G)$	the greatest edge label in graph G.
$\|S\|$	size of the set S.

2.1   From Kite Images to Kite Graphs: Construction of the Dataset

The structure of a Kite can be naturally represented by a graph. On satellite images, Kites appears as flat surfaces delimited by a set of lines. To convert kites' images into attributed graphs, the first stage is to extract the kites from the images by edge detection. Edge (segment or line) detection in images is an intensively studied topic in image analysis [3, 9]. Besides, several recent methods such as [7, 11, 16] give good result on satellite images. The main difficulty with such methods is to find the adequate settings to obtain an acceptable segment detection for a specific application. For the kites, we investigated several solutions with various settings and the LSD algorithm [7] gave us the most satisfactory set of segments (see Figure 1(b)). However, we added a refinement step to delete isolated segments and to merge adjacent ones (see Figure 1(c)). Then, a skeleton is generated by thinning the segments (see Figure 1(d)). The graph is constructed from the skeleton by representing lines by edges and ending points of lines by vertices (Figure1(e)). Each vertex is labeled with a two-dimensional attribute giving its position and an n-dimensional attribute containing the angles between every pair of consecutive incident edges. According to the state of preservation of the kite, a graph obtained by this process can have a single connected component (i.e., the kite is totally preserved) or it can be composed by two or several connected components (i.e., some parts of the kite have been destroyed).

We executed our algorithm on an important number of images containing kites with different state of preservation. We classified the obtained graphs into three preservation levels:

1. State I: the kite is entire and well preserved. The kite graph may be disconnected but the disconnections are neither frequent nor important.
2. State II: the kite graph is very disconnected. Some parts of the kite are not present.
3. State III: The graph is not a kite. These graphs are obtained by executing the algorithm on images that do not contain kites.

Figure 2 depicts some examples in each case. We have actually 200 graphs in each dataset with the characteristics summarized in Table 2.

With the help of the archeologists, we selected from the graphs in State-I dataset, the most preserved kites and modified them manually to obtain prototype kite graphs. Also, to be able to deal with disconnected kite graphs without

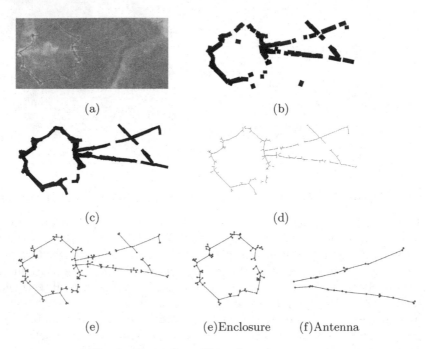

(a)   (b)

(c)   (d)

(e)   (e)Enclosure   (f)Antenna

Fig. 1. Illustration of kite detection

Table 2. Graph Dataset Characteristics

$avg|V|$: average number of vertices. $avg|E|$: average number of edges. $max|V|$: maximum number of vertices. $max|E|$: maximum number of edges. $avgAng$: average value of the angles. $maxAng$ maximum angle value.

avg $\|V\|$	max $\|V\|$	avg $\|E\|$	max $\|E\|$	avg Ang	max Ang
13	622	14	754	124	180

adding significant computing costs, we constructed a prototype graph for each kite component, namely : antenna and enclosure. Figures 1(e) and 1(f) give respectively an exemvale of a kite enclosure and a kite antenna.

2.2   A Multi-level Graph Comparison Approach for Kite Recognition

The proposed kite recognition solution consists of three levels: a global similarity measure, a local similarity measure and a reconstruction process.

The global similarity measure aims to rapidly discard objects that cannot be kites based on the size and the form of the objects compared to the prototypes. It computes graph invariants. The global similarity between two graphs G_1 and G_2 is given by:

$$G\_Sim(G_1, G_2) = w_1 * d_{Convex}(G_1, G_2) + w_2 * d_{Edges}(G_1, G_2) \\ + w_3 * d_{Angles}(G_1, G_2) + w_4 * d_{Vertices}(G_1, G_2)$$

(1)

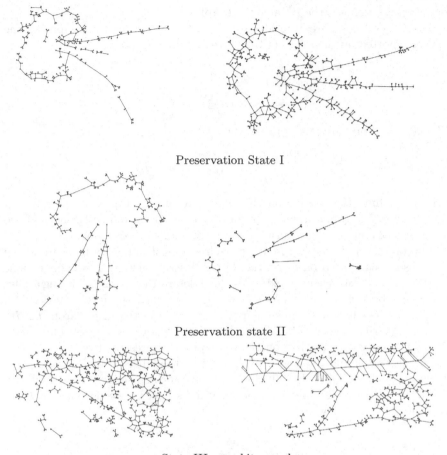

Preservation State I

Preservation state II

State III: non kite graphs

Fig. 2. Dataset

where $w_i, i = 1, 4$ are weighting coefficients with $w_1 + w_2 + w_3 + w_4 = 1$, $d_{Convex}(G_1, G_2)$ and $d_{Angles}(G_1, G_2)$ compare the global geometric forms of the two kites based on the convexity of the angles and the total value of the angles respectively:

$$d_{Convex}(G_1, G_2) = \left| \frac{\|Angles_{G_1} < ConvexityThreshold\|}{\|Angles_{G_1}\|} - \frac{\|Angles_{G_2} < ConvexityThreshold\|}{\|Angles_{G_2}\|} \right| \qquad (2)$$

where $ConvexityThreshold$ is at most equal to 180° but it will be defined according the form of the kites, $Angles_{G_i}$ denotes the set of angles of graph G_i.

$$d_{Angles}(G_1, G_2) = \frac{|\sum Angle_{i, G_1} - \sum Angle_{j, G_2}|}{Max(\sum Angle_{i, G_1}, \sum Angle_{j, G_2})} \qquad (3)$$

where $Angle_{i,G}$ denotes the ith angle of graph G.

$d_{Edges}(G_1, G_2)$ compares the global size of the two kites by comparing the distances reported on the edges of the corresponding graphs.

$$d_{Edges}(G_1, G_2) = \frac{\left| \sum\limits_{i=1}^{\|E(G_1)\|} \ell(e_i) - \sum\limits_{j=1}^{\|E(G_2)\|} \ell(e_j) \right|}{Max\left(\sum\limits_{i=1}^{\|E(G_1)\|} \ell(e_i), \sum\limits_{j=1}^{\|E(G_2)\|} \ell(e_j) \right)} \tag{4}$$

$d_{Vertices}(G_1, G_2)$ compares the order of the two graphs.

$$d_{Vertices}(G_1, G_2) = \frac{\left| \|V(G_1)\| - \|V(G_2)\| \right|}{Max(\|V(G_1)\|, \|V(G_2)\|)} \tag{5}$$

The algorithm takes as inputs the three prototype graphs, i.e., $G_{Antenna}$, $G_{Enclosure}$ and G_{Kite} and a query graph. For each connected component of the query graph, the algorithm returns the most similar kite component.

The second level of the proposed approach is a local similarity measure where the graphs are compared based on their local descriptions using edit operations as in [15] and [14]. However, we extended local descriptions by considering angles degree and labels of the edges. We denote the obtained similarity measure by $GeoL\_Sim$ for Geometric Local similarity measure to distinguish it from L_Sim the local similarity measure of [15] and [14] that do not uses angles. Thus, with $GeoL\_Sim$ each vertex of the compared graphs is represented by a signature, i.e. a vector, containing the degree of the vertex, the labels of its incident edges and the angles between each pair of consecutive incident edges. The similarity between two vertices v_1 and v_2 is computed as a distance between their signatures $S(v_1)$ and $S(v_2)$ as follows:

$$
\begin{aligned}
d(S(v_1), S(v_2)) &= \omega_1 * \frac{|deg(v_1) - deg(v_2)|}{Max(\Delta(G_1), \Delta(G_2))} \\
&+ \omega_2 * \sum_{k=1}^{Max(\Delta(G_1), \Delta(G_2))} \frac{|\ell(e_{1,k}) - \ell(e_{2,k})|}{Max(\Delta(G_1), \Delta(G_2)) * Max(\mathcal{L}(G_1), \mathcal{L}(G_2))} \\
&+ \omega_3 * \sum_{k=1}^{Max(\Delta(G_1), \Delta(G_2))-1} \frac{|Ang_{1,k} - Ang_{2,k}|}{(Max(\Delta(G_1), \Delta(G_2)) - 1) * Max(\mathcal{A}(G_1), \mathcal{A}(G_2))}
\end{aligned}
\tag{6}
$$

where where $\omega_i, i = 1, 4$ are weighting coefficients with $\omega_1 + \omega_2 + \omega_3 = 1$, $Ang_{i,k}$ is the kth angle of vertex v_i.

Using these signatures, we construct a matrix M where $M[i, j] = d(S(v_i), S(v_j))$ for each pair of vertices v_i and v_j such that $v_i \in V(G_1)$ and $v_j \in V(G_2)$. We define the local similarity measure between G_1 and G_2 by:

$$GeoL\_Sim(G_1, G_2) = \min_{h} \sum_{v_i \in V(G_1), h(v_i) \in V(G_2)} d(S(v_i), S(h(v_i))) \tag{7}$$

where $h : V(G_1)) \rightarrow V(G_2))$ is a bijection.

Computing this distance is then equivalent to solve the affectation problem in the matrix $|V(G_1| \times |V(G_2)|$ in which each element represents the distance between the signature of the ith vertex of (G_1)) and the signature of jth vertex of G_2. To do so, we use the Hungarian algorithm [10] to obtain the minimum cost in $O(n^3)$ time.

Similarly to the global similarity algorithm, the local similarity algorithm takes as input a query graph representing an image and determines for each connected component of the query graph the most similar kite component.

When a query graph passes the local similarity measure with more than one connected component classified as a kite part, we need to know if these kite parts are parts of the same kite or belongs to different kites. This is the aim of the reconstruction step that uses the coordinates of the vertices to eventually reconstruct the entire kite from different components: i.e, enclosure and antennas.

3   Evaluation

For evaluation, we used all the available graphs in the real dataset described in Section 2. We undertook three series of experiments to evaluate the robustness and the accuracy of our similarity measure. We compared our approach with the initial framework based on local structures comparison presented in [15] and [14] and described in Section 1. This approach is denoted L_Sim and do not use angles in the local structures of the vertices. In the first experiment, we focused on determining the optimal values for $Form_{threshold}$ and the weighting coefficients of the global and the local similarity measures. Based on these experiments that concerned mainly the well preserved kite graphs (State-I dataset) and the non kite graphs (State-III dataset), we set $Form_{threshold} = 0.5$ and gave an important value to the weight of the labels of the edges, i.e., the distances. $ConvexityThreshold$ was set to $150°$.

In the second series of experiments we evaluated the scalability of our approach over an increasing number of nodes in the query graphs. From the 600 graphs of the dataset, we constructed a set of query groups : $Q_{10},...,Q_{620}$ each of which contains 10 graphs. All the graphs of the same query group Q_i have the same number of vertices, i.e., i. This means that the number of vertices of the query graphs varies from 10 to 620. This variation is obtained by adding 10 graphs till Q_{120} than by by adding 100 graphs. Figure 3 shows the average runtime performance of G_Sim, $GeoL\_Sim$ and L_Sim. The X-axis shows the number of vertices contained in the query graph and the Y-axis the average runtime, in log scale, obtained over the query group of the corresponding graph size when compared to the 4 kite prototype graphs. This figure shows clearly the interest of using the global similarity level which is largely faster than the local similarity measure. The runtime performance of the the two local similarity measures $GeoL\_Sim$ and L_Sim are equivalent and confirm their theoretic polynomial time complexity.

Finally, we evaluated the accuracy of the proposed approach by performing classification. Table 3 depicts our results. We can see that $GeoL\_Sim$ is more accurate than L_Sim on the three datasets. This confirms that considering angles has an important added value for kite recognition. We can also see that

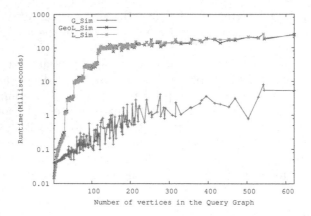

Fig. 3. Runtime Vs number of vertices

Table 3. Classification

Dataset	G_Sim	$GeoL\_Sim$	L_Sim
$State - I$	61	94	90
$State - II$	46	80	60
$State - III$	71	67.5	43

$GeoL\_Sim$ and L_Sim achieve better than G_Sim in the positive datasets (those that contains kites). However G_Sim is better in the negative dataset. So, by using successively G_Sim and $GeoL\_Sim$ we achieve good performance in both cases.

4 Conclusions

In this paper, we make use of the structural information of kites in a graph-based recognition framework, in which we take a query image, retrieve a graph representation of the image, and perform detailed matching to verify the similarity of the query graph with a database of kite prototype graphs. We propose a graph comparison framework that deals with the geometric form of the kite graphs and their disconnection. The key idea is to first discard the non kite graphs using a rapidly computable similarity measure based mainly on the geometric form of the kites. Our preliminary experimentation results are acceptable and encouraging. As a perspective, we are interested by comparing our approach with other graph similarity measures mainly those based on kernels and also with non graph based recognition solutions by implementing a feature and classifier based approach.

Acknowledgement. This work was supported by the LABEX IMU (ANR-10-LABX-0088) of Université de Lyon, within the program "Investissements d'Avenir" (ANR-11-IDEX-0007) operated by the French National Research Agency (ANR).

References

1. Bunke, H., Riesen, K.: Recent advances in graph-based pattern recognition with applications in document analysis. Pattern Recognition 44, 1057–1067 (2011)
2. Chung, Y.C., Han, T., He, Z.: Building recognition using sketch-based representations and spectral graph matching. In: 2009 IEEE 12th International Conference on Computer Vision, pp. 2014–2020 (September 2009)
3. Desolneux, A., Moisan, L., Morel, J.M.: Meaningful alignments. Int. J. Comput. Vision 40(1), 7–23 (2000)
4. Échallier, J.C., Braemer, F.: Nature and function of 'desert kites': new datta and hypothesis. Paléorient 21(1), 35–63 (1995)
5. Foggia, P., Percannella, G., Vento, M.: Graph matching and learning in pattern recognition in the last 10 years. International Journal of Pattern Recognition and Artificial Intelligence 28(01), 1450001 (2014)
6. Gao, X., Xiao, B., Tao, D., Li., X.: A survey of graph edit distance. Pattern Analysis Applications (13), 113–129 (2010)
7. von Gioi, R., Jakubowicz, J., Morel, J.M., Randall, G.: Lsd: A fast line segment detector with a false detection control. IEEE Transactions on Pattern Analysis and Machine Intelligence 32(4), 722–732 (2010)
8. Helms, S., Betts, A.V.G.: The desert 'kites' of the badiyat esh-sham and north arabia. Paléorient 13(1), 41–67 (1987)
9. Illingworth, J., Kittler, J.: A survey of the hough transform. Computer Vision, Graphics, and Image Processing 44(1), 87–116 (1988)
10. Kuhn, H.W.: The Hungarian method for the assignment problem. Naval Research Logistics Quarterly 2, 83–97 (1955)
11. Lacoste, C., Descombes, X., Zerubia, J.: Point processes for unsupervised line network extraction in remote sensing. IEEE Transactions on Pattern Analysis and Machine Intelligence 27(10), 1568–1579 (2005)
12. Lopresti, D.P., Wilfong, G.T.: Comparing Semi-Structured Documents via Graph Probing. In: Workshop on Multimedia Information Systems, pp. 41–50 (2001)
13. McKay, B.: Practical graph isomorphism. Congress Numerantium 87, 30–45 (1981)
14. Raveaux, R., Burie, J.C., Ogier, J.M.: A graph matching method and a graph matching distance based on subgraph assignments. Pattern Recognition Letters 31, 394–406 (2010)
15. Riesen, K., Bunke, H.: Approximate graph edit distance computation by means of bipartite graph matching. Image and Vision Computing 27, 950–959 (2009)
16. Rochery, M., Jermyn, I.H., Zerubia, J.: Higher order active contours. Int. J. Comput. Vision 69(1), 27–42 (2006)
17. Sanfeliu, A., Fu, K.: A distance measure between attributed relational graphs for pattern recognition. IEEE Transactions on Systems, Man, and Cybernetics (Part B) 13(3), 353–363 (1983)
18. Ullmann, J.R.: An Algorithm for Subgraph Isomorphism. J. ACM 23(1), 31–42 (1976), http://dx.doi.org/10.1145/321921.321925
19. Vento, M.: A long trip in the charming world of graphs for pattern recognition. Pattern Recognition 48(2), 291–301 (2015)
20. Wang, Q., Jiang, Z., Yang, J., Zhao, D., Shi, Z.: A hierarchical connection graph algorithm for gable-roof detection in aerial image. IEEE Geoscience and Remote Sensing Letters 8(1), 177–181 (2011)
21. Xiao, Y., Dong, H., Wu, W., Xiong, M., Wang, W., Shi, B.: Structure-based graph distance measures of high degree of precision. Pattern Recognition 41, 3547–3561 (2008)

GEM++: A Tool for Solving Substitution-Tolerant Subgraph Isomorphism

Julien Lerouge[1], Pierre Le Bodic[2], Pierre Héroux[1], and Sébastien Adam[1(✉)]

[1] LITIS EA 4108, BP 12, University of Rouen,
76801 Saint-Etienne du Rouvray, France
[2] H. Milton Stewart School of Industrial and Systems Engineering
Georgia Institute of Technology
765 Ferst Drive NW, Atlanta, 30332, United States
{Julien.Lerouge,Pierre.Heroux,Sebastien.Adam}@litislab.eu,
lebodic@gatech.edu

Abstract. The substitution-tolerant subgraph isomorphism is a particular error-tolerant subgraph matching that allows label substitutions for both vertices and edges. Such a matching is often required in pattern recognition applications since graphs extracted from images are generally labeled with features vectors computed from raw data which are naturally subject to noise. This paper describes an extended version of a Binary Linear Program (BLP) for solving this class of graph matching problem. The paper also presents GEM++, a software framework that implements the BLP and that we have made available for the research community. GEM++ allows the processing of different sub-problems (induced isomorphism or not, directed graphs or not) with complex labelling of vertices and edges. We also present some datasets available for evaluating future contributions in this field.

Keywords: Binary linear programming · Subgraph isomorphism · Graph matching toolkit · Graph datasets

1 Introduction

Given a query and a target graph, the subgraph isomorphism problem consists in deciding whether there exists a subgraph of the target which is isomorphic to the query graph, i.e. there exists a one-to-one mapping between the vertices (respectively the edges) of the query and the vertices (resp. the edges) of a subgraph of the target. This theoretical problem has been the subject of many contributions in the literature [1, 2, 8, 11–13] since it finds applications in various fields such as biosciences, chemistry, knowledge management, social network analysis, scene analysis... In the context of structural pattern recognition for image analysis, algorithms targeting subgraph isomorphism are particularly useful since they can be used to simultaneously consider segmentation and recognition of objects of interest (the query graphs) in a whole image (the target graph).

Beyond its computational complexity (subgraph isomorphism is an NP-complete problem [7]), a main drawback of existing approaches for subgraph

© Springer International Publishing Switzerland 2015
C.-L. Liu et al. (Eds.): GbRPR 2015, LNCS 9069, pp. 128–137, 2015.
DOI: 10.1007/978-3-319-18224-7_13

isomorphism is the requirement of a strict matching between the query graph and a subgraph in the target graph. Now, in pattern recognition applications, graphs to be analyzed are usually affected by distorsions, that can result from the intrinsic variability of patterns in the image, from the digitization procedure, or from the graph construction steps. Among the possible distorsions, we are concerned in this paper with modifications of vertices/edges labels. Hence, we tackle a particular class of matching problem we call substitution-tolerant subgraph isomorphism. A subgraph isomorphism is said to be substitution-tolerant when editing operations on vertices and edges labels are allowed at a given cost. The graph topology remains unchanged.

In [9], we have considered the substitution-tolerant subgraph isomorphism as an optimization problem, modeled with a Binary Linear Program (BLP). A first formulation has been described and some tests have shown the efficiency of the approach for a real world application concerning symbol detection. This paper extends this work with three contributions. First, from a theoretical point of view, we present two extensions to the previous formulation. They concern (i) the handling of both directed or undirected graphs and (ii) the computation of induced subgraph matching. The second contribution is related to the implementation of the formulation in a framework called GEM++. The software implements the BLP with the ability to tackle many graph labelling functions (nominal, real, vectorial...) and to customize vertices and edges edit costs according to these labellings. GEM++ is available online for the research community[1]. Finally, we present some datasets dedicated to the evaluation of substitution-tolerant subgraph isomorphism approaches. These datasets are free to use for evaluating future contributions in this field.

The paper is organized as follows. Section 2 presents the extended formulation, in order to introduce the description of the software toolkit GEM++ which is given in section 3. Then, in section 4, we describe synthetic and real datasets dedicated to substitution-tolerant subgraph isomorphism. Finally, section 5 draws a conclusion of the paper and proposes future directions.

2 Proposed Formulations

2.1 Definitions

Definition 1. A *graph* $\mathcal{G}$ is a couple $\mathcal{G} = (V_{\mathcal{G}}, E_{\mathcal{G}})$, where $V_{\mathcal{G}}$ is a set of vertices and $E_{\mathcal{G}} \subseteq V_{\mathcal{G}} \times V_{\mathcal{G}}$ is a set of edges. The graph $\mathcal{G}$ is said undirected if its edges have no orientation, i.e. $\forall (i, j) \in E_{\mathcal{G}}, (i, j) = (j, i) \in E_{\mathcal{G}}$. Otherwise, $\mathcal{G}$ is a directed graph.

Definition 2. Given $\mathcal{S} = (V_{\mathcal{S}}, E_{\mathcal{S}})$ and $\mathcal{G} = (V_{\mathcal{G}}, E_{\mathcal{G}})$ two graphs verifying $|V_{\mathcal{S}}| \leq |V_{\mathcal{G}}|$, an injective function $f : V_{\mathcal{S}} \to V_{\mathcal{G}}$ is a *subgraph isomorphism* from a graph $\mathcal{S}$ to a graph $\mathcal{G}$ if and only if f verifies $\forall (i, j) \in V_{\mathcal{S}} \times V_{\mathcal{S}}, (i, j) \in E_{\mathcal{S}} \Rightarrow (f(i), f(j)) \in E_{\mathcal{G}}$.

---

[1] http://litis-ilpiso.univ-rouen.fr/

2.2   Binary Linear Programming

Binary linear programming, also known as 0-1 linear programming, is a restriction of linear programming where the variables are binary. Hence, the general form of a BLP is as follows:

$$\min_{x} c^T x \tag{1a}$$

$$\text{subject to } Ax \le b \tag{1b}$$

$$x \in \{0,1\}^n \tag{1c}$$

where $c \in \mathbb{R}^n, A \in \mathbb{R}^{n \times m}$ and $b \in \mathbb{R}^m$ are data of the problem. A solution of this optimization problem is a vector x of n binary variables. A is used to express linear inequality constraints (1b). The objective function $c^T x$ is a linear combination of all variables of x weighted by the components of the vector c. The optimal solution is the one that minimizes the objective function (1a) and respects constraints (1b) and (1c). Once equations (1a) to (1c) are correctly formulated, the second step consists in implementing this model using a mathematical solver. Given an instance of the problem, the solver explores the tree of solutions with the branch-and-cut algorithm, and finds the best feasible solution, in terms of the objective function optimization.

2.3   Solving the Problem with a BLP

As shown in figure 1, the formulation of the substitution-tolerant subgraph isomorphism problem as a BLP implies the definition of two sets of variables:

- for each pair of vertices $i \in V_{\mathcal{S}}$ and $k \in V_{\mathcal{G}}$, there is a variable $x_{i,k}$, such that $x_{i,k} = 1$ if vertices i and k are matched together, 0 otherwise.
- for each pair of edges $ij \in E_{\mathcal{S}}$ and $kl \in E_{\mathcal{G}}$, there is a variable $y_{ij,kl}$, such that $y_{ij,kl} = 1$ if edges ij and kl are matched together, 0 otherwise.

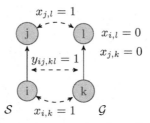

Fig. 1. An example of matching

Using such variables, the objective function, i.e. the global cost for matching $\mathcal{S}$ to a subgraph of $\mathcal{G}$, is defined as follows:

$$\min_{x,y} \left(\sum_{i \in V_{\mathcal{S}}} \sum_{k \in V_{\mathcal{G}}} c_V(i,k) * x_{i,k} + \sum_{ij \in E_{\mathcal{S}}} \sum_{kl \in E_{\mathcal{G}}} c_E(ij,kl) * y_{ij,kl} \right) \tag{2a}$$

where $c_V : V_S \times V_{\mathcal{G}} \to \mathbb{R}^+$ as well as $c_E : E_S \times E_{\mathcal{G}} \to \mathbb{R}^+$ are functions which define the cost of associating vertices and edges.

The following constraints encode the substitution-tolerant subgraph isomorphism problem (see [9] for a justification of these constraints):

$$\sum_{k \in V_{\mathcal{G}}} x_{i,k} = 1 \qquad \forall i \in V_S \tag{2b}$$

$$\sum_{kl \in E_{\mathcal{G}}} y_{ij,kl} = 1 \qquad \forall ij \in E_S \tag{2c}$$

$$\sum_{i \in V_S} x_{i,k} \leq 1 \qquad \forall k \in V_{\mathcal{G}} \tag{2d}$$

$$\sum_{kl \in E_{\mathcal{G}}} y_{ij,kl} \leq x_{i,k} \quad \forall k \in V_{\mathcal{G}}, \forall ij \in E_S \tag{2e}$$

$$\sum_{kl \in E_{\mathcal{G}}} y_{ij,kl} \leq x_{j,l} \quad \forall l \in V_{\mathcal{G}}, \forall ij \in E_S \tag{2f}$$

2.4   Extensions

Induced Subgraph Isomorphism. An *induced subgraph isomorphism* is a more stringent problem, defined by :

Definition 3. Given $S = (V_S, E_S)$ and $\mathcal{G} = (V_{\mathcal{G}}, E_{\mathcal{G}})$ two graphs verifying $|V_S| \leq |V_G|$, an injective function $f : V_S \to V_{\mathcal{G}}$ is an *induced subgraph isomorphism* from a graph S to a graph $\mathcal{G}$ if and only if $\forall (i,j) \in V_S \times V_S, (i,j) \in E_S \Leftrightarrow (f(i), f(j)) \in E_{\mathcal{G}}$.

The equivalence in definition 3 requires an additional set of constraints in the BLP :

$$\sum_{i \in V_S} x_{i,k} + \sum_{j \in V_S} x_{j,l} - \sum_{ij \in E_S} y_{ij,kl} \leq 1 \quad \forall kl \in E_{\mathcal{G}} \tag{2g}$$

Undirected Graphs. If S and $\mathcal{G}$ are undirected graphs, the set of constraints (2e) and (2f) are respectively modified into (2h) and (2i) and, if necessary, the set of constraints (2g) is modified into (2j) :

$$\sum_{kl \in E_{\mathcal{G}}} y_{ij,kl} \leq x_{i,k} + x_{j,k} \quad \forall k \in V_{\mathcal{G}}, \forall ij \in E_S \tag{2h}$$

$$\sum_{kl \in E_{\mathcal{G}}} y_{ij,kl} \leq x_{j,l} + x_{i,l} \quad \forall l \in V_{\mathcal{G}}, \forall ij \in E_S \tag{2i}$$

$$\sum_{i \in V_S} (x_{i,k} + x_{i,l}) + \sum_{j \in V_S} (x_{j,l} + x_{j,k}) - \sum_{ij \in E_S} y_{ij,kl} \leq 1 \quad \forall kl \in E_{\mathcal{G}} \tag{2j}$$

132   J. Lerouge et al.

2.5   Search for Multiple Solutions

Depending on the application context, it may be the case that the query graph that is searched for has many instances in the target graph. As defined in section 2, the BLP model is only able to find the optimal solution. There are multiple ways to search for many solutions [3]. In the context of our study, we have chosen to call iteratively the model and to discard the successive optimal solutions after each call. Such a solution is linear in the number of instances. We implemented three different strategies to discard an optimal solution. After each iteration, given the optimal solution $(\bar{x}\ \bar{y})^T$, our algorithm may perform one of the following cut by adding the corresponding constraint to the BLP:

- cut exactly the optimal solution:

$$\sum_{i \in V_S, k \in V_G} \bar{x}_{i,k} \cdot x_{i,k} \leq |V_S| - 1 \tag{3a}$$

- cut any solution containing a matching between vertices (or edges) of S and G present in the optimal solution:

$$\sum_{i \in V_S, k \in V_G} \bar{x}_{i,k} \cdot x_{i,k} = 0 \tag{3b}$$

- cut any solution involving a vertex (or an edge) of G already matched to a vertex (edge) of S in the optimal solution:

$$\sum_{i \in V_S, k \in V_G} \left(\sum_{j \in V_S} \bar{x}_{j,k} \right) \cdot x_{i,k} = 0 \tag{3c}$$

The new constraint, that cuts the current optimal solution, is added to the model. Hence, this solution becomes infeasible for the next run. The solver can be called again and will be able to find, if it exists, another optimal solution.

3   GEM++

GEM++ is a Graph Extraction and Matching software that we developed in order to implement the subgraph matching approach described in section 2. It is written in C++ and is multiplatform. The framework is composed of shared software libraries, that gathers the common functionalities provided by our framework (graphs and weights handling, integer programming, call to a mathematical solver), and the command-line utility GEM++sub, that solves our subgraph isomorphism problem.

3.1   Graphs and Substitution Costs

GEM++ is able to import directed or undirected, labeled or unlabeled graphs, formatted either in Graph Modeling Language (GML) or in Graph eXchange Language (GXL). Figure 2 shows two examples of directed labeled graphs. The target graph is transcribed in GXL and GML formats in listings 1.1 and 1.2.

```
1   <?xml version="1.0" encoding="UTF-8"?>
2   <gxl>
3   <graph id="query" edgemode="directed">
4   <node id="a"><attr name="x"><float>0.9</float></attr></node>
5   <node id="b"><attr name="x"><float>0.3</float></attr></node>
6   <edge from="a" to="b"><attr name="y"><float>0.4</float></attr></edge>
7   <edge from="b" to="a"><attr name="y"><float>0.2</float></attr></edge>
8   </graph>
9   </gxl>
```

Listing 1.1. GXL example (query.gxl)

```
1   graph [
2     label "target"
3     directed 1
4     node [ id 0 x 0.5 ]
5     node [ id 1 x 0.7 ]
6     node [ id 2 x 0.4 ]
7     edge [ source 0 target 1 y 0.6 ]
8     edge [ source 1 target 2 y 0.1 ]
9     edge [ source 2 target 1 y 0.8 ]
10  ]
```

Listing 1.2. GML example (target.gml)

```
1   nodes_features_weights
2   x 1.0
3   edges_features_weights
4   y 2.0
```

Listing 1.3. Feature weights (weights.fw)

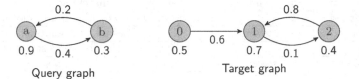

Query graph   Target graph

Fig. 2. An instance of the substitution-tolerant subgraph isomorphism problem

In order to tune the substitution costs c_V and c_E introduced in equation (2a), we defined a weighted euclidean distance on vertices and edges labels. Let us define $\mu : V_S \cup V_G \to L_V$ and $\xi : E_S \cup E_G \to L_E$ the functions that assign labels to each vertex and each edge of S and G. The vertices and edges label spaces L_V and L_E may be composed of any combination of numeric (real) and symbolic attributes. As symbolic attributes are taken from discrete sets, for the sake of simplicity, we transform them into strictly positive integers. Therefore, we have for the vertices $L_V = \mathbb{R}^m \times (\mathbb{N}^*)^n$, with $m \geq 0$ and $n \geq 0$. Let us define $\mu_\mathbb{R} : V_S \cup V_G \to \mathbb{R}^m$ and $\mu_{\mathbb{N}^*} : V_S \cup V_G \to (\mathbb{N}^*)^n$, such that $\forall v \in V_S \cup V_G, \mu(v) = (\mu_\mathbb{R}(v), \mu_{\mathbb{N}^*}(v))$. Let $\mathbf{w} = (w_1, \dots, w_m) \in \mathbb{R}^m$ and $\mathbf{z} = (z_1, \dots, z_n) \in \{0; 1\}^n$ be two vectors of weights, the vertex substitution cost c_V is defined as follows:

$$c_V(i, j) = \begin{cases} \infty & \text{if } \exists k \in [\![1; n]\!], z_k = 1 \text{ and } \mu_{\mathbb{N}^*}(i) \neq \mu_{\mathbb{N}^*}(j), \\ \sqrt{\sum_{k=1}^m w_k^2 (\mu_\mathbb{R}(i)_k - \mu_\mathbb{R}(j)_k)^2} & \text{otherwise.} \end{cases}$$

The binary $\mathbf{z}$ weights are used to ignore some symbolic attributes. The same system is applied to the edge substitution costs. The listing 1.3 gives an example of a feature weights configuration file for vertices and edges.

3.2   Installation, Solvers and Usage

For Ubuntu or Debian-based Linux distributions, we provide pre-compiled packages of GEM++ in .deb format for x86 and x64 architectures. We also provide a Windows installer for 64 bits systems.

In order to solve the formulation, GEM++ relies on mathematical programming solvers. GEM++ is compatible with Gurobi[2] and IBM CPLEX[3], two widespread and performant commercial solvers. Gurobi and IBM provide free academic licenses for their solvers. GEM++ is also compatible with the GNU Linear Programming Toolkit[4] (GLPK), a free-software part of the GNU Project, available in the package repositories of Debian and Ubuntu.

The GEM++ package provides the `GEM++sub` command-line utility. Provided a couple of graphs and label substitution weights, `GEM++sub` is able to solve both substitution-tolerant and exact subgraph isomorphism problems. The available options for the command are summed up in the table 1.

Table 1. Command-line options offered by GEM++

Complete name	Short	Argument	Effect
`--weights`	`-w`	*file*.fw	Configures the weights for substitution costs
`--solution`	`-o`	*file*.sol	Outputs the matching solution(s)
`--induced`	`-i`		Performs induced subgraph isomorphism
`--number`	`-n`	*integer*	Searches up to n solutions
`--cut`	`-c`	*cut_strategy*	Switch to the given cut strategy
`--solver`	`-s`	*solver_name*	Switch to the given solver

3.3   Example

Let us run a solving example of the substitution-tolerant subgraph isomorphism problem, taking the query and target graphs showed in figure 1 as input of the problem. As the topology must be preserved while matching the query graph to a subgraph of the target graph, the only two feasible solutions are $s_1 = \left\{ \begin{matrix} a \to 1 & ab \to 12 \\ b \to 2 & ba \to 21 \end{matrix} \right\}$ and $s_2 = \left\{ \begin{matrix} a \to 2 & ab \to 21 \\ b \to 1 & ba \to 12 \end{matrix} \right\}$.

Let us use the weights defined in the listing 1.3, i.e. $w_x = 1$ and $w_y = 2$. Then, the overall costs of the solutions s_1 and s_2 are 2.1 and 1.9 respectively, and s_2 is the optimal solution. This solution can be discarded using strategy (3a) or (3b), in either case the algorithm will find s_1 as optimal solution at the second iteration. However, the use of strategy (3c) would forbid any new solution, since vertices 1 and 2 were already matched in s_2.

---

[2] http://www.gurobi.com
[3] http://www-01.ibm.com/software/commerce/optimization/cplex-optimizer/
[4] http://www.gnu.org/software/glpk

Here is the command that must be run on a terminal to retrieve the two solutions in a file named `output.sol` (see listing 1.4), using the default (3a) cut strategy :

```
$ GEM++sub -o output.sol -w weights.fw -n 2 query.gxl target.gml
```

```
1   Solution {                      12   Solution {
2       objective value : 1.9       13       objective value : 2.1
3       nodes substitution {        14       nodes substitution {
4           a 2  0.5                 15           a 1  0.2
5           b 1  0.4                 16           b 2  0.1
6       }                           17       }
7       edges substitution {        18       edges substitution {
8           a->b 2->1  0.8          19           a->b 1->2  0.6
9           b->a 1->2  0.2          20           b->a 2->1  1.2
10      }                           21       }
11  }                               22   }
```

Listing 1.4. Output example (output.sol)

If we had used different feature weights, e.g. $w_x = 1$ and $w_y = 1$, the overall costs of the solutions s_1 and s_2 would have been 1.2 and 1.4 respectively, and thus the first optimal solution would have been s_1 instead of s_2. Therefore, the weighting system allows the user to tune the substitution costs in order to match the application needs.

4   Datasets

Although some datasets have been proposed to evaluate exact subgraph isomorphism [4, 6], to the best of our knowledge there are no datasets designed to benchmark the search of substitution-tolerant subgraph isomorphism. In this section, we present two datasets designed for that purpose and made available for the community.

4.1   Synthetic Datasets

The synthetic dataset is composed of four subparts. Each of these datasets contains pairs of graphs. One member of the pair is the query graph and the other is the target graph. For each graph pair, a groundtruth information gives the mapping between each vertex in the query graph to the corresponding vertex in the target graph.

`ILPIso_exact_synth` and `ILPIso_noisy_synth` datasets have been generated using the Erdös-Rényi model [5] which generates random graphs given n, the number of vertices, and p, the probability that two vertices u and v are linked with an edge (u, v). `ILPIso_exact_connected_synth` and `ILPIso_noisy_connected_synth` datasets have been generated with a model, also parameterized with n and p, with the same meaning, but the model used warrants that generated graphs are connected. The procedure used to generate graph pairs was the following:

1. The query graph in generated with a specific graph model generator.
2. A copy of the query is created.
3. In the noisy versions, vertex/edges labels are added a Gaussian noise.
4. The copy is completed to the desired size of the target graph.

Labels for vertices and edges are scalar values randomly drawn in $[-100, 100]$ according to a uniform distribution. For the noisy versions, labels are added a random number drawn from a Gaussian distribution ($m = 0, \sigma^2 = 5$).

Finally, each graph pair is parameterized by p, the probability of existence of an edge, n_p, n_t, which respectively denotes the number of vertices in the query graph and in the target graph. For ILPIso_exact_synth and ILPIso_noisy_synth datasets, parameters are such that $p \in \{0.01, 0.05, 0.1\}$, $n_p \in \{10, 25, 50\}$ and $n_t \in \{50, 100, 250, 500\}$. For ILPIso_exact_connected_synth and ILPIso_noisy_connected_synth datasets, the same values are used for n_p and n_t. p is chosen in $\{0.02, 0.05, 0.1\}$ but according to the value of n_p, the connectivity constraint prevents to select too low values of p.

4.2   Real Datasets

This ILPISO_real dataset is composed of 16 query graphs and 200 target graphs. These graphs respectively describe structural representations of architectural symbols and architectural flooplans from the SESYD dataset[5] in which these symbols appear in their context. The extraction of this graph representation is detailed in [10]. Basically, a vertex is associated to each white connected component of the image of the plan or symbol, and an edge is created between vertices representing adjacent white regions. The features that are used to label vertices are mainly shape descriptors (24 first Zernike moments), whereas edges are labeled with feature vectors that characterize the relationship between adjacent regions (relative scale and distance).

Considering their structural representations, the problem of locating occurrences of symbols on a floorplan turns into the search of subgraphs of the target graphs that are isomorphic with query graphs representing symbols. Each symbol may occur once or several times on a floorplan, or may not occur at all. For each target graph representing a flooplan, the groundtruth information provides the identifiers of vertices that are involved in the symbol occurrence. The whole ILPISO_real dataset contains 5609 symbol occurrences with an average of 28 occurrences per target graph. The graphs corresponding to symbol instances contain 4 vertices and 7 edges on average, whereas the structural representations of the plans contain 121 vertices and 525 edges on average.

5   Conclusion

In this paper, we have presented GEM++, a software which implements a BLP for the search for substitution-tolerant subgraph isomorphism. This work is an

---

[5] http://mathieu.delalandre.free.fr/projects/sesyd/

extension of [9], which now allows to handle undirected graphs, numeric and symbolic attributes, feature weighting, and induced sugraph isomorphism. The tool is available online, has been designed to be easily installed on several operating systems and may be customized at several levels. In particular, it can be used in conjuction with several mathematical solvers. We also provide synthetic and real graph datasets specifically designed for the problem of substitution-tolerant subgraph isomorphism. Our future works concern the tuning of the weights **w** and **z**, the vectors which define the relative weights between attributes in the substitution cost. They could be determined by a learning phase aiming at optimizing the performance on a validation set. Moreover, this work could also be continued by integrating tolerance to topological modifications.

Acknowledgement. This work was partly supported by grants from the ANR-11-JS02-010 Lemon.

References

1. Cordella, L.P., Foggia, P., Sansone, C., Vento, M.: Performance evaluation of the VF graph matching algorithm. In: Proc. of the Int'l Conf. on Image Analys. and Proc., pp. 1172–1177 (1999)
2. Cordella, L.P., Foggia, P., Sansone, C., Vento, M.: A (sub)graph isomorphism algorithm for matching large graphs. IEEE Trans. on PAMI 26(10), 1367–1372 (2004)
3. Danna, E., Fenelon, M., Gu, Z., Wunderling, R.: Generating multiple solutions for mixed integer programming problems. In: Fischetti, M., Williamson, D.P. (eds.) IPCO 2007. LNCS, vol. 4513, pp. 280–294. Springer, Heidelberg (2007)
4. De Santo, M., Foggia, P., Sansone, C., Vento, M.: A large database of graphs and its use for benchmarking graph isomorphism algorithms. Pattern Recogn. Lett. 24(8), 1067–1079 (2003)
5. Erdös, P., Rényi, A.: On random graphs. Public. Mathemat. 6, 290–297 (1959)
6. Foggia, P., Sansone, C., Vento, M.: A database of graphs for isomorphism and subgraph isomorphism benchmarking. In: Proc. Third IAPR TC-15 Int'l Workshop Graph Based Representations, pp. 176–187 (2001)
7. Garey, M.R., Johnson, D.S.: Computers and Intractability: A Guide to the Theory of NP-Completeness. Freeman & Co. (1979)
8. Ghahraman, D.E., Wong, A.K.C., Au, T.: Graph optimal monomorphism algorithms. IEEE Transactions on System, Man and Cybernetics 10, 181–188 (1980)
9. Le Bodic, P., Héroux, P., Adam, S., Lecourtier, Y.: An integer linear program for substitution-tolerant subgraph isomorphism and its use for symbol spotting in technical drawings. Pattern Recognition 45(12), 4214–4224 (2012)
10. Le Bodic, P., Locteau, H., Adam, S., Héroux, P., Lecourtier, Y., Knippel, A.: Symbol detection using region adjacency graphs and integer linear programming. In: Proc. of the Int'l Conf. on Doc. Analys. and Recog., pp. 1320–1324 (2009)
11. Solnon, C.: Alldifferent-based filtering for subgraph isomorphism. Artificial Intelligence 174(12-13), 850–864 (2010)
12. Ullmann, J.R.: An algorithm for subgraph isomorphism. J. ACM 23(1), 31–42 (1976)
13. Wong, A.K.C., You, M., Chan, S.C.: An algorithm for graph optimal monomorphism. IEEE Transactions on System, Man and Cybernetics 20(3), 628–638 (1990)

A Graph Database Repository and Performance Evaluation Metrics for Graph Edit Distance

Zeina Abu-Aisheh$^{(\boxtimes)}$, Romain Raveaux, and Jean-Yves Ramel

Laboratoire d'Informatique (LI), Université François Rabelais,
37200, Tours, France
{zeina.abu-aisheh,romain.raveaux,jean-yves.ramel}@univ-tours.fr
http://www.li.univ-tours.fr/

Abstract. Graph edit distance (GED) is an error tolerant graph matching paradigm whose methods are often evaluated in a classification context and less deeply assessed in terms of the accuracy of the found solution. To evaluate the accuracy of GED methods, low level information is required not only at the classification level but also at the matching level. Most of the publicly available repositories with associated ground truths are dedicated to evaluating graph classification or exact graph matching methods and so the matching correspondences as well as the distance between each pair of graphs are not directly evaluated. This paper consists of two parts. First, we provide a graph database repository annotated with low level information like graph edit distances and their matching correspondences. Second, we propose a set of performance evaluation metrics to assess the performance of GED methods.

Keywords: Graph edit distance · Performance evaluation metrics · Matching correspondence

1 Introduction

Attributed relational graphs are powerful representation structures that have been widely used to represent structural description of objects in pattern recognition (PR), computer vision and other related fields. When objects are represented by graphs, the problem of object comparison is turned into a graph matching (GM) one where an evaluation of structural and attributed similarity of two graphs has to be found [1]. The similarity evaluation, based on GM, consists in finding correspondences between vertices and edges of two graphs that satisfy some constraints ensuring that similar substructures in one graph are mapped to similar substructures in the other [2].

Among *error-tolerant* GM problems, GED is of great interests. GED is a graph matching paradigm whose concept was first reported in [3, 4]. Its basic idea is to find the best set of edit sequences that can transform graph g_1 into graph g_2 by means of edit operations on graph g_1. The allowed operations are inserting, deleting and/or substituting vertices and their corresponding edges.

© Springer International Publishing Switzerland 2015
C.-L. Liu et al. (Eds.): GbRPR 2015, LNCS 9069, pp. 138–147, 2015.
DOI: 10.1007/978-3-319-18224-7_14

In the literature, many exact and approximate approaches have been proposed to solve GED [5–8]. As a first step to evaluate such approaches, one needs to find repositories dedicated to evaluating GM in general or GED in particular.

Lots of graphs repositories have been made publicly available for the community [9, 10]. However, to the best of our knowledge, most of these repositories have been put forward for classification and clustering experiments. Moreover, only high level information has been given to the community such as the class labels of the objects represented by graphs. When evaluating classification, the matching quality is evaluated indirectly through a recognition rate which highly depends on the classifier and does not allow a clear analysis of the matching algorithms. On the other hand, low level information has not been provided. For instance, the matching between vertices and edges of each pair of graphs.

We, authors, believe that providing low level information is of great interest for understanding the behavior of GED methods in terms of accuracy and speed as a function of graph size and attribute types. Hence, instead of proposing yet a completely new graph database repository, we propose adding low level information for well-known and publicly used databases. For that purpose, the GREC, Mutagenicity, Protein, and CMU databases were selected [9, 11]. Added information consists of the best found distance for each pair of graphs as well as their vertex to vertex and edge to edge matching. This information helps at assessing the feasibility of exact and approximate methods.

Our repository aims at making a first step towards a GM repository that is able to assess the accuracy of error-tolerant GM methods. All the graph databases and their added information are publicly available[1]. Moreover, we propose novel performance evaluation metrics that aim at comparing a set of GED approaches based on several significant criteria. For instance, deviation and solution optimality. All the provided criteria are assessed under time and memory constraints.

The remainder of this paper is organized as follows: In Section 2, a focus on the related works is given. In Section 3, the graph set repository is described. In Section 4, the different performance evaluation metrics are put forward to evaluate GED approaches. Section 5 demonstrates a use case of these performance evaluation metrics. Finally, Section 6 is devoted to conclusions and perspectives.

2   Related Works

Table 1 synthesizes the graph databases presented in the literature. One may notice that exact GM has been evaluated at the matching level. However, error-tolerant GM has been evaluated at the classification level rather than the matching one. [11] is devoted to error-tolerant GM with ground truth information. However, graphs have the same number of vertices and thus the scalability measure cannot be assessed. Indeed, there is a lack of performance comparison measures for error-tolerant GM methods, whether exact or approximate ones. Moreover, none of the graph repositories was dedicated to assessing the performance of GED methods.

---

[1] http://www.rfai.li.univ-tours.fr/PagesPerso/zabuaisheh/GED-benchmark.html

Table 1. Synthesis of graph databases

Ref	Problem Type	Graph Type	Database Type	Measure Type	Purpose
[12]	Exact GM	Non-attributed	Synthetic	Accuracy and scalability	Matching
[11]	Error-tolerant GM	Attributed	Real-world	Memory consumption, accuracy and matching quality	Matching
[9]	Error-tolerant GM	Attributed	Real-world	Accuracy and running time	Classification
[13, 14]	Exact GM	Attributed	Synthetic	Accuracy and scalability	Matching
[15]	Exact GM	(Non)attributed	Real-world	Scalability	Matching

In this paper, we focus on GED methods since GED is a flexible paradigm that has been used in different applications in PR. To understand better the behavior of each GED method, we provide new performance metrics.

3   Graph Set Repository

This section is devoted to the description and the justification of the selected graph databases as well as the details of the information added to each database.

3.1   Databases

Several databases are integrated in our repository. This section is devoted to the advantages of each of these databases as well as graphs selection.

We aim at evaluating GED approaches when increasing the number of vertices. To this end, we decomposed each database into disjoint subsets, each of which contains graphs that have the same number of vertices. Moreover, we aim at studying the behavior of each algorithm on different types of attributes.

For each database, two non-negative meta parameters associated to GED are included (τ_{node} and τ_{edge}) where τ_{node} denotes a vertex deletion or insertion whereas τ_{edge} denotes an edge deletion or insertion. A third meta parameter α is integrated to control whether the edit operation cost on the vertices or on the edges is more important. From each graph matching pair, we derive two notions: a distance between each pair of graphs and vertex-vertex and edge-edge matching or so-called edit sequence.

GREC Database: This database consists of a subset of the symbol database underlying the GREC 2005 competition [16].

Database Interest: GREC is composed of undirected graphs of rather small size (*i.e.,* up to 24). In addition, continuous attributes on vertices and edges play an important role in the matching process. Such graphs are representative of pattern recognition problems where graphs are involved in a classification stage.

Cost Function: Additionally to (x, y) coordinates, the vertices of graphs from the GREC database are labeled with a type (ending point, corner, intersection, circle). For edges, two types (line, arc) are employed. The cost functions of vertex and edge substitutions, deletions and insertions are defined as follows:

$$- \; c(u \rightarrow \epsilon) = c(\epsilon \rightarrow v) = \alpha.\tau_{node}$$
$$- \; c(u \rightarrow v) = \begin{cases} \alpha.|\mu(u) - \mu(v)|, & \text{if labels are similar} \\ \alpha.2.\tau_{node}, & \text{otherwise} \end{cases}$$
$$- \; c(p \rightarrow q) = 2.(1 - \alpha).Dirac(\mu(p), \mu(q))$$
$$- \; c(p \rightarrow \epsilon) = c(\epsilon \rightarrow q) = (1 - \alpha).\tau_{edge}$$

where $u \in V_1$, $v \in V_2$, $p \in E_1$ and $q \in E2$. V_* and E_* refer to the sets of vertices and edges of graph g_*, respectively. μ is a function which returns the attribute(s) of each vertex/edge. $(* \rightarrow \epsilon)$ and $(\epsilon \rightarrow *)$ denote vertex/edge deletion and insertion, respectively. In our experiments, we have set τ_{node}, τ_{edge} and α to 90, 15 and 0.5 respectively. The meta parameters' values of the GREC, Mutagenicity and Protein databases are taken from the PhD thesis of Riesen [17].

Database Decomposition: We filtered and decomposed the database into the following subsets: (GREC-5, GREC-10, GREC-15 and GREC-20), each subset has 10 graphs of size 5, 10, 15 and 20, respectively in addition to GREC-mix that has graphs of different sizes taken from the aforementioned subsets. Graphs were chosen from the train set and from different classes to capture, at best, graph variability. 100 pairwise comparisons per subset are conducted.

Mutagenicity Database: The Mutagenicity database was originally prepared by the authors of [18]. For simplicity, we denote this database as MUTA.

Database Interest: MUTA is representative of GM problems where graphs have only symbolic attributed. MUTA gathers large graphs up to 70 vertices.

Cost Function: The cost functions of operations on vertices and edges are defined as follows:

$$- \; c(u \rightarrow \epsilon) = c(\epsilon \rightarrow v) = 2.\alpha.\tau_{node}$$
$$- \; c(u \rightarrow v) = \begin{cases} 0, & \text{if they have the same symbols} \\ \alpha.2.\tau_{node}, & \text{otherwise} \end{cases}$$
$$- \; c(p \rightarrow \epsilon) = c(\epsilon \rightarrow q) = (1 - \alpha).\tau_{edge} \qquad - \; c(p \rightarrow q) = 0$$

where $(\tau_{node}, \tau_{edge}, \alpha)$ values of Mutagenicity are set to (11,1.1,0.25).

Database Decomposition: (MUTA-10, MUTA-20, MUTA-30 ... MUTA-70) were selected and shrunk to 10 graphs per subset. In some subsets of MUTA, the number of train graphs was less than 10. Hence, to complete the subsets, we took some graphs from the test or the validation subsets. We also added another subset, denoted by MUTA-mix, that contains 10 graphs of various number of vertices. As in GREC, 100 comparisons are carried out in each subset.

Protein Database: The protein database was first reported in [18].

Database Interest: This database contains numeric attributes on each vertex as well as a string sequence that is used to represent the amino acid sequence.

Cost Function: The cost functions of matching operations are defined as follows:

- $c(u \rightarrow \epsilon) = c(\epsilon \rightarrow v) = \alpha.\tau_{node}$
- $c(u \rightarrow v) = \begin{cases} SED(\mu(u), \mu(v)), & \text{if labels are similar} \\ \alpha.\tau_{node}, & \text{otherwise} \end{cases}$
- $c(p \rightarrow \epsilon) = c(\epsilon \rightarrow q) = (1 - \alpha).\tau_{edge}$
- $c(p \rightarrow q) = \begin{cases} 0, & \text{if labels are similar} \\ \alpha.\tau_{edge}, & \text{otherwise} \end{cases}$

where $(\tau_{node}, \tau_{edge}, \alpha)$ are set to (11,1,0.75) and SED is string edit distance [19]. Given two amino acid sequences (s_1 and s_2), the corresponding cost of matching symbols in s_1 and s_2 is defined as follows:
c(k $\rightarrow$ m) = c(k $\rightarrow$ ϵ) = c(ϵ $\rightarrow$ m) = 1, c(k $\rightarrow$ m) = 0, s.t. $k, m \in s_1, s_2$.

Database Decomposition: 10 graphs were selected from Protein-20, Protein-30 and Protein-40. In addition, 10 graphs were picked up from the aforementioned Protein subsets and were put in the mixed database referred to as Protein-mix. 100 pairwise matchings per subset are conducted and integrated in the repository.

CMU Database: The CMU model house sequence is made up of a series of images of a toy house that has been captured from different viewpoints. 111 images in total are publicly available. 660 comparisons are carried out.

Database Interest: Unlike the aforementioned databases, the CMU database has a model house sequence that consists of a series of images. Each of which has been manually annotated by a human being providing its key points (corner features). Moreover, the ground truth is attached with each pair of graphs.

Cost Function: We empirically set (τ_{node}, α) to $(\infty, 0.5)$. The cost functions of matching vertices and edges are similar to [20]. We formalize them as follows:

- $c(u \rightarrow \epsilon) = c(\epsilon \rightarrow v) = \alpha.\tau_{node}$ $- c(u \rightarrow v) = 0$
- $c(p \rightarrow \epsilon) = c(\epsilon \rightarrow q) = (1 - \alpha).\mu(q)$ $- c(p \rightarrow q) = (1 - \alpha).|\mu(p) - \mu(q)|$

Graphs Construction: A manual identification of corner features, or points, was done to represent vertices on each of the rotated images. Then, the Delaunay triangulation was applied on the corner-features in order to identify edges and finally transform the images into graphs. Vertices are labeled with (x,y) coordinates while edges are labeled with the distance between vertices. Each graph has 30 vertices and matched with graphs at 10, 20, 30, 40, 50, 60, 70, 80, and 90 in the rotation sequence. Since vertex to vertex matchings are given, the accuracy of the final matching sequence of any method p can be computed by verifying whether or not the matched vertices are correct.

3.2 Added Low Level Information

For each graphs pair (g_i, g_j), the initial content of the databases in addition to the following low level information are provided: the optimal or sub-optimal solution provided by the most accurate GED method for $d(g_i, g_j)$, the name of the most accurate GED method, the solution status (*i.e.*, optimal or sub-optimal) and the edit path sequence. Table 2 illustrates an example of two graphs taken from GREC-5 and their added information in our repository. All these information are available as csv files in our website.

Table 2. Low level information (taken from the file GREC5-lowlevelinfo.csv)

G1Name	G2Name	Method	Distance	Optimal	Matching
					Vertex:0 → 0=37.476/ Vertex:1 → 1=6.519/ Vertex:2 → 2=
					32.070/ Vertex:4 → 4=34.409/ Vertex:3 → 3=24.703/
image3_23	image3_25	*BS−100*	135.178	false	*Edge:2 ↔ 3 → 2 ↔ 3 =0.0/ Edge:0 ↔ 4 → 0 ↔ 4=0.0*

4   Performance Evaluation Metrics

In this section, we define the new metrics used for evaluating any GED method. We propose to analyze the behavior of the different methods under a time constraint (C_T) as well as a memory constraint (C_M):

Deviation: We compute the error committed by each method p over the reference distances. For each pair of graphs matched by method p, we provide the following deviation measure:

$$dev(g_i, g_j)^p = \frac{|d(g_i, g_j)^p - R_{g_i, g_j}|}{R_{g_i, g_j}}, \ \forall (i, j) \in [\![1, m]\!]^2, \forall p \in \mathcal{P} \qquad (1)$$

where m is the number of graphs. $d(g_i, g_j)^p$ is the distance obtained when matching g_i and g_j using method p while R_{g_i, g_j} corresponds to the ground truth provided in our csv files. R_{g_i, g_j} represents the best distance among all methods p for matching graphs g_i and g_j. As all comparisons are evaluated under a small or a big C_T, R_{g_i, g_j} does not necessarily have to be the optimal distance. In other words, R_{g_i, g_j} represents the best solution found under a small or a big C_T. This solution could be optimal when C_T is reasonable to solve the given matching problem. For each method p, the mean $dev(g_i, g_j)^p$ is computed.

Matching Dissimilarity: Let EP refer to vertex to vertex mappings between g_1 into g_2. We aim at finding how dissimilar are two EPs (*i.e.*, EP_p and EP_q) that correspond to matching g_i and g_j using methods p and q. Thus, the idea here is to see how far an EP from the ground truth EP integrated in our repository. To this objective, we count the differences between the two solutions, we exclude edge correspondences in order to only concentrate on vertex correspondences.

Mathematically saying, the distance between EP_p and EP_q is defined as:

$$d(EP_p, EP_q) = \sum_{i=1}^{n} \sum_{j=1}^{m} \delta(EP_p(i), EP_q(j)) \tag{2}$$

where $n = |EP_p|$, $m = |EP_q|$ and $\delta(EP_p(i), EP_q(j))$ is the well-known Kronecker Delta function:

$$\delta(EP_p(i), EP_q(j)) = \begin{cases} 0, & \text{if } EP_p(i) = EP_q(j) \\ 1, & \text{if } EP_p(i) \neq EP_q(j) \end{cases} \tag{3}$$

Number of Explored Nodes: We propose to measure the number of explored nodes in the tree search for each comparison $d(g_i, g_j)$. This number represents the number of moves in the search tree needed to obtain the best found solution. The mean number of explored nodes in the tree search is calculated as follows:

$$\overline{\#explored\_nodes}_p = \frac{1}{m \times m} \sum_{i=1}^{m} \sum_{j=1}^{m} expnd(d(g_i, g_j)^p) \tag{4}$$

where $m \times m$ is the total number of comparisons per subset, m is the number of graphs to be matched and $expnd(d(g_i, g_j)^p)$ is the number of explored nodes obtained when matching graphs g_i and g_j of subset s using method p. Note that p has to be a tree-search based algorithm (*e.g.*, $A*$ [5] and BeamSearch [21]).

Number of Best Found Solutions: For each subset, we count the number of times the best solution is found by method p. This number indicates that method p was able to find the best solution, not necessarily the optimal one. The solution is supposed to be the best one when compared to all the involved methods.

Number of Unfeasible Solutions: For each method p, we measure the number of times method m was not able to find an EP. This case can happen when C_T or C_M is violated before finding a complete solution. (*i.e.*, when there is only incomplete matching correspondences). Lower bound methods [8] always give unfeasible solutions as they only output distances without matching sequences.

Number of Time-Out and Out-Of-Memory Cases: For each subset, we count the number of times method p violates C_T and C_M respectively.

Running Time: We measure the overall time in milliseconds (ms), for each GED computation, including all the inherits costs computations. The mean running time is calculated per subset s and for each method p.

Time-Deviation Scores: To sum up advantages and drawbacks of each method p, a projection of p on a two-dimensional space ($\mathbb{R}^2$) is achieved by using *speed-score* and *deviation-score* features defined in equations 7 and 8 where speed and deviation are two concurrent criteria to be minimized. First, for each database, the mean deviation and the mean time is derived as follows:

$$\overline{dev^p} = \frac{1}{m \times m} \sum_{i=1}^{m} \sum_{j=1}^{m} dev(g_i, g_j)^p \tag{5}$$

$$\overline{time^p} = \frac{1}{m \times m} \sum_{i=1}^{m} \sum_{j=1}^{m} time(g_i, g_j)^p \ and \ (i,j) \in [\![1,m]\!]^2 \tag{6}$$

where $dev(g_i, g_j)$ is the deviation of each $d(g_i, g_j)$ and $time(g_i, g_j)$ is the running time of each $d(g_i, g_j)$. To obtain comparable results between databases, mean deviations and times are normalized between 0 and 1 as follows:

$$\overline{deviation\_score^p} = \frac{1}{\#subsets} \sum_{i=1}^{\#subsets} \frac{\overline{dev_i^p}}{max\_dev_i} \tag{7}$$

$$\overline{speed\_score^p} = \frac{1}{\#subsets} \sum_{i=1}^{\#subsets} \frac{\overline{time_i^p}}{max\_time_i} \tag{8}$$

where $max\_dev_i = \max(\overline{dev_i^p})$ and $max\_time_i = \max(\overline{time_i^p}) \ \forall p \in \mathcal{P}$

5   Use Case

In this section, we demonstrate a use case of the aforementioned metrics. Two GED methods were evaluated. On the exact method side, the A^* algorithm applied to GED problem is a foundation work [5]. It is the most well-known exact method that is often used to evaluate the accuracy of approximate methods. On the approximate method side, the truncated version of A^*, *i.e.*, Beam Search (*BS-x*), was chosen. This method is one of the most accurate heuristics of the literature. In this use case, x in *BS-x* has been set to 1 and 100, respectively.

In this paper, we only provide results on the aforementioned methods. The results on the other databases and other methods are demonstrated in our website. Also, due to the large number of matchings considered and the exponential complexity of the tested methods, C_T has been set to 300 seconds which is large enough to let the methods search deeply into the solution space and to ensure that many search tree nodes will be explored. C_M has been set to 1Gb.

Figure 1(a) represents the deviation from the best distances integrated in our repository. Figure 1(b) points out the matching dissimilarity when comparing $d(g_i, g_j)^p$ with the best matching in our repository. Both figures reveal that *BS-100* had the least deviation. Similarly, Figure 1(c) shows that *BS-100* had the highest number of best found solutions. In average, *BS-100* beat the other methods with 38 more best solutions. A^* suffered from high memory consumption, see Figure 1(g) and thus it outputted unfeasible solutions, as demonstrated in Figure 1(d). In average, on MUTA-40, MUTA-50, MUTA-60 and MUTA-70, *BS-100* explored around 4043 search nodes more than the other methods. Figure 1(h) shows that, in average, *BS-1* was the fastest algorithm. However, according to Figure 1(i), *BS-100* gave the best trade-off between deviation and speed.

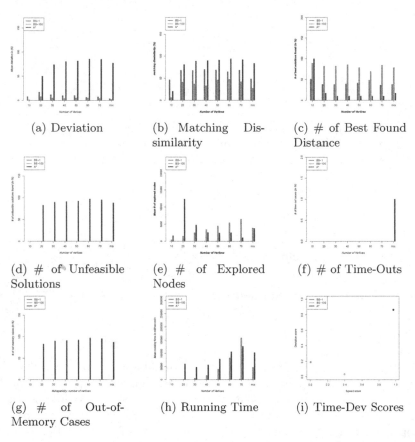

(a) Deviation

(b) Matching Dis-similarity

(c) # of Best Found Distance

(d) # of Unfeasible Solutions

(e) # of Explored Nodes

(f) # of Time-Outs

(g) # of Out-of-Memory Cases

(h) Running Time

(i) Time-Dev Scores

Fig. 1. Results on MUTA and under a reasonable time constraint ($C_T = 300$ seconds)

6 Conclusions

The contribution of this paper is two-fold: First, additional low level annotation has been made publicly available for some representative graph databases. Each graph comparison is coupled with the best solution found by a GED method. Four databases with different characteristics are integrated (GREC, MUTA, Protein and CMU). We have proposed to evaluate the scalability of GED methods on GREC, MUTA and Protein as they have graphs of different number of vertices. Thus, these databases were decomposed into subsets, each of which has graphs whose number of vertices are equal. On the other hand, CMU has geometric graphs of 30 vertices, each of which has been subjected to rotations. Second, this paper has presented performance evaluation metrics that assess GED methods under time and memory constraints. The aim of this paper is to make GED methods better comparable against each other. For that reason, we highly encourage the community not only to use the information provided in the repository, but also to integrate their algorithms' answers when obtaining more accurate results.

In future work, we will further expand this repository by integrating other publicly available databases.

References

1. Vento, M.: A long trip in the charming world of graphs for pattern recognition. Pattern Recognition 48(2), 291–301 (2015)
2. Sorlin, S., Solnon, C., Michel Jolion, J.: A generic graph distance measure based on multivalent matchings (2007)
3. Sanfeliu, A., Fu, K.: A distance measure between attributed relational graphs for pattern recognition. IEEE Transactions on S, M, and C 13, 353–362 (1983)
4. Bunke, H.: Inexact graph matching for structural pattern recognition. Pattern Recognition Letters 1(4), 245–253 (1983)
5. Fankhauser, S., Riesen, K., Bunke, H.: Speeding up graph edit distance computation through fast bipartite matching. In: Jiang, X., Ferrer, M., Torsello, A. (eds.) GbRPR 2011. LNCS, vol. 6658, pp. 102–111. Springer, Heidelberg (2011)
6. Justice, D., Hero, A.: A binary linear programming formulation of the graph edit distance. IEEE Transactions on PA and MI 28(8), 1200–1214 (2006)
7. Riesen, K., Bunke, H.: Approximate graph edit distance computation by means of bipartite graph matching. Image and Vision Computing 28, 950–959 (2009)
8. Fischer, A., Suen, C.Y., Frinken, V., Riesen, K., Bunke, H.: A fast matching algorithm for graph-based handwriting recognition. In: Kropatsch, W.G., Artner, N.M., Haxhimusa, Y., Jiang, X. (eds.) GbRPR 2013. LNCS, vol. 7877, pp. 194–203. Springer, Heidelberg (2013)
9. Riesen, K., Bunke, H.: IAM graph database repository for graph based pattern recognition and machine learning. In: da Vitoria Lobo, N., Kasparis, T., Roli, F., Kwok, J.T., Georgiopoulos, M., Anagnostopoulos, G.C., Loog, M. (eds.) S+SSPR 2008. LNCS, vol. 5342, pp. 287–297. Springer, Heidelberg (2008)
10. Xu, K.B.: Benchmarks with hidden optimum solutions for graph problems, http://www.nlsde.buaa.edu.cn/kexu/benchmarks/graph-benchmarks.htm
11. Cmu house and hotel datasets, http://vasc.ri.cmu.edu/idb/html/motion
12. A large database of graphs and its use for benchmarking graph isomorphism algorithms 24(8), 1067 – 1079 (2003)
13. Conte, D., et al.: Challenging complexity of maximum common subgraph detection algorithms 11(1), 99–143 (2007)
14. Foggia, P., et al.: A performance comparison of five algorithms for graph isomorphism. In: IAPR TC-15, pp. 188–199 (2001)
15. Carletti, V., Foggia, P., Vento, M.: Performance comparison of five exact graph matching algorithms on biological databases. In: Petrosino, A., Maddalena, L., Pala, P. (eds.) ICIAP 2013. LNCS, vol. 8158, pp. 409–417. Springer, Heidelberg (2013)
16. Grec competition, http://symbcontestgrec05.loria.fr/index.php
17. Riesen, K., Bunke, H.: Graph Classification and Clustering Based on Vector Space Embedding (2010)
18. Kazius, J., et al.: Derivation and validation of toxicophores for mutagenicity prediction. Journal of Medicinal Chemistry 48(1), 312–320 (2005)
19. Wagner, et al.: The string-to-string correction problem 21(1), 168–173 (1974)
20. Zhou, F., De la Torre, F.: Factorized graph matching. In: IEEE Conference on Computer Vision and Pattern Recognition (CVPR) (2012)
21. Neuhaus, M., et al.: Fast suboptimal algorithms for the computation of graph edit distance. Structural and Syntactic Pattern Recognition, 163–172 (2006)

Improving Hausdorff Edit Distance
Using Structural Node Context

Andreas Fischer[1,2(✉)], Seiichi Uchida[3], Volkmar Frinken[3],
Kaspar Riesen[4], and Horst Bunke[5]

[1] Department of Informatics, University of Fribourg
Boulevard de Pérolles 90, 1700 Fribourg, Switzerland
`andreas.fischer@unifr.ch`
[2] Institute of Complex Systems, University of Applied Sciences and Arts
Western Switzerland, Boulevard de Pérolles 80, 1700 Fribourg, Switzerland
`andreas.fischer@hefr.ch`
[3] Faculty of Information Science and Electrical Engineering, Kyushu University
744 Motooka, Nishi-ku, Fukuoka 819-0395, Japan
`{uchida,vfrinken}@ait.kyushu-u.ac.jp`
[4] Institute for Informations Systems, University of Applied Sciences and Arts
Northwestern Switzerland, Riggenbachstrasse 16, 4600 Olten, Switzerland
`kaspar.riesen@fhnw.ch`
[5] Institute of Computer Science and Applied Mathematics, University of Bern
Neubrückstrasse 10, 3012 Bern, Switzerland
`bunke@iam.unibe.ch`

Abstract. In order to cope with the exponential time complexity of graph edit distance, several polynomial-time approximation algorithms have been proposed in recent years. The Hausdorff edit distance is a quadratic-time matching procedure for labeled graphs which reduces the edit distance to a correspondence problem between local substructures. In its original formulation, nodes and their adjacent edges have been considered as local substructures. In this paper, we integrate a more general structural node context into the matching procedure based on hierarchical subgraphs. In an experimental evaluation on diverse graph data sets, we demonstrate that the proposed generalization of Hausdorff edit distance can significantly improve the accuracy of graph classification while maintaining low computational complexity.

1 Introduction

Graph-based pattern recognition is characterized by both high flexibility and high computational complexity. On the one hand, labeled graphs offer a high flexibility for object representation by describing parts of an object with nodes and binary relations among the parts with edges. Furthermore, nodes and edges can be labeled with more detailed descriptors such as symbols, numbers, and feature vectors. This flexible representation has found widespread application in the field of pattern recognition, for example in bioinformatics [1], image classification [2], and web content mining [3].

© Springer International Publishing Switzerland 2015
C.-L. Liu et al. (Eds.): GbRPR 2015, LNCS 9069, pp. 148–157, 2015.
DOI: 10.1007/978-3-319-18224-7_15

On the other hand, the computation of a dissimilarity measure for labeled graphs, which is fundamental for pattern recognition systems, involves a high computational complexity. Unlike fixed-size feature vectors, ordered strings, and hierarchical trees, the nodes of a graph are not ordered and can be linked arbitrarily with edges, which dramatically increases the combinatorial possibilities for matching two graphs. Several efficient procedures have been proposed in the past decades [4] that are able to derive a dissimilarity measure in polynomial time with respect to the number of nodes. However, in order to avoid exponential time complexity, the proposed methods typically constrain the graph structure or the graph label domain, thus giving up some of the flexibility for object representation. Examples include spectral methods [5] based on the eigendecomposition of the adjacency matrix which primarily target unlabeled graphs, methods for planar graphs [6], and methods for graphs with unique node labels [7].

Graph edit distance [8,9] is a general dissimilarity measure that is applicable to virtually any kind of graphs. Closely related to string edit distance, also known as Levenshtein distance [10], the method is based on the minimum cost sequence of edit operations that transforms one graph into another considering node and edge substitution, deletion, and insertion. As no constraints are imposed on the graph structure and the graph label domain, the full flexibility for object representation is kept. However, graph edit distance is an NP-complete problem for which no efficient solution exists. The exact solution is usually obtained by means of exhaustive A* search with exponential time complexity, thus limiting the applicability of graph edit distance to rather small graphs in practice.

In a recent approximation framework for graph edit distance [11], a cubic-time matching procedure has been proposed that reduces the edit distance to an assignment problem between local substructures. It aims to combine the high flexibility of graph edit distance with an efficient algorithmic solution, which is provided by the Hungarian algorithm [12], and has shown a high performance for graph classification when compared with the exact edit distance. In addition, the approximation method provides upper and lower bounds for graph edit distance [13]. Runtime improvements include the method proposed in [14], which is able to reduce the number of possible assignments if the underlying cost function satisfies certain metric properties.

The approximation framework has been recently extended with a quadratic-time matching procedure, namely the Hausdorff edit distance [15]. The proposed method reduces the assignment problem to a Hausdorff matching problem [16] between two sets of local substructures, which can be resolved in quadratic rather than cubic time and provides a lower bound for graph edit distance. Although a certain loss in accuracy has to be taken into account, promising results for graph classification could be reported.

The stepwise reduction from exponential to quadratic time complexity has been achieved by ignoring the global graph structure. Instead, only local substructures are matched and the approximation methods rely strongly on the node and edge labels in order to find the best correspondence between the substructures. Indeed, the procedures are not well-suited for matching unlabeled, purely structural graphs. For cubic-time approximation based on the assignment problem, the methods proposed in [17,18] address this issue by including more structural information in the assignment cost. In [17], subgraphs consisting of a central node, its adjacent edges, as well as the end nodes of the edges have been taken into account as local substructures. In [18], walks of length k starting at the central node have been integrated into the matching procedure.

In this paper, we aim to strengthen the importance of structural integrity for the Hausdorff edit distance by extending the local substructures. In its original formulation [15], we have considered only nodes and their adjacent edges as local substructures, which has allowed us to derive a lower bound for graph edit distance. The proposed extension considers a larger structural context around the nodes by means of hierarchical subgraphs. In an experimental evaluation, we demonstrate that this generalization of Hausdorff edit distance can significantly improve the accuracy of graph classification while maintaining low computational complexity.

The remainder of this paper is organized as follows. First, the Hausdorff edit distance is discussed in Section 2. Then, the integration of structural node context is presented in Section 3. Afterwards, experimental results are provided in Section 4 and conclusions are drawn in Section 5.

2   Hausdorff Edit Distance

In the following, we provide some basic definitions for graph edit distance and review the Hausdorff edit distance. For more details, we refer to [15].

2.1   Definition of Graph Edit Distance

A *graph* g is a four-tuple $g = (V, E, \mu, \nu)$ where V is the finite set of nodes, $E \subseteq V \times V$ is the set of edges, $\mu : V \to L_V$ is the node labeling function, and $\nu : E \to L_E$ is the edge labeling function. L_V and L_E are label domains, for example symbols, numbers, or the vector space $\mathbb{R}^n$.

Considering two graphs $g_1 = (V_1, E_1, \mu_1, \nu_1)$ and $g_2 = (V_2, E_2, \mu_2, \nu_2)$, *edit operations* transform nodes and edges of g_1 into nodes and edges of g_2. We take three edit operations into account for $u \in V_1$ and $v \in V_2$, namely *substitutions* $(u \to v)$, *deletions* $(u \to \epsilon)$, and *insertions* $(\epsilon \to v)$. The same edit operations are used for edges $p \in E_1$ and $q \in E_2$, that is substitutions $(p \to q)$, deletions $(p \to \epsilon)$, and insertions $(\epsilon \to q)$. An *edit path* is a sequence of edit operations $(e_1, \ldots, e_k)$ that transform g_1 into g_2.

A *cost function* $\mathcal{C}$ assigns non-negative costs to node and edge edit operations. An example for Euclidean labels is the Euclidean cost function with substitution cost $\mathcal{C}(u \to v) = ||\mu_1(u) - \mu_2(v)||$ and $\mathcal{C}(p \to q) = ||\nu_1(p) - \nu_2(q)||$. The cost for deletion and insertion is usually set to a constant value. Without loss of generality, we assume $\mathcal{C}(u \to \epsilon) = \mathcal{C}(\epsilon \to v) = C_n$ and $\mathcal{C}(p \to \epsilon) = \mathcal{C}(\epsilon \to q) = C_e$ for all types of cost functions. Finally, the graph edit distance is defined as the minimum cost edit path between g_1 and g_2. Formally,

$$GED(g_1, g_2) = \min_{(e_1,\dots,e_k)\in\Upsilon(g_1,g_2)} \sum_{i=1}^{k} \mathcal{C}(e_i) \tag{1}$$

where $\Upsilon(g_1, g_2)$ denotes the set of all edit paths. $\Upsilon(g_1, g_2)$ is usually computed by means of exhaustive A* search over all possible node edit operations.

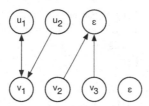

Fig. 1. Local substructure correspondence

2.2 Approximation Based on Hausdorff Matching

Instead of finding the best global correspondence between nodes and edges, the Hausdorff edit distance reduces the problem to finding the best correspondence between local substructures, that is nodes and their adjacent edges. An example is illustrated in Figure 1 for $V_1 = \{u_1, u_2\}$ and $V_2 = \{v_1, v_2, v_3\}$. Multiple assignments to the same substructure are allowed and two special ϵ nodes represent deletions and insertions.

Formally, the Hausdorff edit distance $HED(g_1, g_2)$ between two graphs g_1 and g_2 is defined as:

$$HED(g_1, g_2) = \sum_{u\in V_1} \min_{v\in V_2\cup\{\epsilon\}} f(u, v) + \sum_{v\in V_2} \min_{u\in V_1\cup\{\epsilon\}} f(u, v) \tag{2}$$

The two summation terms calculate nearest neighbor distances between the node sets similar to the Hausdorff distance between finite subsets of a metric space. Nearest neighbors are found with respect to the node function

$$f(u, v) = \begin{cases} C_n + \sum_{i=1}^{|P|} \frac{C_e}{2} & \text{for node deletion } (u \to \epsilon) \\ C_n + \sum_{i=1}^{|Q|} \frac{C_e}{2} & \text{for node insertion } (\epsilon \to v) \\ \frac{C(u\to v)+\frac{HED(P,Q)}{2}}{2} & \text{for node substitution } (u \to v) \end{cases} \tag{3}$$

where P is the set of edges adjacent to u and Q is the set of edges adjacent to v. Half of the implied edge cost is added to the node cost. In the case of substitution, only half of the total cost is considered.

In order to estimate the edge cost implied by node substitution, the edge sets P and Q are matched in the same manner as the node sets, that is based on a Hausdorff edit distance

$$HED(P, Q) = \sum_{p\in P} \min_{q\in Q\cup\{\epsilon\}} g(p, q) + \sum_{q\in Q} \min_{p\in P\cup\{\epsilon\}} g(p, q) \tag{4}$$

with the corresponding edge function

$$g(p, q) = \begin{cases} C_e & \text{for edge deletion } (p \to \epsilon) \\ C_e & \text{for edge insertion } (\epsilon \to q) \\ \frac{c(p \to q)}{2} & \text{for edge substitution } (p \to q) \end{cases} \tag{5}$$

The divisions by 2 in the node function $f(u, v)$ and the edge function $g(p, q)$ ensure that the Hausdorff edit distance is a lower bound of graph edit distance [15]. In order to limit the underestimation, a minimum edit cost is returned. For $HED(g_1, g_2)$ the minimum edit cost is $||V_1| - |V_2|| \cdot C_n$ and for $HED(P, Q)$ the minimum edit cost is $||P| - |Q|| \cdot C_e$.

3   Structural Node Context

In its original formulation, the Hausdorff edit distance considers only nodes and adjacent edges as local substructures. In this section, we present a generalized Hausdorff edit distance which takes a structural node context into account in order to strengthen the structural integrity of the matching procedure.

3.1   Definition

Let $g = (V, E, \mu, \nu)$ be a labeled graph, $u \in V$, and $n \in \mathbb{N}$. We define the *structural node context* with degree n as a hierarchical subgraph

$$c_n(u, g) = (L_1, \ldots, L_n) \tag{6}$$

where $L_i = (V_i, E_i, \mu_i, \nu_i)$ are subgraphs of g with

- $V_i = \{v \in V \mid \text{the shortest path length between } u \text{ and } v \text{ is } (i - 1) \}$
- $E_i = \{(v_1, v_2) \in E \mid v_1 \in V_i \lor v_2 \in V_i\}$
- $\mu_i(v) = \mu(v) \ \forall v \in V_i$
- $\nu_i(e) = \mu(e) \ \forall e \in E_i$.

For the shortest path, neither the direction nor the labels of the edges are taken into account. The structural node context $c_n(u, g)$ can be constructed iteratively for $i = 1, \ldots, n$ as illustrated in Figure 2.

3.2   Modified Hausdorff Edit Distance

In order to integrate the structural node context into the Hausdorff edit distance, we first define a distance between two node contexts $c_n(u, g_1) = (L_1, \ldots, L_n)$ and $c_n(v, g_2) = (M_1, \ldots, M_n)$

$$d(c_n(u, g_1), c_n(v, g_2)) = \sum_{i=1}^{n} HED(L_i, M_i) \tag{7}$$

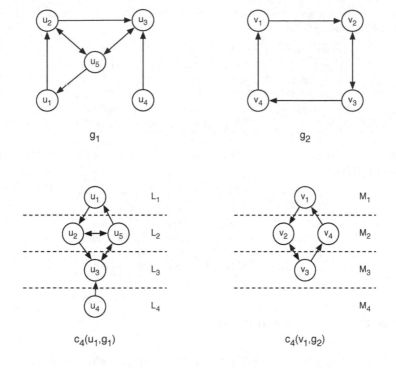

Fig. 2. Structural node context with degree $n = 4$ for nodes u_1 and v_1

Based on this distance, we find the nearest neighbor $b_n(u, g_2)$ of u in g_2 and the nearest neighbor $b_n(v, g_1)$ of v in g_1

$$b_n(u, g_2) = \underset{v \in V_2}{\operatorname{argmin}} \, d(c_n(u, g_1), c_n(v, g_2)) \tag{8}$$

$$b_n(v, g_1) = \underset{u \in V_1}{\operatorname{argmin}} \, d(c_n(u, g_1), c_n(v, g_2)) \tag{9}$$

Finally, we compute the Hausdorff edit distance with respect to these nearest neighbors

$$HED_n(g_1, g_2) = \sum_{u \in V_1} \min_{v \in \{b_n(u, g_2), \epsilon\}} f(u, v) + \sum_{v \in V_2} \min_{u \in \{b_n(v, g_1), \epsilon\}} f(u, v) \tag{10}$$

The case $n = 1$ corresponds with the original Hausdorff edit distance considering only nodes and adjacent edges as the local substructures. The case $n > 1$ considers an extended structural context for selecting the nearest neighbor correspondence.

In the example shown in Figure 2, the computation of Equation 7 is illustrated for structural node context with degree $n = 4$. To determine whether or not v_1 is the nearest neighbor of u_1, the Hausdorff edit distance is computed between all subgraphs L_i and M_i for $i = 1, 2, 3, 4$.

Table 1. Recognition accuracy results on the test set. The difference Δ is calculated between HED_n and HED_1.

Database	AED	HED_1	HED_n	Δ	n
Letters I	99.7	97.9	99.5	+1.6	5
Letters II	94.3	86.9	92.4	+5.5	3
Letters III	89.9	79.2	83.7	+4.5	2
Fingerprints	80.3	82.8	82.5	−0.3	2
Molecules	99.6	99.2	99.5	+0.3	5

Table 2. Runtime results on the test set. The overhead factor × is calculated between HED_6 and HED_1.

Database	AED	HED_1	HED_6	×
Letters I	35.9	10.5	16.2	1.5
Letters II	37.1	10.8	16.9	1.6
Letters III	49.1	11.6	21.5	1.9
Fingerprints	410.7	91.0	130.3	1.4
Molecules	1050.0	46.0	281.2	6.1

4   Experimental Evaluation

We have evaluated the proposed generalization of Hausdorff edit distance in a series of graph classification experiments. Several data sets from the IAM graph database repository [19] have been selected that differ in graph size, graph structure, and label domains.

The *Letters I-III* data sets contain line drawings of 15 letters in three distortion degrees. The graphs have less than 10 nodes, the nodes are labeled with Cartesian coordinates, and the edges are unlabeled. The *Fingerprints* data set contains images of 4 different types of fingerprints. The graphs have up to 26 nodes, the nodes are unlabeled, and the edges are labeled with an angular value. Finally, the *Molecules* data set contains active and inactive molecular compounds from an AIDS screening database. The graphs have up to 95 nodes, the nodes are labeled with a chemical symbol, and the edges are unlabeled. For every data set, a training, validation, and test set is defined to perform graph classification experiments. For more details on the data sets and the classification task, we refer to [15,19].

The classification accuracy results are reported in Table 1 in percentage. HED_1 is the original Hausdorff edit distance with context degree $n = 1$. For the proposed generalization HED_n the context degree n has been optimized on the validation set over a range of $n \in \{1, 2, \ldots, 6\}$. The results are put into context with the related cubic-time approximation of graph edit distance by means of assignment edit distance AED [11]. First, we can observe that in four out of five cases the use of structural node context has improved the classification accuracy. The improvements are statistically significant (t-test, $\alpha = 0.05$). In one case, the accuracy has decreased but without statistical significance.

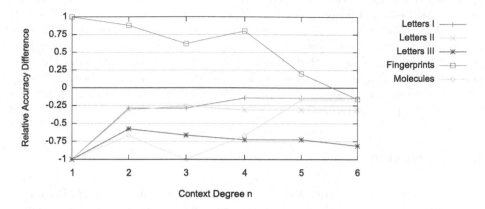

Fig. 3. Relative accuracy difference between HED_n and AED on the test set for context degree n

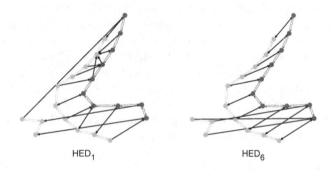

HED$_1$ HED$_6$

Fig. 4. Node correspondence for two graphs from the *Fingerprints* data set

Runtime results are reported in Table 2 in seconds for recognizing the whole test set. Comparisons have been performed on a 2GHz processor with respect to a Java implementation based on [20].[1] When extending the context degree from $n = 1$ to $n = 6$, overhead factors between 1.4 and 6.1 are observed. When compared with AED, the overall runtimes are still significantly below.

A more detailed analysis of the accuracy results is provided in Figure 3. It reports the relative accuracy difference

$$\frac{acc(HED_n) - acc(AED)}{|acc(HED_1) - acc(AED)|}$$

between HED_n and AED on the test set for all context degrees considered in the experiments. In all cases, the accuracy of HED_n approaches the accuracy of

---

[1] http://www.fhnw.ch/wirtschaft/iwi/gmt

AED when using structural node context. Accordingly, the accuracy increases on four data sets and decreases on the *Fingerprints* data set.

Finally, a qualitative example of the node correspondence is shown in Figure 4 for two graphs from the *Fingerprints* data set. One graph is visualized with light gray nodes and the other with dark gray nodes. Black arrows indicate the node correspondence, that is the nearest neighbor selection according to Equation 10. Clearly, the use of context degree $n = 6$ leads to a higher structural integrity of the matching procedure than $n = 1$.

5   Conclusions

Hausdorff edit distance is a quadratic-time matching procedure for labeled graphs which approximates graph edit distance. In this paper, we propose a generalization of the method by means of structural node context in order to improve the structural integrity of the solution.

In an experimental evaluation on diverse graph data sets, we demonstrate that the use of structural node context can significantly increase the recognition accuracy while maintaining low computational complexity. With the proposed method, the gap between quadratic-time and cubic-time approximation of graph edit distance in terms of accuracy could be significantly reduced.

Future lines of research include a more comprehensive evaluation of the generalized Hausdorff edit distance on data sets with very large graphs. Furthermore, an integration of the structural node context into other related approximations like the assignment edit distance seems promising. Finally, a label context could be taken into account instead of a structural context in order to strengthen the impact of the label information.

Acknowledgement. This work has been supported by the *Swiss National Science Foundation* fellowship project P300P2_151279 and the *Hasler Foundation* Switzerland.

References

1. Borgwardt, K., Ong, C., Schönauer, S., Vishwanathan, S., Smola, A., Kriegel, H.P.: Protein function prediction via graph kernels. Bioinformatics 21(1), 47–56 (2005)
2. Harchaoui, Z., Bach, F.: Image classification with segmentation graph kernels. In: Proc. Int. Conf. on Computer Vision and Pattern Recognition, pp. 1–8 (2007)
3. Schenker, A., Last, M., Bunke, H., Kandel, A.: Classification of web documents using graph matching. Int. Journal of Pattern Recognition and Artificial Intelligence 18(3), 475–496 (2004)
4. Conte, D., Foggia, P., Sansone, C., Vento, M.: Thirty years of graph matching in pattern recognition. Int. Journal of Pattern Recognition and Artificial Intelligence 18(3), 265–298 (2004)
5. Wilson, R., Hancock, E., Luo, B.: Pattern vectors from algebraic graph theory. IEEE Trans. on Pattern Analysis and Machine Intelligence 27(7), 1112–1124 (2005)

6. Neuhaus, M., Bunke, H.: An error-tolerant approximate matching algorithm for attributed planar graphs and its application to fingerprint classification. In: Fred, A., Caelli, T.M., Duin, R.P.W., Campilho, A.C., de Ridder, D. (eds.) SSPR&SPR 2004. LNCS, vol. 3138, pp. 180–189. Springer, Heidelberg (2004)
7. Dickinson, P., Bunke, H., Dadej, A., Kraetzl, M.: Matching graphs with unique node labels. Pattern Analysis and Applications 7(3), 243–254 (2004)
8. Sanfeliu, A., Fu, K.S.: A distance measure between attributed relational graphs for pattern recognition. IEEE Trans. on Systems, Man, and Cybernetics 13(3), 353–363 (1983)
9. Bunke, H., Allermann, G.: Inexact graph matching for structural pattern recognition. Pattern Recognition Letters 1(4), 245–253 (1983)
10. Wagner, R.A., Fischer, M.J.: The string-to-string correction problem. Journal of the Association for Computing Machinery 21(1), 168–173 (1974)
11. Riesen, K., Bunke, H.: Approximate graph edit distance computation by means of bipartite graph matching. Image and Vision Computing 27(4), 950–959 (2009)
12. Kuhn, H.: The Hungarian method for the assignment problem. Naval Research Logistic Quarterly 2, 83–97 (1955)
13. Riesen, K., Fischer, A., Bunke, H.: Estimating graph edit distance using lower and upper bounds of bipartite approximations. Int. Journal of Pattern Recognition and Artificial Intelligence 29(2), 1–27 (2015)
14. Serratosa, F.: Fast computation of bipartite graph matching. Pattern Recognition Letters 45, 244–250 (2014)
15. Fischer, A., Suen, C., Frinken, V., Riesen, K., Bunke, H.: Approximation of graph edit distance based on Hausdorff matching. Pattern Recognition 48(2), 331–343 (2015)
16. Huttenlocher, D.P., Klanderman, G.A., Kl, G.A., Rucklidge, W.J.: Comparing images using the Hausdorff distance. IEEE Trans. on Pattern Analysis and Machine Intelligence 15, 850–863 (1993)
17. Raveaux, R., Burie, J., Ogier, J.: A graph matching method and a graph matching distance based on subgraph assignments. Pattern Recognition Letters 31(5), 394–406 (2010)
18. Gaüzère, B., Bougleux, S., Riesen, K., Brun, L.: Approximate graph edit distance guided by bipartite matching of bags of walks. In: Fränti, P., Brown, G., Loog, M., Escolano, F., Pelillo, M. (eds.) S+SSPR 2014. LNCS, vol. 8621, pp. 73–82. Springer, Heidelberg (2014)
19. Riesen, K., Bunke, H.: IAM graph database repository for graph based pattern recognition and machine learning. In: da Vitoria Lobo, N., Kasparis, T., Roli, F., Kwok, J.T., Georgiopoulos, M., Anagnostopoulos, G.C., Loog, M. (eds.) S+SSPR 2008. LNCS, vol. 5342, pp. 287–297. Springer, Heidelberg (2008)
20. Riesen, K., Emmenegger, S., Bunke, H.: A novel software toolkit for graph edit distance computation. In: Kropatsch, W.G., Artner, N.M., Haxhimusa, Y., Jiang, X. (eds.) GbRPR 2013. LNCS, vol. 7877, pp. 142–151. Springer, Heidelberg (2013)

Learning Graph Model
for Different Dimensions Image Matching

Haoyi Zhou[1(✉)], Xiao Bai[1], Jun Zhou[2], Haichuan Yang[1], and Yun Liu[1]

[1] School of Automation Science and Electrical Engineering,
Beihang University XueYuan Road No.37, Beijing, HaiDian District, China
`{Haoyi,baixiao,yanghaichuan,Yunliu}@buaa.edu.cn`
[2] School of Information and Communication Technology,
Griffith University Nathan, 4111 QLD, Brisbane, Australia
`jun.zhou@griffith.edu.au`

Abstract. Hyperspectral imagery has been widely used in real applications such as remote sensing, agriculture, surveillance, and geological analysis. Matching hyperspectral images is a challenge task due to the high dimensional nature of the data. The matching task becomes more difficult when images with different dimensions, such as a hyperspectral image and an RGB image, have to be matched. In this paper, we address this problem by investigating structured support vector machine to learn graph model for each type of image. The graph model incorporates both low-level features and stable correspondences within images. The inherent characteristics are depicted by using graph matching algorithm on weighted graph models. We validate the effectiveness of our method through experiments on matching hyperspectral images to RGB images, and hyperspectral images with different dimensions.

Keywords: Matching · Graph Model · Hyperspectral Image

1 Introduction

With the development of imaging technology, images can be acquired in different modalities, such as RGB image, radar image, IR image, UV image, hyperspectral image, and so on. Robust matching algorithms are needed for the purpose of fusion and registration between these images which have disparity in numbers of dimensions. The difficulties come from distinctive representation of image content in multi-modal images.

In this paper, we take hyperspectral image as an example of multi-dimensional data and investigate matching methods in two settings, i.e., matching hyperspectral image and RGB image, and matching hyperspectral images with different dimensions.

© Springer International Publishing Switzerland 2015
C.-L. Liu et al. (Eds.): GbRPR 2015, LNCS 9069, pp. 158–167, 2015.
DOI: 10.1007/978-3-319-18224-7_16

1.1 Motivation

A hyperspectral image consists of grayscale bands each of which covers a small range of the light wavelengths[1]. We study the hyperspectral image matching problem from three aspects:

(1) **Correspondence.** Our method is proposed based on the observation that local invariant features are interrelated among different hyperspectral/band images. That is to say, steady points and spatial features or structures [19] within or across bands of a hyperspectral image shall also be observable in other hyperspectral images, their spectral bands, or RGB images of the same scene.

(2) **Graph Model.** Image matching is defined as finding the correct relationships between two point sets. To this end, many methods based on image keypoints have been proposed [15,16]. Because the internal structure of objects turns out to be important, the matching problem can be formulated as a graph matching problem by building and matching graphs from keypoints in each image [6]. Current graph methods cannot solve the matching problem effectively when images are in different dimensions. In this paper, we develop a uniform graph model that preserves stable correspondences between two images.

(3) **Learning Method.** With a complete attributed graph [18] defined as the uniform model, the distinctive characteristic of each band affects the attribute of every vertex and influences the connections represented by edges. It is reasonable to adopt a learning method on weighted attributed graph, aiming to select the most discriminative components and to boost the performance of matching function.

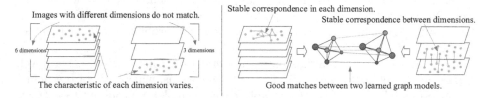

Fig. 1. Left Panel: difficulty in matching different dimensions. Right panel: our motivation.

1.2 Related Work

We firstly review some recent work on matching or registration of multidimensional hyperspectral images. One active area that adopt this technique is painting conservation, which only permits "seeing" without "touching". New close range hyperspectral sensors have been developed for this purpose [1,9]. While some approaches do not require image matching or calibration [11], methods based on

---

[1] Hyperspectral images can be divided into two types, i.e., remote sensing and close range, according to the distance from the imaging device to the objects. We focus on the latter in this paper.

local weighted mean measurements have achieved better performance in image registration [7]. However, these methods can only deal with image matching problem in constrained conditions.

Keypoint description methods, such SIFT[15] and SURF[2], have been used to match multi-dimensional images. However, a series of experiments show that they do not bring much advantages when matching across bands in wide wavelength range [19]. To get better results, Mukherjee et al proposed a method which merges the first few principal components of hyperspectral image in a Gaussian scale-space representation [17]. Meanwhile, a nonlinear combing function is used to get a scalar image along the spectral dimension using a scalar SIFT operator. Image matching performance can be further improved by developing better the keypoint detection approaches. For example, vector SIFT [8] operator extends SIFT based upon the multiple Gaussian scale space. Modified SIFT [20] descriptor preserves local edges with local contrast and differential excitation function. However, directly extending traditional image matching method into the spectral domain can not solve the complex hyperspectral image matching problem. In particular, the spatial correspondence among feature points can not be fully described in point-to-point matching prototype. Therefore, it is worthwhile to bring graph model into this problem so as to characterize the relationship between keypoints.

Our method requires the building of a complete graph and imposing edge-preserving constraints. Some basic and powerful matching algorithms have been proposed to serve for this purpose, such as Spectral Matching [13], Integer Projected Fixed Point Method [14], and Reweighted Random Walks [5]. Recently, Cho et al [3] parameterized the objective function of graph matching. This approach focuses on improving the matching results on similarity function of vertices. Our graph model defined weights on both edges and vertices. We followed the idea in [4] which formulates the graph matching problem into a structured output setting [21].

The main contribution of this paper are threefold. First, we present a novel method for matching images with different dimensions, which is far from being well investigated. Second, a novel graph learning model is proposed to extract stable correspondences among multi-dimensional images. Third, a formulation of structured output learning is redefined to deal with various correspondences.

2   Methodology

This section contains formulas and definition of three key steps of our method, i.e., graph construction, parameterization, and learning.

2.1   Graph Construction

In this step, we construct a complete graph. Two important issues need to be addressed here, sort out vital feature points as vertices, and build finite but effective correspondences.

(1) **Limited Feature Points.** We use the SIFT [15] or Hessian[16] detectors to detect local feature points in each band image. Known from [19], only repeatedly detected feature points are robust for cross band matching. Moreover, the graph will be too complex to match if too many keypoints are selected. Therefore, a k-means clustering method is performed on all points according to locations, and only k keypoints that are closest to the clustering centers in each dimension are selected. An undirected graph G_i, where i indexes the band number, is constructed from the selected keypoints in each dimension. As a consequence, a D-band hyperspectral image generates a graph set $\mathbf{G} = \{G_1, G_2, \ldots, G_D\}$.

(2) **Various Correspondence.** The relevance between different bands can be captured by directly matching the spatial positions of the keypoints based on their feature descriptor using robust matching approach such as RANSAC [10]. We store the pair-wise points matching results between adjacent bands in C_i. Thus, various correspondences $C_1, C_2, \ldots, C_{D-1}, C_D$ form a set $\mathbf{C}$, where C_1 defines the initial order of matching sequence.

(3) **Landmark Points.** Manually labelled landmark points can be treated as another type of correspondence. A manually selected subset of feature points are collected on each band of hyperspectral images. Then, we carefully mark the pair-wise matching keypoints between adjacent bands, generating a set $\mathbf{L}$, which is similar to $\mathbf{C}$.

2.2   Graph Matching Formulation

To learn a graph model on exact undirected graph matching, we firstly review the graph formulation. An attributed graph [18] is represented as $G = (V, E, A)$ where V contains nodes v_i in the graph, E consists of edges e_{ij} between vertices. An attribute a_{ii} is assigned to each node and a_{ij} to each edge, which are real numbers or vectors, respectively.

Considering the matching problem between two attributed graphs $G = (V, E, A)$ and $G' = (V', E', A')$, an assignment matrix $\mathbf{X} \in \{0,1\}^{N \times N'}$ defines one of the possible matching correspondences, where $N = |V|$ and $N' = |V'|$ are the number of vertices in each graph. $x_{ij} \in \mathbf{X}$ should satisfy the following condition

$$x_{ij} = \begin{cases} 1, & \text{if } v_i \in V \text{ matches } v_j \in V' \\ 0, & \text{otherwise} \end{cases} \tag{1}$$

In this paper, we follow [5,6,3,4] and denote $x \in \{0,1\}^{NN'}$ as a column-wise vectorized replica of $\mathbf{X}$. A score function $S(G, G', x)$ represents the similarity of graph attributes. So the graph matching problem can be formulated using this maximizing score function $S(G, G', x)$ for possible assignment vector x:

$$x^* = \underset{x}{\operatorname{argmax}} S(G, G', x)$$

$$s.t. \ x \in \{0,1\}^{NN'}, \ \sum_{i=1}^{N} x_{ij} \leq 1, \ \sum_{j'=1}^{N'} x_{i'j'} \leq 1 \tag{2}$$

where constraints in Eq. (2) is required by the one-to-one matching between G and G'. Then the score function $S(G, G', x)$ is re-written to transform Eq. (1) into a manageable form. A similarity matrix $\mathbf{M}$ of $NN' \times NN'$ is introduced where the non-diagonal element $\mathbf{M}_{ij;pq} = s_E(a_{ip}, a'_{jq})$ preserves the edge similarity between edge e_{ip} in G and e_{jq} in G'. The diagonal term $\mathbf{M}_{ij;ij} = s_V(a_i, a'_j)$ contains the node similarity between vertices v_i in G and v_j in G'. As a result, the score function is updated as:

$$S(G, G', x) = \sum_{x_{ij}=1} s_V(a_i, a'_j) + \sum_{x_{ij}=1, x_{pq}=1} s_E(a_{ip}, a_{jq})$$
$$= x^T \mathbf{M} x$$
(3)

The graph matching problem in Eq. (1) is formulated as Integer Quadratic Program (IQP) problem known to be NP hard in Eq. (3).

2.3 Parameterized Graph Model

Given a multi-band image, an attributed graph G_i is constructed on each band using the above steps. We can treat each G_i as a weaker sample of the potentially optimized graph model. The correspondences $C_i \in \mathbf{C}$ with respect to G_i are crucial clues in finding such a model.

To address the problem of learning graphs, we assumed that there exist a graph model G^* that fits various correspondences. The matching between G^* and graph G_i of each band can be recovered by generating an assignment vector x which reflects the correspondence maps G^* to $\mathbf{G}$. Therefore, inspired by the structured output framework [21] and graph learning [3,12], we use the score function $S(G, G^*, x)$ as a compatibility function, getting scores of all possible assignment vectors from correspondence set $\mathbf{C}$. The problem of finding the best graph match from Eq.(2) can be expressed as:

$$x^*(G^*, \beta; G) = \underset{x \in \S_1(C_G)}{\mathrm{argmax}}\ S(G^*, G, x; \beta)$$
(4)

where β is a weighting vector defined in [4], which indicates the importance of vertices and edges. $\S_1(\cdot)$ denotes the assignment vector generated from the first matching results in correspondence set $\mathbf{C}$. The learning process is performed on the pair-wise examples $\mathbf{Y} = \{(G_1, x_1), \ldots, (G_d, x_d)\}$, where $x_i = \S_1(C_{G_i})$ is ith correspondence of band i.

We parameterized both G^* and β in a vectorized form to gain a linear form of S. Following [4], the similarity functions s_V and s_E are dot products of two attribute vectors $s_V(a_i, a'_j) = a_i \bullet a_j$ and $s_E(a_{ip}, a'_{jq}) = a_{ip} \bullet a_{jq}$. Moreover, the attribute vector $\Theta(G^*) = [\cdots; a_i^*; \cdots; a_{ij}^*; \cdots]$. Feature map $\Psi(G, x) = [\cdots; a_{\pi(i)}^*; \cdots; a_{\pi(i)\pi(j)}^*; \cdots]$ are also defined in vectorized form, where $\pi(i) = j$ denotes a matching of v_i^* in G^* to v_j in G according to assignment vector x. With the requirement of max-margin framework [21], the scoring function in Eq. (4) can be factorized in a dot product form[4] $S(G^*, G, x; \beta) = \mathbf{w} \bullet \Psi(G, x)$, where

$\mathbf{w} \equiv \beta \odot \Theta(G^*)$ combines the weights β and attributes $\Theta(G^*)$ ($\odot$ denotes the Hadamard product). Then Eq. (4) can be rewritten in a linear form as follows

$$x^*(G^*, \beta; G) = \underset{x \in \S_1(G)}{\operatorname{argmax}} \mathbf{w} \cdot \Psi(G, x) \tag{5}$$

Attributes Definition. We represent edge attributes by a histogram of log-polar bins [4] to match the dot product assumption. The polar-angle of each edge is measured from the mean characteristic angle of all vertices in case there is no such angle. This will enhance the robustness of the method against rotation. Then, the SIFT descriptor [15] is adopted as the attribute of vertices for describing the local appearance.

2.4   Graph Model Learning

Now, with the labeled examples $\mathbf{Y} = \{(G_1, x_1), \ldots, (G_d, x_d)\}$, a structured output maximum margin framework [21] can handle the learning of $\mathbf{w}$. It can be applied to Eq. (5) and equivalently expressed in a margin re-scaling maximization formulation

$$\min_{\mathbf{w}, \xi} \frac{1}{2}\|\mathbf{w}\|^2 + \frac{\lambda}{d}\sum_{i=1}^{d} \xi_i$$
$$s.t. \begin{cases} \forall i : \xi_i \geq 0 \\ \forall i, \ \forall x \in \S_1(C_G)\backslash x_i : \langle \mathbf{w}, \delta\Psi(G, x)\rangle \geq \Delta(x_i, x) - \xi_i \end{cases} \tag{6}$$

where $\delta\Psi(G, x) \equiv \Psi(G, x_i) - \Psi(G, x)$, $\Delta(x_i, x)$ stands for penalty of loss function on choosing x instead of the reliable assignment vector x_i, λ is a parameter controlling the trade-off between regularization and loss terms.

Noticed that only the first correspondence C_1 in $\mathbf{C}$ is brought into the Eq. (5). However, there is always more than one correspondence besides the landmarks. Despite the fact that manually landmark labelling is time-consuming, we show that combining landmarks with various correspondences is an efficient way of improving the performance of the graph model. Thus, we add an additional set of constraints and modify Eq. (6) into

$$\min_{\mathbf{w}, \xi, \nu} \frac{1}{2}\|\mathbf{w}\|^2 + \frac{\lambda_1}{d}\sum_{i=1}^{d} (\xi_i + \lambda_2\nu_i)$$
$$s.t. \begin{cases} \forall i : \xi_i, \ \nu_i \geq 0 \\ \forall i, \ \forall x \in \S_1(C_G)\backslash x_i : \langle \mathbf{w}, \delta\Psi(G, x)\rangle \geq \Delta(x_i, x) - \xi_i \\ \forall i, \ \forall x' \in \S_2(C_G)\backslash x_i : \langle \mathbf{w}, \delta\Psi(G, x')\rangle \geq \Delta(x_i, x') - \nu_i \end{cases} \tag{7}$$

where λ_1 is the trade-off parameter, λ_2 controls the effect of second correspondence. More correspondences or manually labelled landmark can also be imported in this way. The structured SVM [21] allows us to maximize the margin of the constraints in Eq.(7). The graph model contained in $\mathbf{w}$ is learned along with structured output. The initial graph model is set at median band as it matches other bands quite well in the experiments [19].

Loss Function. A loss function is required in the objective function of Eq. (7) to measure the sensitivity of a predicted assignment vector x^* against the input correspondences in assignment vector x. We define the normalized Hamming loss as $\Delta(x_i, x) = 1 - \frac{x_i \cdot x}{\|x_i\|_F^2}$, where $\| \cdot \|_F$ denotes the Frobenius norm, on mismatch rate. This has been proved to be effective in graph matching method [3,4].

2.5 Matching Method

Given an attribute graph G_1^* with weights β_1 is being generated on a multi-band image, we can perform the same steps on the target (different) band image to obtain an attribute graph G_2^* and weights β_2. The graph matching problem formulated in Eq. (2) and Eq. (3) can be parameterized with weights as:

$$x^* = \operatorname*{argmax}_{x} x^T \mathbf{M_w} x \ \ s.t. \ \ x \in \{0,1\}^{NN'}, \ \sum_{i=1}^{N} x_{ij} \leq 1, \ \sum_{j'=1}^{N'} x_{i'j'} \leq 1 \ \ (8)$$

where $\mathbf{M_w} = (\beta_1 \odot \beta_2) \cdot \mathbf{M}$. So the traditional matching methods mentioned in Section 1.2 are elegant options to address this standard graph matching problem.

3 Experiments

We have applied our algorithm on multi-dimensional images in the Multispectral Image Database [22]. Fig. 2 shows two examples[2].

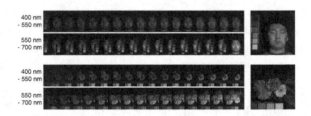

Fig. 2. Hyperspectral images (left) in the range from 400nm to 700nm with a step size of 10nm and the corresponding RGB image (right)

Firstly, we generated two image sets from the 14 selected hyperspectral images. In set 1, the first 25 bands of each image form a 25-dimensional image and the matching is performed on the rest 6 bands of the same scene. Then, we matched each 31-dimensional image with the corresponding RGB image in set 2. A tolerance $t = 10$, approximately 2% of the band image size (512 pixels), is defined to determine correct matchings within acceptable position. The matching accuracy is calculated by $\frac{\sharp\{correct\ matches\}}{\sharp\{all\ matches\}}$, where $\sharp\{\cdot\}$ denotes the number of pairs. Landmarks were manually marked on 10 selected points for adjacent bands.

---

[2] The dataset contains 32 scenes. Every scene contains a 31 band hyperspectral image and a composite RGB image. We selected 14 of them in various classes, including face, paints, toys, flowers and etc.

Table 1. Matching accuracy on set 1 (25 band images to 6 band images)

Methods	face	face^t	watercolors	watercolors^t	flowers	flowers^t	toys	toys^t	Average
SIFT(average)	73.8	72.1	80.9	80.5	89.4	89.3	87.5	85.6	73.7
SIFT(worst)	35.7	30.8	50.1	44.4	79.9	78.2	78.0	73.0	-
Vector SIFT	81.3	74.2	82.3	78.9	93.1	**90.4**	85.1	81.4	78.1
our+SM	83.3	80.3	**96.7**	**93.3**	86.7	86.7	90.0	87.7	87.9
our+RRWM	**86.7**	**81.1**	**96.7**	90.0	**93.3**	90.2	**93.3**	**92.8**	**88.8**

Table 2. Matching accuracy on set 2 (31 band images to RGB images)

Methods	face	face^t	watercolors	watercolors^t	flowers	flowers^t	toys	toys^t	Average
SIFT(average)	83.1	78.1	82.1	82.2	91.5	91.3	75.9	74.1	79.6
SIFT(worst)	55.5	41.2	73.8	73.2	81.5	80.8	69.1	34.3	-
Vector SIFT	78.2	76.6	84.2	83.2	92.7	90.4	76.2	74.1	81.9
our+SM	80.0	73.3	90.0	86.9	90.0	89.1	86.7	83.3	85.0
our+RRWM	80.0	76.9	**93.3**	86.7	**96.7**	90.0	88.1	86.7	85.7
our^R+RRWM	83.3	79.7	86.7	86.7	93.3	**92.2**	90.0	90.0	89.5
our^{RL}+RRWM	**93.3**	**90.2**	90.0	**89.7**	90.0	86.7	**93.3**	91.4	**91.3**

The SIFT [15] and vector SIFT[8] matching algorithms were applied between every band of images pairs. We calculated the mean accuracy and record the worst case.

The experimental results on set 1 are shown in Table 1. The accuracy of keypoint based method various significantly. The worst case is observed in matching bands that are apart to each other, as reported in [19]. Setting $k = 30$ and $\lambda = 0.25$, we obtain novel graph models for each scene. Using matching method in [13] and [5], our proposed methods outperform the traditional matching methods on the average results and avoid the worst case. Accuracy varies in less extent on our matching methods. In addition, the *face^t* represents 12 rotated scenes from 0° to 360° and reveals decrease in accuracy.

The experimental results on set 2 are presented in Table 2. Compared with previous experiment, the accuracies decrease slightly due to the bigger gap between RGB images and hyperspectral images. But our method still demonstrate improvement over the alternative methods.

Moreover, we applied random rotations $\mathbb{R}_i$ in the range of $[0°, 360°]$ both on G and $\mathbf{C}$ to demonstrate the anti-rotation ability of graph models. The results are shown on the penultimate row in Table 2. The rotation also generates some negative influences on matching outcomes, e.g., in *watercolors* and *flowers*.

The method our^{RL} introduces landmark correspondences $\mathbf{L}$ as the second correspondence into the structured output max-margin framework in Eq. (7) with $\lambda_2 = 0.15$ and $\lambda_3 = 0.35$. The landmarks in *face* turn to be highly effective rather than in *flowers* and *watercolors*. The latter contains insignificant correspondences and complicated pieces, like petals, which are hard to recognize correctly.

Some visualized band matching results are shown in Fig. 3.

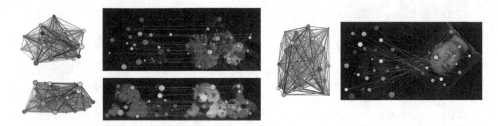

Fig. 3. Weighted Graph model for hyperspectral image(left) and matching between different band images (right)

4   Conclusions

Matching of images with different dimensions is a non-trivial problem. In this paper, we have propose a learning weighted graph model to catch the most discriminative feature points and correspondences among each band image. A structured output framework with multiple constraints is applied to address the *NP*-Hard graph matching problem. The experiments show significant improvement on the matching performance between different bands, especially with few landmarks, when compared against several alternative methods.

Acknowledgement. This research is supported by the National Natural Science Foundation of China (NSFC) projects No.61370123 and No.61402389.

References

1. Balas, C., Papadakis, V., Papadakis, N., Papadakis, A., Vazgiouraki, E., Themelis, G.: A novel hyper-spectral imaging apparatus for the non-destructive analysis of objects of artistic and historic value. Journal of Cultural Heritage 4, 330–337 (2003)
2. Bay, H., Ess, A., Tuytelaars, T., Van Gool, L.: Speeded-up robust features (SURF). Computer Vision and Image Understanding 110(3), 346–359 (2008)
3. Caetano, T.S., McAuley, J.J., Cheng, L., Le, Q.V., Smola, A.J.: Learning graph matching. IEEE Transactions on Pattern Analysis and Machine Intelligence 31(6), 1048–1058 (2009)
4. Cho, M., Alahari, K., Ponce, J.: Learning graphs to match. In: IEEE International Conference on Computer Vision, pp. 25–32. IEEE (2013)
5. Cho, M., Lee, J., Lee, K.M.: Reweighted random walks for graph matching. In: Daniilidis, K., Maragos, P., Paragios, N. (eds.) ECCV 2010, Part V. LNCS, vol. 6315, pp. 492–505. Springer, Heidelberg (2010)
6. Cour, T., Srinivasan, P., Shi, J.: Balanced graph matching. Advances in Neural Information Processing Systems 19, 313 (2007)
7. Diem, M., Lettner, M., Sablatnig, R.: Multi-spectral image acquisition and registration of ancient manuscripts 224, 129–136 (2007)
8. Dorado-Muñoz, L.P., Velez-Reyes, M., Mukherjee, A., Roysam, B.: A vector SIFT operator for interest point detection in hyperspectral imagery. In: 2nd Workshop on Hyperspectral Image and Signal Processing: Evolution in Remote Sensing, pp. 1–4. IEEE (2010)

9. Easton Jr., R.L., Knox, K.T., Christens-Barry, W.A.: Multispectral imaging of the Archimedes palimpsest. In: IEEE Applied Imagery Pattern Recognition Workshop, pp. 111–111. IEEE Computer Society (2003)
10. Fischler, M.A., Bolles, R.C.: Random sample consensus: a paradigm for model fitting with applications to image analysis and automated cartography. Communications of the ACM 24(6), 381–395 (1981)
11. Fontana, R., Gambino, M.C., Greco, M., Marras, L., Pampaloni, E.M., Pelagotti, A., Pezzati, L., Poggi, P.: 2D imaging and 3D sensing data acquisition and mutual registration for painting conservation. In: Electronic Imaging 2005, pp. 51–58. International Society for Optics and Photonics (2005)
12. Hare, S., Saffari, A., Torr, P.H.: Efficient online structured output learning for keypoint-based object tracking. In: IEEE Conference on Computer Vision and Pattern Recognition, pp. 1894–1901. IEEE (2012)
13. Leordeanu, M., Hebert, M.: A spectral technique for correspondence problems using pairwise constraints. In: IEEE International Conference on Computer Vision, vol. 2, pp. 1482–1489. IEEE (2005)
14. Leordeanu, M., Hebert, M., Sukthankar, R.: An integer projected fixed point method for graph matching and map inference. In: Advances in Neural Information Processing Systems, vol. 22, pp. 1114–1122 (2009)
15. Lowe, D.G.: Distinctive image features from scale-invariant keypoints. International Journal of Computer Vision 60(2), 91–110 (2004)
16. Mikolajczyk, K., Schmid, C.: Scale & affine invariant interest point detectors. International Journal of Computer Vision 60(1), 63–86 (2004)
17. Mukherjee, A., Velez-Reyes, M., Roysam, B.: Interest points for hyperspectral image data. IEEE Transactions on Geoscience and Remote Sensing 47(3), 748–760 (2009)
18. Pelillo, M.: A unifying framework for relational structure matching. In: Fourteenth International Conference on Pattern Recognition, vol. 2, pp. 1316–1319. IEEE (1998)
19. Saleem, S., Bais, A., Sablatnig, R.: A performance evaluation of SIFT and SURF for multispectral image matching. In: Campilho, A., Kamel, M. (eds.) ICIAR 2012, Part I. LNCS, vol. 7324, pp. 166–173. Springer, Heidelberg (2012)
20. Saleem, S., Sablatnig, R.: A modified SIFT descriptor for image matching under spectral variations. In: Petrosino, A. (ed.) ICIAP 2013, Part I. LNCS, vol. 8156, pp. 652–661. Springer, Heidelberg (2013)
21. Tsochantaridis, I., Joachims, T., Hofmann, T., Altun, Y.: Large margin methods for structured and interdependent output variables. Journal of Machine Learning Research, 1453–1484 (2005)
22. Yasuma, F., Mitsunaga, T., Iso, D., Nayar, S.K.: Generalized assorted pixel camera: postcapture control of resolution, dynamic range, and spectrum. IEEE Transactions on Image Processing 19(9), 2241–2253 (2010)

VF2 Plus: An Improved version of VF2 for Biological Graphs

Vincenzo Carletti[✉], Pasquale Foggia, and Mario Vento

DIEM, Department of Information Engineering,
Electrical Engineering and Applied Mathematics,
University of Salerno, Salerno, Italy
{vcarletti,pfoggia,mvento}@unisa.it
http://mivia.unisa.it

Abstract. Subgraph isomorphism is a common problem in several application fields where graphs are the best suited data representation, but it is known to be an NP-Complete problem. However, several algorithms exist that are fast enough on commonly encountered graphs so as to be practically usable; among them, for more than a decade VF2 has been the state of the art algorithm used to solve this problem and it is still the reference algorithm for many applications. Nevertheless, VF2 has been designed and implemented ten years ago when the structural features of the commonly used graphs were considerably different. Hence a renovation is required to make the algorithm able to compete in the challenges arisen in the last years, such as the use of graph matching on the very large graphs coming from bioinformatics applications. In this paper we propose a significant set of enhancements to the original VF2 algorithm that enable it to compete with more recently proposed graph matching techniques. Finally, we evaluate the effectiveness of these enhancement by comparing the matching performance both with the original VF2 and with several recent algorithms, using both the widely known MIVIA graph database and another public graph dataset containing real-world graphs from bioinformatics applications.

1 Introduction

In several application fields graphs are the best suited data structure to represent entities and relationships composing the problem domain. Think, as examples, to bioinformatics or chemoinformatics [2,9,10], where proteins and drugs structures, metabolic networks, interaction networks and so on, are naturally represented using graphs. Other noteworthy examples are social network analysis, where the social interactions are represented by graphs of people, and the semantic web, where the Resource Description Framework (RDF) defines a standard graph format to represent the relationships existing among a set of web resources. In many of these applications, a relevant problem is the search for a pattern graph within a target graph, namely the subgraph isomorphism problem [3,6,18]. This problem is known to be NP-Complete in the worst case, i.e. when the two graphs

© Springer International Publishing Switzerland 2015
C.-L. Liu et al. (Eds.): GbRPR 2015, LNCS 9069, pp. 168–177, 2015.
DOI: 10.1007/978-3-319-18224-7_17

are complete or in general symmetric. Thus, the only way to solve the subgraph isomorphism problem in a reasonable time is to exploit the knowledge on the structure of the graphs. Due to this fact, many algorithms have been proposed during the last decade. In order to provide a complete overview of the current state of the art we identified three main trends: tree search based algorithms, constraint programming algorithms and index based algorithms. The first trend comprises algorithms coming from the field of artificial intelligence; these algorithms deal with the problem by searching the solution inside a state space, generally using a depth-first search Ullmann [16] and VF/VF2 [4] are two very popular algorithms adopting this paradigm. Recently, new algorithms tried to refine this approach for specific kinds of graphs, as instance RI, RIDS [1] and L2G. The constraint programming approach [20] deals with subgraph isomorphism by using a diametrically opposite method, with respect to the previous one. Indeed, the rationale is to filter, among all the possible couples of nodes, those that are surely not contained in a complete matching solution. The algorithm by McGregor [12] is an early example of this trend, that has been followed during the last decade by Larrosa and Valiente[11], ILF [20], LAD [15] and the last algorithm by Ullmann [17]. The last trend comprises algorithms that extend the pure database approach, i.e. graph indexing, where the main problem is just to determine if a pattern graph is inside a target graph, without identifying which subgraphs in the target graph are isomorphic with the pattern. These algorithms exploit a set of features to define an index for the pattern graph structure, used for driving the search for possible solutions on the target. Recent algorithms following this trend are: GraphQL [8], QuickSI [14], GADDI [21], SPath [22] and TurboIso [7]. The great variety of algorithms is due to the wide diversity of problems, and so of graph representations, in which the subgraph isomorphism is employed. Among these it is important to note that VF2 is one of the most employed algorithm for graph and subgraph isomorphism due to its flexibility [2] in addressing different exact graph matching problems. However, VF2 has been proposed more than ten years ago, when the main applications of exact graph matching were fingerprint recognition, character recognition and molecular structure comparison; thus, the structural features of the graphs employed then were considerably different from the current one, think as example to the size. In this paper, we propose a set of improvements to the VF2 algorithm for large biological graphs and we call the improved version VF2 Plus. Finally, in order to provide an assessment of the improvements, we compared VF2 Plus with VF2, RI, L2G and LAD on the subgraph isomorphism problem.

2   Preliminaries

Before starting the description of the improvements in VF2 Plus, it is important to provide an overview of VF2. As introduced in Section 1, VF2 [4] is a well known state of art algorithm able to deal with different exact graph matching problems, based on a state space representation: each state is a partial mapping between the two given graphs, while goal states are complete mappings consistent with the constraints of the problem we are facing. Hence, the search space

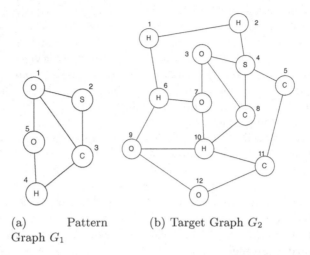

(a)   Pattern   (b) Target Graph G_2
Graph G_1

Fig. 1. Graphs used as example to detail how VF2 and VF2 Plus work

is explored using a depth-first strategy with backtracking, driven by a set of feasibility rules to prune, in advance, unfruitful search paths.

In order to provide more details, let us introduce some notations that will be used in the following. Let be $G_1 = (V_1, E_1)$ the pattern graph (Figure 1(a)), and $G_2 = (V_2, E_2)$ the target graph (Figure 1(b)), being V_1, V_2 and E_1, E_2 respectively the vertices and the edges sets of each graph. Let us define the size of the two graphs as the number of the nodes inside the sets V_1 and V_2, so that $|G_1| = |V_1|$ and $|G_2| = |V_2|$.

The algorithm has to keep track, at each state, of the vertex pairs inside the current mapping and of the set of nodes where to search for the next candidate pairs. As shown in Figure 3, each state s stores mapped couples in a set $M(s)$, namely the core set, and uses vertices in two separated sets $M_1(s)$ and $M_2(s)$, respectively for first and second graph. Then, for each vertex in $M_1(s)$ and $M_2(s)$, neighbors not yet in the core set are kept into two sets $T_1(s)$ and $T_2(s)$, called terminal sets. Obviously, in the case of directed graph, each states uses two different pairs of terminal sets: $T_1^{in}(s)$, $T_2^{in}(s)$ for neighbors connected by incoming edges and $T_1^{out}(s)$, $T_2^{out}(s)$ for these connected by outgoing edges. Despite this representation seems to need a space complexity quadratic with respect to the number of the states, VF2 exploits an optimized representation able to reduces this complexity to linear.

Before generating a new state, VF2 selects the next candidate pair (u, v) by picking u from $T_1(s)$ and v from $T_2(s)$ if $T_1(s) \neq 0$ and $T_2(s) \neq 0$, or from the sets $\widetilde{N}_1(s)$, $\widetilde{N}_2(s)$ that contain remaining vertices of G_1 and G_2, not in the core or in the terminal sets:

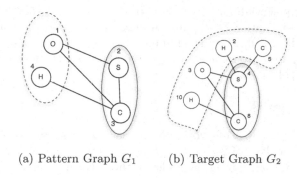

(a) Pattern Graph G_1 (b) Target Graph G_2

Fig. 2. Core sets (solid line) and terminal sets (dotted line) used by VF2 to represent the state of the current matching and the next candidates set. In the example we can identify the following sets: $M(s) \equiv \{(3,8),(2,4)\}$, $M_1(s) \equiv \{3,2\}$, $M_2(s) \equiv \{8,4\}$, $T_1(s) \equiv \{1,4\}$, $T_2(s) \equiv \{2,3,5,10\}$.

$$P_1(s) = \begin{cases} T_1(s), & \text{if } |T_1(s)| \neq 0 \wedge |T_2(s)| \neq 0 , \\ \widetilde{N}_{1p}(s), & \text{otherwise} \end{cases} \tag{1a}$$

$$P_2(s) = \begin{cases} T_2(s), & \text{if } |T_1(s)| \neq 0 \wedge |T_2(s)| \neq 0 , \\ \widetilde{N}_{2p}(s), & \text{otherwise} \end{cases} \tag{1b}$$

After that the candidate pair (u,v) has been selected, VF2 evaluates if it is certainly not moving towards a solution by using a set of feasibility rules:

Core Rule. The main rule needed to solve the specific problem we are facing; it guarantees to generate only consistent states, i.e. states wherein the partial mapping can possibly be part of a solution.

Look-Ahead Rules. Secondary rules used to search for possible inconsistencies, in order to detect early states without any goal descendants.

In order to understand how feasibility rules work, let us consider the problem of subgraph isomorphism on undirected graphs G_1 and G_2, as shown in Figure 1; VF2 uses the following rules:

$$R_{core}(s,u,v) \iff$$
$$\forall u' \in adj(u) \cap M_1(s) \exists! v' \in adj(v) \cap M_2(s) : (u',v') \in M(s) \tag{2a}$$
$$\wedge \forall v' \in adj(v) \cap M_2(s) \exists! u' \in adj(u) \cap M_1(s) : (u',v') \in M(s)$$

$$R_{term}(s,u,v) \iff |adj(u) \cap T_1(s)| = |adj(v) \cap T_2(s)| \tag{2b}$$

$$R_{new}(s,u,v) \iff |adj(u) \cap \widetilde{N}_1(s)| = |adj(v) \cap \widetilde{N}_2(s)| \tag{2c}$$

The effectiveness of the previous rules in terms of search space reduction and complexity has been well discussed in a recent paper by N. Dahm and H. Bunke [5].

However, the exponential nature of the subgraph isomorphism problem and the wide variety of application fields raise the need for an algorithm that is generic with respect to the specific contexts, but easy to specialize, by using simple heuristics if possible, in order to reduce the explored search space and consequently the time to find the solutions.

3   Improvements

Once we described how VF2 works, it is easier to discuss the improvements provided. In particular, VF2 Plus improves two important weaknesses of VF2: the total order relationships on the nodes of the pattern graph and the structure of the terminal sets.

3.1   Total Order Relationship

The first weakness of VF2 lies in the total order relationship employed. Such order relationship, defined on the nodes of the pattern graph, is very simple and provides no real advantage to the algorithm other than producing a search space structured as a tree. In fact, if the search space were a graph then a state with k different pairs could be reached from $k!$ different paths, making the search unfeasible. However, if well defined, a total order relationship can give an additional advantage by providing a sequence of nodes of the first graph that makes the algorithm able to explore the search space more efficiently. Thus, VF2 Plus uses a sorting procedure that prefers nodes, on the pattern graph, with the lowest probability to find a candidate on the target graph and the highest number of connections with the nodes already used by the algorithm. The procedure provides a sorted node sequence that imposes to the algorithm how to explore the pattern graph. For this reason, once the sequence is given the algorithm already knows which node of the pattern graph it is going to select as candidate at each level of the search space. Indeed, it is trivial to understand that all of the states at the same level share the same candidate node on the pattern graph. In this way, VF2 Plus is able to compute the state of the terminal sets related to the pattern graph at each level before staring the matching.

Evaluate the Probability. On subgraph isomorphism a node $v \in G_2$ is feasible with a node $u \in G_1$ if they share the same label, $lab(u) = lab(v)$, and their degree is compatible, $deg(u) \le deg(v)$. Thus, both semantic and structural node features, i.e. the label and the degree, can be employed to evaluate the probability to find a candidate $v \in G_2$ for $u \in G_1$.

Firstly, the procedure computes the frequency of each label and degree on G_2. So, let us define:

- $P_{lab}(L)$: a priori probability to find a node with label L on G_2.
- $P_{deg}(d)$: a priori probability to find a node with degree d on the G_2.

For each pattern node $u \in G_1$ the probability is computed as follows:

$$P(u) = P_{lab}(L) * \cup_{d' \ge d} P_{deg}(d')$$ (3)

Order Relationships. Once the probability is computed, the algorithm begins the ordering procedure. Let's define the set of already sorted nodes M, T the set of nodes candidate to be selected and *degreeM*, of a node in $u \in T$, as the number of edges that are connected to nodes in M. Then the algorithm proceeds as follows:

1. Select the node with the lowest probability.
   (a) If more nodes share the same probability, select the one with maximum degree.
   (b) If more nodes share the same probability and they have the maximum degree, select the first.
2. Put the selected node in the set M and all of its neighboors in the set T.
3. From the set T select the node with the maximum *degreeM*.
   (a) If more nodes share the same *degreeM*, select the one with the lowest probability.
   (b) If more nodes share the same probability and the same *degreeM*, select the first.
4. Go to 2

3.2   Terminal Sets

The second critical point lies in the terminal sets, T_1 and T_2. As shown in the Equations (1) (2), these sets are fundamentals both for generating new candidate pairs and for applying the look-ahead rules. But, while for the latter VF2 analyzes just the neighborhood of both the nodes in the candidate pair (u, v), for the former it could explore the terminal sets completely. Thus, since the search for a new candidate pair is performed repeatedly at each state, it is clear that the size of the terminal sets affects the efficiency of VF2. The growth of the sets depends on the average degree, the clustering coefficient and the size of the analyzed graphs. As example, a very bad case is represented by large sparse graphs, such as protein graphs. In these situations VF2 explores more useless candidate nodes inside the terminal sets, while the part of the set that contains feasible candidates is very small. VF2 Plus reduces significantly the explored part of the terminal set by using a new procedure of candidate selection based on two main improvements. First, in order to find a node in G_2 to map to the node $u \in G_1$, the new procedure analyzes just the neighborhood of the node $v' \in G_2$ that is already mapped to a neighbor u' of u. If there are no neighbors of u in the core set, then the pair will be searched outside the terminal sets. Second, before starting the matching, VF2 Plus applies a node classification procedure on the nodes of both the graphs. The aim of this classification is to divide the terminal sets in different subsets, one for each class, and so reduce the number of feasible candidate pair and make the look-ahead rules stronger as applied to each class subset. Indeed, given a set C of n non overlapped class C_i with $i \in [0, n] \subset \mathbb{N}$, the terminal sets can be structured as:

$$\begin{aligned} T_1(s) &\equiv T_1^{C1}(s) \cup T_1^{C2}(s) \cup \cdots \cup T_1^{Cn}(s) \\ T_2(s) &\equiv T_2^{C1}(s) \cup T_2^{C2}(s) \cup \cdots \cup T_2^{Cn}(s) \end{aligned} \tag{4}$$

And the look-ahead rule can be rewritten as follows:

$$R_{ahead}(u, v, s) \iff |adj(u) \cap T_1^{C1}(s)| = |adj(u) \cap T_2^{C1}(s)|$$
$$\wedge |adj(u) \cap T_1^{C2}(s)| = |adj(u) \cap T_2^{C2}(s)| \wedge |adj(u) \cap T_1^{C3}(s)| = |adj(u) \cap T_2^{C3}(s)| \quad (5)$$
$$\wedge \cdots \wedge |adj(u) \cap T_1^{Cn}(s)| = |adj(u) \cap T_2^{Cn}(s)|$$

In many real application is possible to identify a suited class function that uses semantic and structural features to classify the nodes. It is important to note that if two nodes of G_1 and G_2 are feasible for the isomorphism they must be in the same class, while the inverse is not true. Thus, once selected a candidate pair from the same class set in T_1 and T_2 is necessary to evaluate the feasibility.

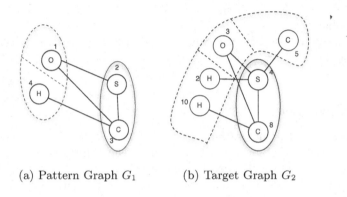

(a) Pattern Graph G_1 (b) Target Graph G_2

Fig. 3. Core sets (solid line) and terminal sets (dotted line) partitioned in different classes. In this example each node with the same label belongs to the same class.

4   Experiments

In order to prove the effectiveness of the proposed algorithm, we compared VF3 with VF2 and other state of the art algorithms: RI, LAD, L2G. The experiments have been conducted on a linux based system (kernel version 3.13) with an Intel Xeon Twelve Core 3.20 Ghz and 32 GBytes of RAM. We have considered two different parameters: the time to find the first matching solution and the time to find all the matching solutions. For each parameter, the measure has been performed several times on the same pair of graphs to avoid caching effects. The comparison has been performed employing the biological datasets proposed by a recent International Contest on Pattern Search in Biological Databases [19]. The datasets are composed of Contact Map and Protein graphs extracted from the Protein Data Bank [13]. Moreover, we have also considered a modified version of the previous datasets wherein the number of solutions, for each pair of graphs, has been increased by reducing the variability of the labels; we called this dataset *Reduced Labels*. The protein dataset is composed of very large and sparse graphs from 500 to 10000 nodes, with average degree of 4; while the contact maps dataset contains medium size and dense graphs from 150 to 800 nodes and average degree of 20. In the original version the two datasets presents respectively 6 and 20 different labels, but in the modified version the variability of the labels has been reduced to 4 and 7.

As shown in Figure 4, VF2 Plus always outperforms VF2, LAD and L2G. Especially on very large and sparse graphs the difference, between VF2 and VF2 Plus, is more than two order of magnitude on the time to find all the solutions. Furthermore, the Table 1 shows that VF2 Plus most often equals RI, the algorithm winner of the International Contest on Pattern Search in Biological Databases. Indeed, in several cases, the gap between VF2 Plus and RI is less then the 10% on the matching time (Table 1). However, VF2 Plus performs better than RI for very large and sparse graphs when the number of solutions increases (Figure 4(b)) and for dense medium graphs when the number of label is higher (Figure 4(c)). In both cases, the advantage is mainly given by the class-based terminal sets and the new candidate selection procedure that significantly reduce the number of unfeasible pairs explored.

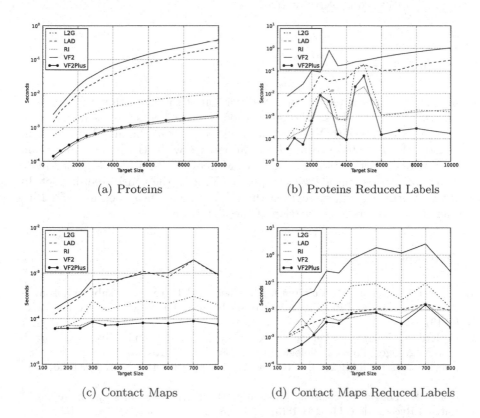

(a) Proteins   (b) Proteins Reduced Labels

(c) Contact Maps   (d) Contact Maps Reduced Labels

Fig. 4. Time in seconds to find all matchings on the four considered datasets

Table 1. The Tables show the results of the experiments on four biological datasets: Proteins Original, Proteins with reduced labels, Contact Maps Original and Contact Maps with reduced labels. For each dataset two different parameters have been considered: the time to found the first matching (First) and the time to found all the matchings (All). Each cell contains the algorithm with the best time, if the gap between two algorithms is less than the 10% both are provided.

Proteins				
	Original		Reduced Labels	
Size	First	All	First	All
600	VF2P/RI	VF2P/RI	VF2P	VF2P
1000	VF2P/RI	VF2P/RI	VF2P	VF2P/RI
1500	VF2P/RI	VF2P/RI	VF2P	VF2P
2000	VF2P/RI	VF2P/RI	VF2P	VF2P/RI
2500	VF2P/RI	VF2P/RI	VF2P	VF2P/RI
3000	VF2P/RI	VF2P/RI	VF2P	VF2P/RI
3500	VF2P/RI	VF2P/RI	VF2P	VF2P
4000	VF2P/RI	VF2P/RI	VF2P	VF2P
4500	VF2P/RI	VF2P/RI	VF2P	RI
5000	VF2P/RI	VF2P/RI	VF2P	VF2P
6000	VF2P/RI	VF2P/RI	VF2P	VF2P
7000	VF2P/RI	VF2P/RI	VF2P	VF2P
10000	VF2P/RI	VF2P/RI	VF2P	VF2P

Contact Maps				
	Original		Reduced Labels	
Size	First	All	First	All
150	VF2P/RI	VF2P/RI	VF2P	VF2P
200	VF2P/RI	VF2P/RI	VF2P	VF2P
250	VF2P/RI	VF2P/RI	VF2P/RI	VF2P/RI
300	VF2P/RI	VF2P/RI	VF2P/RI	VF2P/RI
350	VF2P/RI	VF2P	VF2P/RI	VF2P/RI
400	VF2P/RI	VF2P	VF2P/RI	VF2P/RI
500	VF2P/RI	VF2P	VF2P/RI	VF2P/RI
600	VF2P	VF2P	VF2P	VF2P/RI
700	VF2P	VF2P	VF2P	VF2P/RI
800	VF2P	VF2P	VF2P	VF2P/RI

5   Conclusions

In this paper we presented an improved version of the algorithm VF2, namely VF2 Plus. The effectiveness of the proposed improvements has been shown by comparing VF2 Plus with VF2 and other three state of the art algorithms. In particular, VF2 Plus has shown to be competitive with RI, the best algorithm in the state of the art on biological graphs. Moreover, VF2 Plus has gained more than one order of magnitude with respect to VF2 on the considered dataset. In the future works, we will strengthen the improvements proposed for VF2 Plus in order to obtain a renewed version of the algorithm.

References

1. Bonnici, V., Giugno, R., Pulvirenti, A., Shasha, D., Ferro, A.: A subgraph isomorphism algorithm and its application to biochemical data. BMC Bioinformatics 14 (2013)
2. Carletti, V., Foggia, P., Vento, M.: Performance Comparison of Five Exact Graph Matching Algorithms on Biological Databases. In: Petrosino, A., Maddalena, L., Pala, P. (eds.) ICIAP 2013. LNCS, vol. 8158, pp. 409–417. Springer, Heidelberg (2013)
3. Conte, D., Foggia, P., Sansone, C., Vento, M.: Thirty years of graph matching in Pattern Recognition. IJPRAI 18(3), 265–298 (2004)
4. Cordella, L., Foggia, P., Sansone, C., Vento, M.: A (sub)graph isomorphism algorithm for matching large graphs. IEEE Transactions on Pattern Analysis and Machine Intelligence 26, 1367–1372 (2004)

 5. Dahm, N., Bunke, H., Caelli, T., Gao, Y.: Efficient subgraph matching using topo-
    logical node feature constraints. Pattern Recognition (June 2014)
 6. Foggia, P., Percannella, G., Vento, M.: Graph Matching And Learning In Pattern
    Recognition On The Last Ten Years. ... Journal of Pattern Recognition ... (2014)
 7. Han, W.S., Lee, J.H., Lee, J.H.: Turbo Iso: Towards Ultrafast And Robust Sub-
    graph Isomorphism Search In Large Graph Databases. In: ... of the 2013 Interna-
    tional Conference on ..., pp. 337–348 (2013)
 8. He, H., Singh, A.K.: Graphs-At-A-Time: Query Language And Access Methods
    For Graph Databases. In: Proceedings of the 2008 ACM SIGMOD International
    ..., pp. 405–417 (2008)
 9. Huan, J., et al: Comparing graph representations of protein structure for min-
    ing family-specific residue-based packing motif. Journal of Computational Biology
    (2005)
10. Lacroix, V., Fernandez, C., Sagot, M.: Motif search in graphs: Application to
    metabolic networks. Transactions on computational biology and bioinformatics
    (Dicember 2006)
11. Larrosa, J., Valiente, G.: Constraint satisfaction algorithms for graph pattern
    matching. Mathematical Structures in Computer Science 12, 403–422 (2002)
12. McGregor, J.: Relational consistency algorithms and their application in finding
    subgraph and graph isomorphisms. Information Sciences 19(3), 229–250 (1979)
13. RCSB: Protein data bank web site (June 2015), http://www.rcsb.org/pdb
14. Shang, H., Zhang, Y., Lin, X., Yu, J.X.: Taming Verification Hardness: An Efficient
    Algorithm for Testing Subgraph Isomorphism, pp. 364–375 (2008)
15. Solnon, C.: Alldifferent-based filtering for subgraph isomorphism. Artificial Intel-
    ligence 174(12-13), 850–864 (2010)
16. Ullman, J.R.: An algorithm for subgraph isomorphism. J. Assoc. Comput.
    Mach. 23, 31–42 (1976)
17. Ullmann, J.: Bit-Vector Algorithms For Binary Constraint Satisfaction And Sub-
    graph Isomorphism. Journal of Experimental Algorithmics (JEA) 15(1) (2010)
18. Vento, M.: A Long Tri. In: The Charming World Of Graphs For Pattern Recogni-
    tion. Pattern Recognition, 1–11 (January 2014)
19. Vento, M., Jiang, X., Foggia, P.: International contest on pattern search in biolog-
    ical databases (June 2015), http://biograph2014.unisa.it
20. Zampelli, S., Deville, Y., Solnon, C.: Solving subgraph isomorphism problems with
    constraint programming. Constraints 15(3), 327–353 (2010)
21. Zhang, S., Li, S., Yang, J.: GADDI: Distance Index Based Subgraph Matching In
    Biological Networks. In: ... of the 12th International Conference on ... (2009)
22. Zhao, P., Han, J.: On Graph Query Optimization In Large Networks. Proceedings
    of the VLDB Endowment 3(1-2), 340–351 (2010)

Report on the First Contest on Graph Matching Algorithms for Pattern Search in Biological Databases

Vincenzo Carletti,[1] Pasquale Foggia[1(✉)], Mario Vento[1], and Xiaoyi Jiang[2]

[1] Department of Information Engineering, Electrical Engineering and
Applied Mathematics University of Salerno, Fisciano, Italy
{mvento,pfoggia,vcarletti}@unisa.it
[2] Department of Computer Science
University of Münster, Münster, Germany
xjiang@uni-muenster.de

Abstract. Graphs are a powerful data structure that can be applied to several problems in bioinformatics, and efficient graph matching is often a tool required for several applications that try to extract useful information from large databases of graphs. While graph matching is in general a NP-complete problem, several algorithms exist that can be fast enough on practical graphs. However, there is no single algorithm that is able to outperform the others on every kind of graphs, and so it is of paramount importance to assess the algorithms on graphs coming from the actual problem domain. To this aim, we have organized the first edition of the Contest on Graph Matching Algorithms for Pattern Search in Biological Databases, hosted by the ICPR2014 Conference, so as to provide an opportunity for comparing state-of-the-art matching algorithms on a new graph database built using several kinds of real-world graphs found in bioinformatics applications. The participating algorithms were evaluated with respect to both their computation time and their memory usage. This paper will describe the contest task and databases, will provide a brief outline of the participating algorithms, and will present the results of the contest.

1 Introduction

Graphs are a powerful data structure that can be applied to several problems in bioinformatics. For instance, a lot of biological data can be represented as graphs: molecule structures, protein secondary structures, protein interaction networks. Since all these data are stored in huge biological data banks, one of the most important challenges for bioinformatics is the development of efficient Pattern Recognition tools to retrieve relevant information. Among these, the search for patterns inside a biological database can be formulated as a graph matching problem [1,9,12].

Graph matching within a biological database is a very computationally intensive problem, especially because of the size of the graphs involved. For instance,

© Springer International Publishing Switzerland 2015
C.-L. Liu et al. (Eds.): GbRPR 2015, LNCS 9069, pp. 178–187, 2015.
DOI: 10.1007/978-3-319-18224-7_18

if we consider a generic protein database, while a pattern can be a small graph of ten nodes, a complete protein graph, where the pattern is to be searched, can have several thousands of nodes and edges. So the number of candidate solutions for all the proteins can be huge.

To reduce the computational complexity, which in the worst case is exponential [5], some researchers have put their efforts in the development of more or less complex heuristics, that can decrease the matching time in the average case, at least for some classes of graphs, but still retain the exponential worst case; others have devoted their attention to inexact, suboptimal matching algorithms, that achieve a polynomial time, but do not ensure the accuracy of the found solutions, which may or may not be acceptable depending on the application. Besides the time complexity, space requirements of the algorithms also are an important concern when working with the large graphs of bioinformatics applications.

For this reason, it is extremely useful, for the researchers applying graph matching to bioinformatics, to obtain a benchmarking of matching algorithms that is not based on performance measurements taken on generic graphs, but exploits real data from bioinformatics databases, and at the same time attempts to highlight which are the conditions that give an advantage or a disadvantage to each of the algorithms.

This is the motivation behind the organization First International Contest on Graph Matching Algorithms for Pattern Search in Biological Databases (Biograph2014), hosted in August 2014 by the International Conference on Pattern Recogniton (ICPR2014). We, as the organizers of the contest, prepared a large dataset containing three different kinds of graphs extracted from bioinformatics databases. The participants to the contest were invited to submit their subgraph isomorphism algorithms, that we have tested measuring both the running time and the used memory. This initiative was endorsed by the Technical Committee #15 (Graph-based Representations in Pattern Recognition) of the International Association for Pattern Recognition (IAPR), since this Technical Committee has a long-standing tradition of promoting and supporting benchmarking initiatives for graph-based algorithms.

This report provides a description of the contest and of the dataset, and presents the results obtained by participant algorithms.

2 The Biograph2014 Contest

2.1 Description of the Contest Task

The competition task was to find the occurrences of a smaller *pattern graph* within a larger *target graph*. Such a task is often part of bioinformatics applications; for instance, in *motif discovery* [8,10] a repeating substructure within a set of larger structures is searched for, under the assumption that a commonly occurring substructure is likely to have some sort of biological significance.

More specifically, the algorithms were required to find all the *subgraph isomorphisms* between the pattern and the target graphs. A subgraph isomorphism is a mapping between the nodes of the pattern and the nodes of the target that is injective, is consistent with the node/edge labels, and preserves the edge structure of the graphs (for a formal definition of subgraph isomorphism see [5]).

Subgraph isomorphism is, as previously said, an NP-complete problem. For this reason, the contest was open also to inexact algorithms. However, only exact algorithms were submitted.

2.2   Performance Measures

We have evaluated the algorithms submitted to the contest according to three criteria:

1. CPU time needed for solving the problem;
2. used memory;
3. accuracy of the found solutions.

Regarding the time, we have performed the measurement by instrumenting the source code of the programs, so as to be able to consider only the time spent during the matching, and not the loading of the input graphs. For each pattern/target pair we have measured both the time required to find the first solution, and the time for all the solutions. The time needed for finding the first solution reflects the needs of applications where it is only required whether the searched substructure is present or not in the target; the time for all the solutions instead is a more appropriate criterion for applications where it is also important to find the position of the substructure within the target.

For the memory, we have used the Valgrind dynamic analysis tool [11] in order to measure the peak of the memory usage for each algorithm. This is an important factor that has to be considered when the application has to work with very large graphs, such as protein structures; while in the literature there are matching algorithms requiring a memory space that grows quadratically with respect to the number of nodes, their usage is not practical for many bioinformatics applications.

The accuracy measurements were introduced to take into account inexact algorithms. In particular, we have measured the number of missed solutions. Since all the algorithms submitted to the contest were exact, we have used these accuracy measures only to check for errors in their implementations. In the end, all the tested implementations have provided correct solutions, and so we will not discuss further these measures.

2.3   The Dataset

We have prepared three sets of graphs representing different structures used in biological problems:

- *Molecules dataset*, containing the chemical structures of different small organic compounds taken from the PubChem database hosted by the National Center for Biotecnology Information (NCBI) [3]; these are small graphs (up to 99 nodes), and the edge density is very low (about 2 edges per node). In this dataset, nodes represent atoms, and edges represent chemical bonds.

- *Proteins dataset*, containing the chemical structures of proteins and protein backbones taken from the Protein DataBank (PDB) [2]; these are very large graphs (up to 10081 nodes), and the edge density is very low (about 2 edges per node). Also in this dataset, nodes represent atoms, and edges represent chemical bonds.
- *Contact maps dataset*, containing the contact maps extracted from proteins in the PDB, by means of the CMView tool [14]. A contact map represents the adjacency relation in the 3D structure of a protein between the aminoacids belonging to different chains. The corresponding graphs are medium sized (up to 733 nodes), but the density is somewhat higher (about 20 edges per node). In this case, nodes represent amino acids in the protein chains, and edges represent the spatial adjacency relationship.

For each of the three datasets, some pattern graphs have been extracted by randomly chosing connected subgraphs of the target graphs. The characteristics of the datasets are summarized in Table 1.

Table 1. Characteristics of the three datasets of the contest

dataset	target graphs	nodes	edges/node	node labels	pattern graphs	pattern nodes
Molecules	10000	8–99	≈ 2	4–5	50	4–64
Proteins	300	535–10081	≈ 2	4–5	60	8–256
Contact maps	300	99–733	≈ 20	21	60	8–256

The dataset has been made publicly available at the link [7].

3   Contest Participants

Six algorithms were submitted to the Biograph2014 contest. In the following we will discuss briefly the characteristics of each of them. Table 2 contains a summary of these algorithms.

Regarding the overall structure of the presented algorithms, they can be roughly divided into two categories: algorithms based on backtracking, and algorithms based on constraint propagation. Backtracking algorithms incrementally build a solution, adding at each step a pair of nodes to the current mapping, after checking if the addition is consistent with the problem constraints and with algorithm-specific heuristics; if they reach a point where no other pair can be added, they *backtrack*, undoing the previous assignment and trying a different one. Algorithms based on constraint propagation, instead, are based on a formulation of graph matching as a Constraint Satisfaction Problem (CSP); they start by computing the *domain* for each node of the pattern, i.e. the set of target nodes that are compatible with it; then the domains are iteratively reduced by propagating constraints on the structure of the mapping, until only a few candidate matchings remain, that can be easily enumerated. Of course, the two approaches can be combined, by first applying constraint propagation to reduce the number of possible matchings, and then, if there is still more than one possible solution, by using backtracking to explore the reduced search space.

Table 2. The algorithms submitted to the contest

name	authors	ref.	structure	heuristics
L2G	I. Almasri		Backtracking	Dynamic node ordering, Minimum Remaining Value
LAD	C. Solnon	[13]	Constraint propagation	AllDifferent constraint
FC	C. Solnon	[13]	Constraint propagation	ForwardChecking(Edges), ForwardChecking(Diff)
PJ	J.-C. Janodet, F. Papadopoulos		Backtracking	Static node ordering on smallest domain, k-successors and k-predecessors
RI	V. Bonnici, R. Giugno	[4]	Backtracking	Static node ordering on connections to matched nodes
RI-DS	V. Bonnici, R. Giugno	[4]	Backtracking/ Constraint propagation	Static node ordering on connections to matched nodes, Domain precomputation based on labels and degrees

4   Contest Results

In the following we will present and discuss the results obtained by the contest participants. In order to provide a baseline value for the performance, we have added the time and memory measurements for the well known VF2 algorithm [6], which was considered very fast when it was introduced, and is still very commonly used.

The experiments have been conducted on a Linux-based system having twelve Intel Xeon 3.20Ghz cores. The system RAM was 32 GBytes.

Figure 1 shows the average matching times for finding all the possible matchings between a pattern, with respect to the number of target nodes. On small and medium graphs (the Molecules and the Contact Maps datasets), RI and RI-DS are the fastest algorithms, with a slight advantage for RI-DS when the graphs are small. After these two algorithms come FC, LAD and L2G, with L2G at an advantage on medium graphs. Finally, PJ is the slowest, with a similar performance to the older VF2 algorithm.

Passing to the very large graphs of the Proteins dataset there is a slight change in the ranking. In this case, RI is definitely able to outperform the others by almost an order of magnitude to the closest competitors. L2G and RI-DS come second, with very similar performance. After them, at more than one order of magnitude, follow the remaining algorithms, with similar times.

This behavior is confirmed by Table 3, which reports for each dataset the total time spent by each algorithm for finding all the solutions.

Thus, for these kinds of graphs, a safe choice would be the RI algorithm, which is the fastest on very large graphs, but is not very far from the fastest (RI-DS) on small and medium graphs.

We have also analyzed the times for finding only the first solution for each pattern/target pair. Table 4 reports this information. While the fastest algorithms

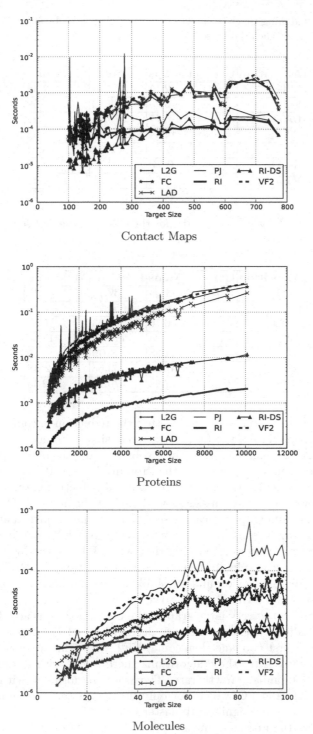

Contact Maps

Proteins

Molecules

Fig. 1. Matching time versus target size for finding all the solutions

Table 3. Total matching times (seconds) for the search of all solutions. Boldface values are the best times for each database.

Algorithm	Contact Maps	Proteins	Molecules
L2G	2.14	58.34	7.38
FC	6.25	908.43	5.59
LAD	6.77	579.21	7.11
PJ	14.54	1169.06	16.02
RI	1.41	**11.62**	3.64
RI-DS	**0.79**	54.51	**2.59**
VF2	7.98	892.66	14.57

Table 4. Total matching times (seconds) for the search of the first solution. Boldface values are the best times for each database.

Algorithm	Contact Maps	Proteins	Molecules
L2G	1.94	37.74	6.14
FC	6.23	651.29	5.16
LAD	6.76	569.03	6.87
PJ	13.24	711.35	13.07
RI	1.38	**8.09**	3.51
RI-DS	**0.78**	53.66	**2.48**
VF2	7.62	570.19	12.43

remain the same, it is interesting to note that the reduction in computation time depends not only on the dataset (as it is expected, since different datasets have on the average a different number of matching solutions for each pattern/target), but also on the algorithm: for instance, on the Proteins dataset, PJ has a reduction of 40% of the total matching time, while RI-DS has a reduction of only 3%.

Figure 2 shows the memory usage plotted against the number of target nodes. For small and medium graphs, it can be seen that the most convenient algorithms are RI and RI-DS; if graphs are medium sized, also L2G has a competitive memory usage. It is also worth noting that while PJ has the highest memory occupation, on these graphs its trend is almost constant. LAD, FC and VF2 have an intermediate behavior, and are similar to each other, although it can be seen that the first two have a higher growth rate.

When we pass to the very large graphs, the best algorithms are RI and L2G, which have an almost identical memory occupation. RI-DS uses slightly more memory than the first two, and then comes VF2, with a higher curve but a very similar growth rate. After these algorithms there is PJ, with a larger memory occupation but a similar growth rate. The worst performers, with respect to memory, are LAD and FC, which not only use two orders of magnitude more memory, but also have a significantly faster growth.

Table 5 shows the maximum memory usage for each algorithm for matching an entire dataset, that confirms the previous observations. However it is worth

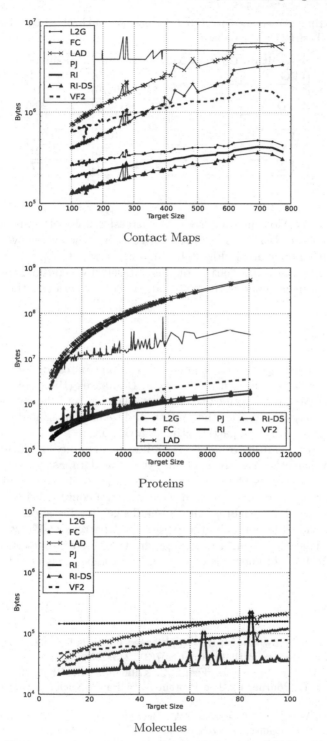

Contact Maps

Proteins

Molecules

Fig. 2. Used memory versus target size for the search of all the solutions

Table 5. Maximum used memory (kbytes) for the search of all solutions. Boldface values are the best for each database.

Algorithm	Contact Maps	Proteins	Molecules
L2G	483	**1679**	154
FC	3318	509270	116
LAD	5602	543550	209
PJ	6613	80388	3737
RI	409	1685	219
RI-DS	**354**	1975	221
VF2	1756	3527	**77**

noting that on the Molecules dataset (the smallest graphs), the maximum memory usage of RI and RI-DS is significantly higher than the average value that can be observed from the chart in Fig. 2, because of a few pattern/target pairs (that can be seen as peaks in the chart). For this dataset, if we consider the maximum memory requirement, none of the newer algorithm outperforms the old VF2.

5   Conclusions

In this paper we have presented the first Contest on Contest on Graph Matching Algorithms for Pattern Search in Biological Databases (Biograph2014), hosted at ICPR2014. We have described the contest challenge and the dataset prepared and made publicly available for the contest, and we have presented and discussed the results obtained by the participating algorithms.

We hope that this initiative will have a following in immediate future. First, we wish to encourage other researchers to use the same database [7] for their benchmarking activities, so as to make the results more easily comparable. Second, we are planning to organize a second edition of the contest, using an extended database and possibly a more complex challenge (for instance, the algorithm in the first edition where required to perform a 1-pattern-vs-1-target matching, while it could be interesting to measure the time for a 1-pattern-vs-n-targets matching, which could allow the algorithm to do some kinds of optimizations).

References

1. Aittokallio, T., Schwikowski, B.: Graph-based methods for analysing networks in cell biology. Briefings in Bioinformatics 7(3), 243–255 (2006)
2. Berman, H., Westbrook, J., Feng, Z., Gilliland, G., Bhat, T., Weissig, H., Shindyalov, I., Bourne, P.: The Protein Data Bank. Nucleic Acids Research 28, 235–242 (2000)
3. Bolton, E., Wang, Y., Thyessen, P.A., Bryant, S.H.: PubChem: Integrated platform of small molecules and biological activities. Annual Reports in Computational Chemistry 4(12) (2008)

4. Bonnici, V., Giugno, R., Pulvirenti, A., Shasha, D., Ferro, A.: A subgraph isomorphism algorithm and its application to biochemical data. BMC Bioinformatics 14 (2013)
5. Conte, D., Foggia, P., Sansone, C., Vento, M.: Thirty years of graph matching in Pattern Recognition. IJPRAI 18(3), 265–298 (2004)
6. Cordella, L., Foggia, P., Sansone, C., Vento, M.: A (sub)graph isomorphism algorithm for matching large graphs. IEEE Transactions on Pattern Analysis and Machine Intelligence 26, 1367–1372 (2004)
7. Foggia, P., Vento, M., Jiang, X.: The biograph2014 contest dataset, http://biograph2014.unisa.it
8. Huan, J., Bandyopadhyay, D., Wang, W., Snoeyink, J., Prins, J., Tropsha, A.: Comparing graph representations of protein structure for mining family-specific residue-based packing motif. Journal of Computational Biology 12(6), 657–671 (2005)
9. Kuhl, F.S., Crippen, G.M., Friesen, D.K.: A combinatorial algorithm for calculating ligand binding. Journal of Computational Chemistry 5(1), 24–34 (1984)
10. Lacroix, V., Fernandez, C., Sagot, M.: Motif search in graphs: Application to metabolic networks. Transactions on Computational Biology and Bioinformatics (2006)
11. Nethercote, N., Seward, J.: Valgrind: A framework for heavyweight dynamic binary instrumentation. In: Proceedings of the 2007 ACM SIGPLAN Conference on Programming Language Design and Implementation, PLDI 2007, pp. 89–100. ACM (2007)
12. Raymond, J., Willett, P.: Maximum common subgraph isomorphism algorithms for the matching of chemical structures. Journal of Computer-Aided Molecular Design 16(7), 521–533 (2002)
13. Solnon, C.: Alldifferent-based filtering for subgraph isomorphism. Artificial Intelligence 174(12-13), 850–864 (2010)
14. Vehlow, C., Stehr, H., Winkelmann, M., Duarte, J.M., Petzold, L., Dinse, J., Lappe, M.: CMView: Interactive contact map visualization and analysis. Bioinformatics (2011), doi:10.1093/bioinformatics/btr163

Approximate Graph Edit Distance Computation Combining Bipartite Matching and Exact Neighborhood Substructure Distance

Vincenzo Carletti[1], Benoit Gaüzère[1(✉)], Luc Brun[2], and Mario Vento[1]

[1] DIEM, Department of Information Engineering, Electrical Engineering and Applied Mathematics, University of Salerno, Fisciano, Italy
{mvento,vcarletti}@unisa.it
[2] GREYC CNRS UMR 6072, ENSICAEN, Caen, France
luc.brun@ensicaen.fr

Abstract. Graph edit distance corresponds to a flexible graph dissimilarity measure. Unfortunately, its computation requires an exponential complexity according to the number of nodes of both graphs being compared. Some heuristics based on bipartite assignment algorithms have been proposed in order to approximate the graph edit distance. However, these heuristics lack of accuracy since they are based either on small patterns providing a too local information or walks whose tottering induce some bias in the edit distance calculus. In this work, we propose to extend previous heuristics by considering both less local and more accurate patterns using subgraphs defined around each node.

1 Introduction

Graph based representations allow to model a wide variety of data [1,2]. However, since graphs do not lie in euclidean spaces, definition of a graph similarity measure is not a trivial problem. Nonetheless, different approaches have been explored to define similarity or dissimilarity measures between graphs [1,2]. A first family of approaches is based on graph theory, using for example the size of the maximum common subgraph of two graphs as their similarity measure. A second family of approaches aims to embed graphs into euclidean spaces by extracting a set of predefined features from either graph structures [4] or spectrum of adjacency matrices [5]. In particular, vectorial representations of graphs can be processed using the large set of well known machine learning and statistical pattern recognition methods defined on vectorial space. However, a drawback of this approach is that graphs encode complex objects using nodes and relationships between nodes and a large amount of information is lost when graphs are transformed in vectors. Hence, instead of defining an explicit embedding of graphs, an alternative approach consists in using graph kernels [8] which correspond to a scalar product between two implicit embedding of graphs. Any graph kernel, which can be seen as a similarity measure between graphs, can then be used in any machine learning method which access to data only through scalar

© Springer International Publishing Switzerland 2015
C.-L. Liu et al. (Eds.): GbRPR 2015, LNCS 9069, pp. 188–197, 2015.
DOI: 10.1007/978-3-319-18224-7_19

products. Another interesting and widely used approach is the graph edit distance. Unlike most of the existing graph similarity measures, this approach aims to define a dissimilarity measure between two graphs directly in the graph space.

The graph edit distance corresponds to a measure of the distortion required to transform one graph into another. The distortion between two graphs G and G' can be encoded by an edit path defined as a sequence of operations transforming G into G'. Such a sequence may include node or edge insertions, removals and substitutions. Given a non-negative cost function c(.) associated to each edit operation, the sum of elementary operation's costs composing the edit path defines its cost. The optimal edit path is then defined as the one having the minimal cost among all possible edit paths transforming G into G'. This minimal cost corresponds then to the graph edit distance between G and G'. More formally, the graph edit distance is defined by the following equation:

$$d_{edit}(G, G') = \min_{(e_1,...,e_k) \in \mathcal{P}(G,G')} \sum_{i=1}^{k} c(e_i).$$ (1)

where $\mathcal{P}(G, G')$ corresponds to all possible edit paths, each edit path consisting in a sequence of edit operations $(e_1, \ldots, e_k)$. Therefore, computing graph edit distance relies on finding an optimal edit path among all possible ones. A common approach consists in traversing the space of all possible edit paths by using an heuristic such as A^* search. This approach, detailed in Section 2, allows to find an exact graph edit distance with the cost of an exponential complexity which restricts it to rather small graphs, typically composed of up to ten nodes.

In order to tackle this complexity, Riesen and Bunke have proposed in [9] a method to compute an approximation of the graph edit distance in a polynomial time. This method exploits the close relationship between node mapping and edit distance to reduce the complexity. Indeed, any mapping between the set of nodes and edges of two graphs induces an edit path which substitutes all mapped nodes and edges, and inserts or removes the non mapped nodes and edges of the two graphs. Conversely, given an edit path between two graphs such that each node and each edge is substituted only once, one can define a mapping between the substituted nodes and edges of both graphs. The heuristic of Riesen and Bunke [9] builds a mapping between the sets of nodes of two graphs using a bipartite assignment algorithm, and deduces an edit path from this mapping. The cost associated to this possibly non optimal edit path corresponds to an overestimation of the exact edit distance. Obviously, the better the assignment is, the lower the overestimation is and thus more accurate is the approximation. The optimal bipartite assignment algorithm is based on a cost function defined between the neighborhoods of each pair of nodes of the two graphs. The idea behind this heuristic being that a mapping between nodes with similar neighborhoods should induce an edit path associated to a low cost and thus close to the optimal one. However, this heuristic may work poorly when the direct neighbourhood does not allow to easily differentiate the nodes such as when considering unlabeled graphs.

Considering this, we can distinguish two approaches which aim to improve the approximation of graph edit distance: A first one consists in starting from the edit path induced by the mapping computed by the original method of Riesen and Bunke, and improve this edit path by slightly modifying it using for example genetic algorithms [11]. A second approach consists in improving the initial node mapping [3]. The approach proposed in [3] associates to each node of the graph a bag of k-walks starting from this node. The mapping cost of two nodes is then computed by an approximation of the cost of mapping their bag of walks. This approach allows to compute a better approximation than the original approach defined in [9], but this gain may be altered by a low accuracy induced both by the use of walks and an approximation of mapping costs. Therefore, in order to compute a more accurate approximation of graph edit distance, we propose to associate each node of the graph to a subgraph including all nodes within a radius of k edges, such graphs being denoted as k-subgraphs. These patterns may provide a more accurate description about the surroundings of a node than a bag of walks. In addition, we propose to evaluate mapping costs between k-subgraphs using either an exact or an approximated graph edit distance.

Our paper is structured as follows. First, in Section 2, we review the computation of exact graph edit distance using the A^* algorithm. We also discuss about the Beam heuristic which allows to compute an approximate graph edit distance by restricting the search space. Second, in Section 3, we review in details the approximation algorithm proposed by Riesen and Bunke in [9]. Then, in Section 4, we present our proposition to improve the approximation of graph edit distance by computing a mapping cost based on k-subgraphs instead of direct neighborhood or k-walks edit distance. Finally, we present some experiments in Section 5 showing the accuracy gain obtained using our approach.

2   Graph Edit Distance Computation: Exact Approach

The problem of computing the exact graph edit distance between two graphs can be formulated as a search problem inside an appropriate state space, as well as for other graph matching problems [1,2]. Considering graph edit distance problem, the state space corresponds to the set of all complete and incomplete edit paths transforming the source graph into the target graph. This search is generally performed using A^* algorithm.

A^* is a well known search algorithm, which uses an heuristic function to conduct the search towards an optimal solution inside a search space. It is proven to be complete, i.e. it always find an optimal solution if it exists, and to be the best suited algorithm to perform an heuristic search. A^* algorithm consists in finding an optimal path starting from an initial state s_0 to a goal state s_g in a search space S. A^* begins thus by exploring the search space from s_0. Then, for each step corresponding to a state $s \in S$, the cost corresponding to the path from s_0 to s is encoded by a past-path cost function summing the cost of each previous step, while an estimation of the cost from s to s_g is provided by an heuristic function. The sum of these two functions provides an approximate cost of the path from s_0 to s_g that passes through s. So, given a current state s, it generates a set of successor states s' and puts non explored ones into a frontier set, namely the *Opening* set. Then, the state s' having the lowest cost estimation to reach the goal state s_g is extracted from the *Opening* set and is chosen as the next state. This search process ends when a goal state s_g is reached or the *Opening* set is empty. It is worth noticing that under certain conditions on heuristic function, A^* always finds an optimal solution.

Time complexity of A^* depends on the specific heuristic, but in the worst case, it is exponential with respect to the length of the shortest path. In order to reduce the complexity of A^* search, a limit on the size of the *Opening* set can be imposed. Clearly, in this way, the search algorithm does not guarantee to always find the optimal solution, but only a sub optimal one. Obviously, the probability of finding the optimal solution decreases as the limit size of *Opening* set decreases. This adaptation of A^* is called Beam Search.

A^* algorithm can be used to find an optimal edit path between two graphs, and thus to compute an exact graph edit distance. To this purpose, we have to clarify the structure of the search space and how the heuristic cost function is defined. On one hand, each state $s \in S$, $s \neq s_g$, corresponds to a partial solution, i.e. a partial edit path which transforms a subgraph H of G into a subgraph H' of G'. On the other hand, a goal state s_g corresponds to an optimal and complete edit path transforming G into G'.

The heuristic cost function encodes, for each state $s \in S$, an estimation of the cost required to reach s_g from s. In this paper, we used the heuristic function defined by Riesen and Bunke in [10]. This heuristic allows to find an optimal mapping between nodes and edges of G and G' which have not been previously mapped. This optimal assignment provides a minimal mapping cost of unprocessed sub graphs of the two graphs $G - H$ and $G' - H'$.

As described previously, we can use Beam Search in order to reduce the complexity of A^*. However, as shown in [7], using a limitation on the size of *Opening* set, we may not find an optimal edit path. Therefore, the computed graph edit distance may correspond to an overestimation of the exact graph edit distance.

3   Graph Edit Distance Computation: Approximate Approach

The graph edit distance approximation framework introduced in [9] reduces the search problem associated to exact graph edit distance computation to a Linear Sum Assignment Problem (LSAP) which can be solved in polynomial time. Considering two graphs G and G', the approach proposed by [9] consists in three steps. First, G and G' are subdivided into two sets of elements, where each element is defined as a bag of patterns encoding the local environment of a node of G or G'. Given these two sets, we can define a cost matrix $\mathbf{C}_\epsilon$ which encodes the cost of mapping two elements between the two sets. Second, we resolve the LSAP according to $\mathbf{C}_\epsilon$ using Munkres' algorithm [6]. This algorithm allows to compute an optimal assignment between the two sets associated to each graph which corresponds to a mapping of nodes of the first graph onto nodes of the second graph. Third, we can deduce an edit path from this optimal mapping by inferring node and edge edit operations. The cost associated to this edit path, which may not be optimal, corresponds to an approximation of the graph edit distance.

More formally, let us first consider an input labeled graph $G = (V, E, \mu, \nu)$, where V encodes the set of nodes, E the set of edges, μ the labeling function defined on nodes and ν the labeling function defined on edges. This graph is associated with a set of bags of patterns $B = \{B_i\}_{i=1,...,|V|}$. Every bag B_i is associated to a node $u_i \in V$ and characterizes the local structure of G around node u_i. The target graph $G' = (V', E', \mu', \nu')$ and its corresponding bags of structural patterns $B' = \{B'_i\}_{i=1,...,|V'|}$ are given analogously. We define a cost $c(B_i \to B'_j)$ for the substitution of two bags of patterns. In order to define cost for inserting or removing bags of patterns, we introduce an empty bag of patterns ε. Then, costs $c(B_i \to \varepsilon)$ and $c(\varepsilon \to B'_j)$ encode respectively removal and insertion of a bag. Given this cost definition, a cost matrix $\mathbf{C}_\varepsilon(B, B')$ encoding costs of substitutions, insertions, and removals of bags of structural patterns is defined as:

$$\mathbf{C}_\varepsilon(B, B') = \begin{bmatrix} \mathbf{C}(B, B') & \mathbf{C}_\varepsilon(B \to \varepsilon) \\ \mathbf{C}_\varepsilon(\varepsilon \to B') & \mathbf{0} \end{bmatrix} \in [0, +\infty]^{(|V|+|V'|) \times (|V'|+|V|)}, \quad (2)$$

where $\mathbf{C}(B, B')_{i,j} = c(B_i \to B'_j)$, $\mathbf{C}_\varepsilon(B \to \epsilon)_{i,j} = c(B_i \to \varepsilon)$ if $i = j$ and $+\infty$ elsewhere. Similarly, $\mathbf{C}_\varepsilon(\varepsilon \to B')_{i,j} = c(\varepsilon \to B'_i)$ if $i = j$ and $+\infty$ elsewhere.

Given the cost matrix $\mathbf{C}_\varepsilon(B, B')$, we can compute an optimal assignment between the sets B and B' in $O((|V| + |V'|)^3)$ time complexity thanks to the use of Munkres' algorithm [6]. This algorithm allows to compute a mapping between the two sets B and B', which may not be unique, having the lowest cost according to $\mathbf{C}_\varepsilon(B, B')$. Since each bag B_i is associated to a node u_i, the optimal assignment provides an optimal assignment between the nodes of both graphs with respect to the associated bags of patterns. That is, the optimal assignment corresponds to a mapping $\psi : V \cup \{\varepsilon\}^{|V'|} \to V' \cup \{\varepsilon\}^{|V|}$ of the nodes V of G to the nodes V' of G'. Due to the definition of the cost matrix, which allows both insertions and removals of elements, the mapping ψ is composed of different forms of node assignments: $u_i \to u'_j$, $u_i \to \varepsilon$, $\varepsilon \to u'_j$, and $\varepsilon \to \varepsilon$. The mapping ψ can be interpreted as a partial edit path between the graphs G and G' which only includes edit operations on nodes. The complete edit path is obtained by completing this partial edit path with edit operations on edges. Using simple graphs, this set of edit operations can be directly inferred from node edit operations since edit operations performed on nodes induces substitution, insertion or deletion of some edges in order to retrieve the target graph. Hence, the set of edges operations required to transform G into G' is obtained from the set of node operations induced by ψ. The cost of the complete edit path is finally returned as an approximate graph edit distance between graphs G and G'.

This approach proposed by Riesen and Bunke allows to compute an approximate edit distance in a polynomial time complexity with respect to the number of nodes. The low complexity of this approach allows then to use the graph edit distance in pattern recognition problems, and reaches good prediction accuracies [9]. However, the approach proposed by [9] associates to each node a bag of patterns defined as the node itself and its direct neighbourhood, i.e. its incident edges and its adjacent nodes. Using this kind of bags of patterns, this approach lacks of accuracy in some applications where the direct neighbourhood of graphs is not sufficiently discriminant. When considering such graphs, the mapping costs associated to each pair of nodes do not differ sufficiently and the optimal mapping depends more on the initial order of nodes than on the graph's structure. Therefore, in order to improve the accuracy of the approximation of graph edit distance, Gaüzère et al. proposed in [3] to enhance the information associated to each node by considering a bag of walks up to a length k instead of only the direct neighbourhood. This approach follows the same scheme as the one used by [9] and described at the beginning of this section, except that the set of bags of patterns associated to a node is defined as the set of walks starting from this node and having a particular length k. Considering such a bag of patterns allows to extend the amount of information associated to each node by taking into account less local structures. The bag of patterns associated to each node is then more discriminant and leads to a better approximation of the graph edit distance. However, the use of bags of walks induces some drawbacks. First, the set of computed walks suffers from the tottering phenomenon which leads to consider irrelevant patterns. These irrelevant patterns affects the mapping cost and thus the approximation of the graph edit distance. In addition, the mapping cost between two bags of walks is only approximated, which induces another loss of accuracy. Therefore, we propose in the next section to tackle these drawbacks by considering k-subgraphs associated to each node.

4   Approximate GED Using K-Graphs

First, let us introduce the concept of k-subgraph (Figure 1). Given a node $u \in G$ and a radius $k \in \mathbb{N}$, we define k-subgraph(u) as the subgraph of G defined by the subset of nodes of G which can be reached from u by a path composed of maximum k edges. Given this concept, each node of the graph can be associated to a sub structure of the graph which encodes a more or less local information, depending on the value of k. However, unlike bags of walks used in [3], k-subgraphs do not suffer from tottering phenomenon.

Following the graph edit distance approximation scheme described in section 3, we propose to define the bag of patterns associated to each node $u \in V$ of a graph $G = (V, E, \mu, \nu)$ as k-subgraph(u). Then, in order to compute the optimal assignment between two sets of patterns, we propose to define matching cost $c(u \rightarrow v)$ as the exact graph edit distance between k-subgraph(u) and k-subgraph(v). However, as explained in previous paragraph, k-subgraph(u) and k-subgraph(v) are respectively centered around nodes u and v. In addition, the matching cost $c(u \rightarrow v)$ must encode the matching ability of u and v and not only the similarity of k-subgraph(u) with k-subgraph(v). Therefore, we propose to restrict the set of possible mappings in such a way that the two central nodes

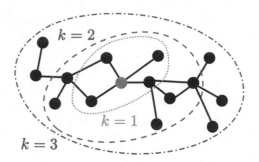

Fig. 1. Examples of k-graphs associated to a central node (in light grey)

u and v are mapped together. Using such a constraint, we force a substitution operation $(u \rightarrow v)$ to be the only node edit operation performed on u and v. This restriction allows also to slightly reduce the complexity required to compute the exact graph edit distance by pruning a part of possible edit paths. Therefore, given two nodes $u \in G$ and $v \in G'$, the cost of the substitution $c(u \rightarrow v)$ is defined as: $c(u \rightarrow v) = d_{(u,v)}(\text{k-subgraph}(u), \text{k-subgraph}(v))$ where $d_{(u,v)}$ corresponds to the graph edit distance between k-subgraph(u) and k-subgraph(v) with restriction on node edit operations involving u and v. The enumeration of k-subgraphs requires to perform a depth first search from each node which is associated to a complexity in $\mathcal{O}(nd^k)$ where n is the number of nodes of the graph, d its maximum degree and k the radius associated to k-subgraph. Therefore, the computational effort is polynomial with the maximum degree of graphs and is linear with the size of the graphs. It is worth noticing that this complexity is only linear for graphs having a bounded degree. In addition, our proposition induces to compute a graph edit distance for each pair of nodes of the graphs to be compared. Considering two graphs having n nodes, these operations induce n^2 graph edit distance computations. Hopefully, these graph edit distances are only computed between graphs of a limited size for reasonable values of k, which limits computational time. However, in order to reduce this computational time, we may use the Beam search algorithm (Section 2) and limit the queue size.

5   Experiments

Following the same test protocol as in [3], we tested our heuristic on 4 graph datasets[1] encoding molecular graphs. For all these experiments and as in [3], insertion/removal costs have been arbitrarily set to 3 for both edges and nodes and substitution cost to 1 for edges and nodes, regardless of node's or edge's labels. Graphs included within the 4 datasets have different characteristics: Alkane and PAH are only composed of unlabeled graphs whereas MAO and Acyclic correspond to labeled graphs. In addition, Alkane and Acyclic correspond to acyclic graphs having a low number of nodes (8 to 9 nodes in average) whereas MAO and PAH correspond to larger cyclic graphs (about 20 nodes in average).

---

[1] These datasets are available at http://iapr-tc15.greyc.fr/links.html

Tables 1 and 2 show a comparison of the accuracy of our proposition with two state of the art methods: the original one from Riesen [9] and an improvement of this approach using k-walks [3], both detailed in Section 3. As in [3], chosen k is the one obtaining the most accurate results. First, Table 1 shows the percentage of entries of edit distance's matrix corresponding to accuracy gain (i.e. computed edit distance is lower), accuracy loss or equivalent accuracy obtained by our approach versus the ones obtained by [9] and [3]. These percentages are displayed for different ways of computing graph edit distance between k-subgraphs: exact graph edit distance (A^*) and Beam search using a queue limit of 1000 (Beam, 1000) or 100 states (Beam, 100). On one hand, we can note that our approach provides always a better approximation of graph edit distance for 63% to 99% of molecules' pairs when compared to the approach of Riesen and 40% to 76% when compared to the approach based on k-walks. These observations are still valid even if we use Beam search algorithm in order to reduce the computational time required by our approximation. On the other hand, we observe an accuracy loss for only $< 1\%$ to 15% when compared to the approach of Riesen and 8% to 41% when compared to the second approach. Note however that the comparison on PAH dataset suffers from the fact that k is limited to 2 for A^* algorithm, versus walks composed of up to 5 nodes in [3]. This limitation is induced by the high computational time required by A^* algorithm. However, we can note that using a faster algorithm which allows to consider larger k-subgraphs, we obtain a better approximation of our graph edit distance. Finally, these two first experiments shows a clear gain on the accuracy of our approximation compared to both state of the art approaches.

Same conclusions can be drawn from Table 2 which shows, for each dataset and each method, the average edit distance ($\bar{d}$) together with the average error of our approximation with respect to the exact graph edit distance and the average time required to compute graph edit distance for a pair of graphs. The exact graph edit distance, and thus the average error, is not available for MAO and PAH datasets since it requires too much time to be computed. First, we can note that our approaches require higher computational times with respect to other approximation frameworks. This observation is coherent with the fact that we

Table 1. Accuracy comparison of approach versus [3] and [9]

Dataset	A^*			Beam, 1000			Beam, 100		
	Gain	Loss	=	Gain	Loss	=	Gain	Loss	=
[9] Alkane	63%	17%	20%	60%	20%	20%	56%	23%	21%
PAH	68%	17%	15%	81%	9%	9 %	81%	10%	9%
MAO	99%	< 1%	< 1%	98%	2%	< 1%	93%	7%	< 1%
Acyclic	75%	15%	10%	72%	18%	11%	68%	21%	11%
[3] Alkane	63%	17%	21%	59%	19%	22%	54%	21%	26%
PAH	40%	41%	18%	63%	23%	14%	61%	25%	14%
MAO	76%	8%	16%	69%	16%	16%	47%	43%	10%
Acyclic	59%	22%	18%	55%	25%	19%	51%	30%	19%

Table 2. Average edit distance ($\bar{d}$), average error ($\bar{e}$) and average time in seconds ($\bar{t}$) for each method and each dataset (KG = our method)

Method	Alkane			Acyclic			MAO		PAH	
	$\bar{d}$	$\bar{e}$	$\bar{t}$	$\bar{d}$	$\bar{e}$	$\bar{t}$	$\bar{d}$	$\bar{t}$	$\bar{d}$	$\bar{t}$
A^*	15		1.29	17		6.02				
[9]	35	18	10^{-2}	35	18	10^{-3}	105	10^{-3}	138	10^{-3}
[3]	33	18	10^{-3}	31	14	10^{-2}	49	10^{-2}	120	10^{-2}
KG, A^*	26	11	2.27	28	9	0.73	44	6.16	129	2.01
KG, Beam 1000	27.3	12.6	0.46	28.6	9.9	0.13	47	5.74	113	19.39
KG, Beam 500	27.6	12.1	0.58	28.7	9.9	0.21	54	6.43	113	19.74
KG, Beam 100	28.4	12.9	0.22	29.7	10.8	0.12	60	1.84	115	4.79
KG, Beam 50	28.8	13.2	0.15	30	11.2	0.09	76	1.10	115	2.73

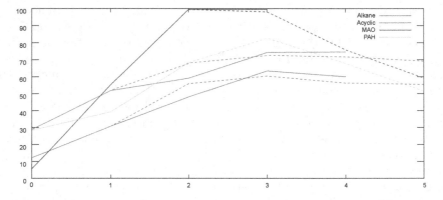

Fig. 2. Gain in accuracy compared to Riesen [9] obtained for each dataset versus the size of considered paths (k). Smashed lines represent the gain using beam search algorithm instead of A^* algorithm.

have to compute, for each pair of nodes to be matched, a graph edit distance between rather large graphs when k is equals to 3 or 4. In addition, computation times obtained for line 3 corresponds to a Matlab/C++ implementation whereas other lines have been computed using a Java implementation [10]. The results shown in these two tables show the gain in accuracy induced by using k-subgraphs to compute a matching cost between nodes instead of using direct neighbourhood [9] or k-walks [3]. In addition, we can note that taking into account a large radius for k-subgraphs increases the accuracy of our edit distance approximations (Figure 2). Conversely to the observation stated in [3], in our framework, considering a large radius does not induce irrelevant patterns and thus the accuracy does not decrease. The decrease in gain accuracy observed for smashed lines for larger k is due to the fact that the limit on the number of states does not allow to find an optimal edit path. This may occur more often as the number of possible edit paths increases.

6   Conclusion

In this article, we have proposed a new heuristic to enhance the approximation of graph edit distance using the framework based on optimal bipartite graph matching. Our proposition aims to use less local and more discriminant sub structures, called k-subgraphs, associated to each node. Despite the high computational time induced by our proposition, our approach is still less time consuming than computing an exact graph edit distance and we obtain a better approximation accuracy than previous methods, hence showing the relevancy of considering larger and exact patterns with respect to previous propositions.

References

1. Conte, D., Foggia, P., Sansone, C., Vento, M.: Thirty years of graph matching in Pattern Recognition. International Journal of Pattern Recognition and Artificial Intelligence 18(3), 265–298 (2004)
2. Foggia, P., Percannella, G., Vento, M.: Graph Matching and Learning in Pattern Recognition on the last ten years. Journal of Pattern Recognition and Artificial Intelligence 28(1) (2014)
3. Gaüzère, B., Bougleux, S., Riesen, K., Brun, L.: Approximate Graph Edit Distance Guided by Bipartite Matching of Bags of Walks. In: Fränti, P., Brown, G., Loog, M., Escolano, F., Pelillo, M. (eds.) S+SSPR 2014. LNCS, vol. 8621, pp. 73–82. Springer, Heidelberg (2014)
4. Gibert, J., Valveny, E., Bunke, H.: Graph embedding in vector spaces by node attribute statistics. Pattern Recognition 45(9), 3072–3083 (2012)
5. Luo, B., Wilson, R.C., Hancock, E.R.: Spectral embedding of graphs. Pattern Recognition 36(10), 2213–2230 (2003)
6. Munkres, J.: Algorithms for the assignment and transportation problems. Journal of the Society for Industrial and Applied Mathematics 5(1), 32–38 (1957)
7. Neuhaus, M., Riesen, K., Bunke, H.: Fast suboptimal algorithms for the computation of graph edit distance. In: Yeung, D.-Y., Kwok, J.T., Fred, A., Roli, F., de Ridder, D. (eds.) SSPR 2006 and SPR 2006. LNCS, vol. 4109, pp. 163–172. Springer, Heidelberg (2006)
8. Ramon, J., Gärtner, T.: Expressivity versus efficiency of graph kernels. In: First International Workshop on Mining Graphs, Trees and Sequences, pp. 65–74 (2003)
9. Riesen, K., Bunke, H.: Approximate graph edit distance computation by means of bipartite graph matching. Image and Vision Computing 27, 950–959 (2009)
10. Riesen, K., Emmenegger, S., Bunke, H.: A novel software toolkit for graph edit distance computation. In: Kropatsch, W.G., Artner, N.M., Haxhimusa, Y., Jiang, X. (eds.) GbRPR 2013. LNCS, vol. 7877, pp. 142–151. Springer, Heidelberg (2013)
11. Riesen, K., Fischer, A., Bunke, H.: Improving Approximate Graph Edit Distance Using Genetic Algorithms. In: Fränti, P., Brown, G., Loog, M., Escolano, F., Pelillo, M. (eds.) S+SSPR 2014. LNCS, vol. 8621, pp. 63–72. Springer, Heidelberg (2014)

Multi-layer Tree Matching Using HSTs

Yusuf Osmanlıoğlu$^{(\boxtimes)}$ and Ali Shokoufandeh

Department of Computer Science, Drexel University
Philadelphia, USA
{osmanlioglu,ashokouf}@cs.drexel.edu

Abstract. Matching two images by mapping image features play a fundamental role in many computer vision task. Due to noisy nature of feature extraction, establishing a one-to-one matching of features may not always be possible. Although many-to-many matching techniques establishes the desired multi map between features, they ignore the spatial structure of the nodes. In this paper, we propose a novel technique that utilizes both the individual node features and the clustering information of nodes for image matching where image features are represented as hierarchically well-separated trees (HSTs). Our method uses the fact that non-leaf nodes of an HST represent a constellation of nodes in the original image and obtains a matching by finding a mapping between non-leaf nodes among the two HSTs. Empirical evaluation of the method on an extensive set of recognition tests shows the robustness and efficiency of the overall approach.

Keywords: Hierarchically well-separated tree · HST · Metric embedding · Graph matching · Tree matching

1 Introduction

Graph matching is a fundamental problem in computer science which has numerous real life applications such as image matching. Given two graphs G and H, the goal of matching is to find correspondences among vertices and edges in of G and H. Exact graph matching is known to be computationally intractable since determining whether G contains an induced subgraph H' that is isomorphic to H, *subgraph isomorphism problem*, is known to be NP-complete [9]. Even a simpler special case of subgraph isomorphism, *graph isomorphism problem*, is neither known to be polynomial time solvable nor NP-complete. Consequently, a large body of work has been devoted to *inexact matching* problem. One of the earliest formulations of this latter problem is due to Shapiro and Haralick [17]. Over the course of three decades, several methods were proposed for tackling the problem including tabu search [19], error-correcting graph matching [6], and convex optimization formulations [1]. Inexact graph matching is also motivated by the need for solving the correspondence problem when the underlying data is noisy. For example, features of two images representing the same object might have minor differences due to factors such as occlusion, noise, or scaling. The dissimilarity can be result of lack of a feature in one of the graphs or a feature

© Springer International Publishing Switzerland 2015
C.-L. Liu et al. (Eds.): GbRPR 2015, LNCS 9069, pp. 198–207, 2015.
DOI: 10.1007/978-3-319-18224-7_20

in one of the graphs can map to several features in the other graph. Therefore, inexact matching may have to deal with mapping sets of features amongst two graphs rather than establishing one-to-one matching among features.

A popular approach for tackling noisy many-to-many matching is based on graph edit distances (e.g., [2,5]). Here the idea is to make local searches over the graph and transform (part of) one graph into the other by minimal cost modifications such as inserting, deleting, substituting, or merging of nodes and edges. Zaslavskiy et al. [20] formulated many-to-many graph matching as a discrete optimization problem and proposed an approximation algorithm based on continuous relaxation. Using spectral graph theory, Caelli and Kosinov [7] approached the many-to-many matching problem as eigendecomposition of the graph's adjacency matrix. Inexact graph matching can also be handled using metric embedding methods. This technique consists of mapping data from a source space to a "simpler" target space while preserving the distances. Demirci et al. [10] devised an isometric embedding,which first represents the graph representations of images as metric trees and then embeds them isometrically in the geometric space under the l_1-norm for solving the many-to-many image matching problem. They used *earth movers distance* for establishing correspondences among subsets of distributions corresponding to the original graphs in embedded space.

Another approach for inexact graph matching based on embedding uses approximation of original graphs by trees (*tree embedding*) followed by an assignment step. It is well known that approximating the solution for many NP-hard problems in general metrics can be done in polynomial time once data is embedded into tree metrics. However, such embeddings tend to introduce distortion. A common technique for overcoming such high distortion is *probabilistic approximation* method of Karp [12]. Utilizing probabilistic embedding, Bartal [3] introduced the notion of *hierarchically well-separated trees (HST)* where edge weights on a path from root to leaves decrease by a constant factor in successive levels. Embedding into HSTs improves the representational power of graphs especially in domains such as image matching since the internal nodes of the tree represents constellations of nodes of the original graph. Thus, tree structure captures the segmentation information at its internal nodes along with the node features at its leaves. Following Bartal's seminal work, there have been several studies on HSTs which improved the upper bound of distortion and introduced deterministic embedding algorithms [4,8,14]. Finally, Fakcharoenphol et al. [11] devised a deterministic algorithm that achieved embedding arbitrary metrics into HSTs with a tight $\Theta(\log n)$ distortion.

It is an underlying assumption that many-to-many matching algorithms use the features associated with individual nodes to establish the correspondences, while ignoring the underlying regions containing the spatial distribution of the matched features. In this paper, we propose a novel matching method that utilizes the spatial distribution of nodes within image regions as well as features to tackle the matching problem. Our method use optimization over HSTs, similar to Kleinberg and Tardos [13] approach, for solving the metric labeling problem. In metric labeling problem, we are given an object graph and a label graph, and the goal is to find a many-to-one mapping between the nodes of the object graph and label graph. Kleinberg and Tardos first embedded the label graph into an HST and then the nodes of the object graph are assigned to the leaves of the HST using a linear programming formulation. Instead of matching a graph with a tree, we propose to first embed both (object and label) graphs into two separate HSTs and then match the nodes of the resulting trees at fixed layers within the trees. Our method takes advantage of the layered structure of HSTs to map nodes at non-leaf levels. As a result, the clustering structure of features that correspond to their spatial distribution with the original image features are taken into account during the matching process. Although embedding of both graphs into HSTs introduces an additional $O(\log n)$ distortion, the overall asymptotic distortion rate stays intact. We demonstrate the utility of our method for image matching problem. Our preliminary results show that the proposed method performs well for standard datasets.

The rest of the document is organized as follows: Section §2 gives an overview of notations and definitions. In Section §3, we present the details of multi-layer tree matching which is followed by its application to image matching problem in Section §4. Finally, Section §5 concludes the document.

2   Notations and Definitions

The term embedding refers to a mapping between two spaces. From a computational point of view, a major goal of embedding is to find approximate solutions to NP-hard problems. Another important use of embedding is to achieve performance gains in algorithms by decreasing the space or time complexity of a polynomial-time solvable problem. Given a set of points P, a mapping $d : P \times P \to R^+$ is called distance function if $\forall\ p, q, r \in P$ the following four conditions are satisfied: $d(p, q) = 0$ iff $p = q$, $d(p, q) \geq 0$, $d(p, q) = d(q, p)$, and $d(p, q) + d(q, r) \geq d(p, r)$. The pair (P, d) is called a *metric space* or a *metric*. A finite metric space (P, d) can be represented as a weighted graph $G = (V, E)$ with shortest path as the distance measure where points in P form the vertex set V and pairwise distances between points become the edge weights. Thus, problems on embeddings to graphs are of particular interest where the size of graph will determine the complexity of the associated algorithm. A commonly used approach for decreasing this size is based on changing the structure of the graph by removing edges that change the distance metric of the graph, removing or adding vertices, or changing weights of edges. This approach, however, introduces *distortion* on distances in the graph which is defined as the product of the maximum factors by which the distances in the graph are stretched and shrunk.

In general, it is hard to find an isometric embedding between two arbitrary metric spaces. Therefore, it is important to find an embedding in which the distances between vertices of the destination metric are as close as possible to their counterparts in the source metric space. In reducing the size of a graph by removing its vertices and edges, one would be interested in culminating the pruning process with a tree since for many problems solving them on trees is much easier compared to arbitrary graphs. Embedding of graphs into trees is a very challenging problem; even for the simple case of embedding an n-cycle into a tree. Karp [12] introduced the idea of *probabilistic embedding* for overcoming this difficulty where given a metric d defined over a finite space P, the main idea is to find a set S of simpler metrics defined over P which dominates d and guarantees the expected distortion of any edge to be small.

Uniform metrics are among the simplest tessellation spaces where all distances are regularly distributed across cells. Such metrics are important from a computational point of view since one can easily apply divide-and-conquer approach to problems under uniform metrics. Motivated by these observations, Bartal [3] defined the notion of *hierarchically well separated trees (HST)* for viewing finite metric spaces as a uniform metric. A k-HST is defined as a rooted weighted tree where edge weights from a node to each of its children are the same and decrease by a factor of at least k along any root to leaf path. Assuming that the maximum distance between any pair of points (diameter) in source space is Δ, source space is separated into clusters (sub-metrics) of diameter $\frac{\Delta}{k}$. The resulting clusters are then linked to the root as child nodes with edges of weight $\frac{\Delta}{2}$. The relation between parent and child nodes continues recursively until the children nodes consist of single data elements. Bartal has shown the lower bound for distortion of embedding into HSTs to be $\Omega(\log n)$. He also provided a randomized embedding algorithm that utilizes probabilistic partitioning with a distortion rate of $O(\log^2 n)$. In sequel works, both Bartal [4] and Charikar et al. [8] introduced deterministic algorithms with smaller distortion ($O(\log n \log \log n)$). Konjevod et al. [14] were the first to improve the upper bound on distortion to $O(\log n)$ for the case of planar graphs. Fakcharoenphol et al. [11] closed the gap for arbitrary graphs by introducing a deterministic algorithm with a tight distortion rate ($\Theta(\log n)$). The deterministic nature of their algorithm made this result of great practical value.

One of the earliest algorithmic application of HST in general family of classification problems is known as *metric labeling* [13]. Given a set of objects P and a set of labels L with pairwise relationships defined among the elements of both sets, the goal is to assign a label to each object by minimizing a cost function involving both separation and assignment costs. Separation cost penalizes assigning loosely related labels to closely related objects while assignment cost penalizes labeling an object with an unrelated label. The cost function $Q(f)$ can be stated as follows:

$$Q(f) = \sum_{p \in P} c(p, f(p)) + \sum_{e=(p,q) \in E} w_e \cdot d(f(p), f(q))$$

where, $c(p, l)$ represents the cost of assigning an object $p \in P$ to a label $l \in L$, w_e is the nonnegative weight associated with edge $e = (p, q)$ indicating the strength of the relation between p and q, and $d(\cdot, \cdot)$ is a similarity measure on the set L of labels. Although, there has been ample studies on solving classification problems using labeling methods, their work was the first that provided a polynomial-time approximation algorithm with a nontrivial performance guarantee.

A natural quadratic programming (QP) formulation for the general metric labeling problem is as follows:

$$
\begin{aligned}
\min \quad & \sum_{p\in P}\sum_{a\in L} c_{p,a}\cdot x_{p,a} + \sum_{p\in P}\sum_{q\in P} w_{p,q} \sum_{a\in L}\sum_{b\in L} d_{a,b}\cdot x_{p,a}\cdot x_{q,b} \\
\text{s.t.} \quad & \sum_{a\in L} x_{p,a} = 1, & \forall p\in P \\
& x_{p,a}\in\{0,1\}, & p\in P, a\in L
\end{aligned}
\tag{1}
$$

To eliminate the quadratic term in the formulation (1), Kleinberg and Tardos [13] used embedding of label graph into an HST and measured the distances between labels over the tree. This made it possible to replace the quadratic term in the objective function with an absolute value of a subtraction. They formulated the problem as a linear program (LP) and solving it, obtained a fractional assignment. Finally, a many-to-one matching from object set to label set is achieved using a deterministic rounding procedure. It is shown that this solution has $O(\log n)$ distortion due to using HSTs both in solving the linear program and in rounding.

We recently extended the original metric labeling to a complicated formulation in the domain of user association in celular networks [18]. The aim in here is to assign users to base stations which offers best downlink rate while keeping the user distribution over base stations balanced to optimize congestion.

3   Multi-layer Tree Matching

In this section we will present a novel technique for solving many-to-many graph matching problem. Our objective is to create correspondence between set of nodes by utilizing their spatial distribution in addition to individual node features. We obtain the underlying structure of nodes by using HSTs. Our technique extends the metric labeling formulation of Kleinberg and Tardos [13]. Specifically, given an object graph $G_{\mathcal{O}}$ and a label graph $G_{\mathcal{L}}$, we embed both of them into separate HSTs $\mathcal{T}_{\mathcal{O}}$ and $\mathcal{T}_{\mathcal{L}}$ using the deterministic embedding algorithm [11] where internal nodes of these HSTs correspond to subsets of features in the original graph. Our goal is to match clusters of $G_{\mathcal{O}}$ to those of $G_{\mathcal{L}}$ as demonstrated in the first row of the figure. We perform this over the layers of HSTs as shown in second row of Fig. 1. For example, matching the clusters in the second layer of the left-tree with the clusters in the third layer of the right-tree in the figure will map non-leaf nodes $a1$ and $a3$ to $b8$ and $b9$, respectively. Using an extended version of metric labeling, we match the set of (non-leaf) nodes $\mathcal{T}_{\mathcal{O}}^{i}$ at some fixed layer i of $\mathcal{T}_{\mathcal{O}}$ with the set of (non-leaf) nodes $\mathcal{T}_{\mathcal{L}}^{j}$ at some fixed layer j of $\mathcal{T}_{\mathcal{L}}$. Intuitively, leaves that are descendants of matched non-leaf nodes will be assigned to each other in a many-to-many fashion. Extending the metric labeling, we formulate the layered tree matching problem as the following integer linear program:

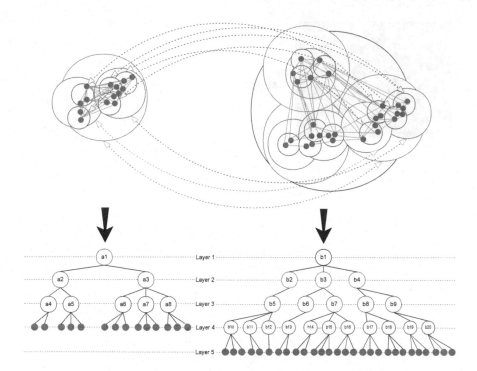

Fig. 1. Given the two graphs in the first row, our method first embeds them into separate HSTs using Fakcharoenphol et al.'s [11] algorithm, resulting with the k-HSTs shown in the second row. Then we match non-leaf nodes of both trees using an extended version of metric labeling.

$$
\begin{aligned}
\min \quad & \sum_{A \in \mathcal{T}_\mathcal{O}^i} \sum_{B \in \mathcal{T}_\mathcal{L}^j} C(A,B) \cdot x_{A,B} + \sum_{P \in \mathcal{T}_\mathcal{O}^i} \sum_{Q \in \mathcal{T}_\mathcal{O}^i} W_{P,Q} \cdot D_{P,Q} \\
\text{s.t.} \quad & \sum_{B \in \mathcal{T}_\mathcal{L}^j} x_{A,B} = 1, && \forall A \in \mathcal{T}_\mathcal{O}^i \\
& D_{P,Q} = \sum_T l_T \cdot \bar{x}_{P,Q,T}, && \forall P,Q \in \mathcal{T}_\mathcal{O}^i \\
& \bar{x}_{P,Q,T} \geq \tilde{x}_{P,T} - \tilde{x}_{Q,T}, && P,Q \in \mathcal{T}_\mathcal{O}^i, \mathrm{T} \in \mathcal{T}_\mathcal{L}^k, k = \{1, \cdots, j\} \\
& \bar{x}_{P,Q,T} \geq \tilde{x}_{Q,T} - \tilde{x}_{P,T}, && P,Q \in \mathcal{T}_\mathcal{O}^i, \mathrm{T} \in \mathcal{T}_\mathcal{L}^k, k = \{1, \cdots, j\} \\
& \tilde{x}_{A,T} = \sum_{B \in T} x_{A,B}, && \forall A \in \mathcal{T}_\mathcal{O}^i \\
& x_{A,B} \in \{0,1\}, && A \in \mathcal{T}_\mathcal{O}^i, B \in \mathcal{T}_\mathcal{L}^j
\end{aligned}
\tag{2}
$$

where $C(A,B)$ is the cost of assigning object node $A \in \mathcal{T}_\mathcal{O}^i$ to label node $B \in \mathcal{T}_\mathcal{L}^j$, $x_{A,B}$ is the probability of node $A \in \mathcal{T}_\mathcal{O}$ to be assigned to node $B \in \mathcal{T}_\mathcal{L}$, $W_{P,Q}$ is the weight of the path between nodes P,Q in $\mathcal{T}_\mathcal{O}$, and $\tilde{x}_{A,T}$ is the joint probability of objects in set A to be assigned to the labels included in subtree T of $\mathcal{T}_\mathcal{L}$. Since matching two clusters requires comparing sets of nodes in each cluster, we use normalized Hausdorff distance[1] for calculating the assignment cost $C(A,B)$ between sets $A \in \mathcal{T}_\mathcal{O}^i$ and $B \in \mathcal{T}_\mathcal{L}^j$.

---

[1] Given a metric space (M,d) and two subsets $X,Y \in M$, the symmetric Hausdorff distance $d_H(X,Y)$ between X and Y is defined as $d_H(X,Y) = \max\{\max_{x \in X} \min_{y \in Y} d(x,y), \max_{y \in Y} \min_{x \in X} d(x,y)\}$.

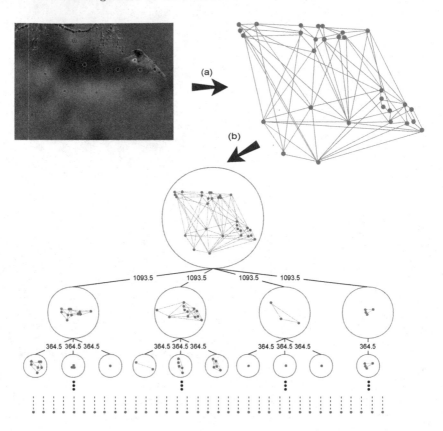

Fig. 2. (a) Graph representation of image features (b) Embedding the graph representation into 3-HST

4   Application to Image Matching

We apply the multi-layer tree matching formulation of §3 to image matching problem. Given an image, our method starts with extracting SIFT [15] features of the image. Then, we represent the image as a complete undirected graph where features are represented with nodes and the relation between features are represented by edges. Next, we embed this complete graph into an HSTs as shown in Fig. 2. Having two such HSTs, we aim to find a mapping between the internal nodes of the HSTs which represent a constellation of features of the images. We match non-leaf nodes at a fixed layer of the first image with non-leaf nodes at some (other) fixed layer of the second image. Comparing layers that are closer to leaf layer yields a finer level of matching whereas matching layers closer to root runs faster albeit resulting in a coarse matching. This tradeoff between robustness and running time can be optimized by progressively altering the layers to match. Assume that given a dataset of images and a query object, we are asked to find the image that resembles the query object most. Since metric labeling takes exponential time in number of nodes of object and label

Table 1. Success rate and average running time for experiments

Method	Success Rate	Average Time
Layer 1	30.07	4
Layer 2	51.6	8
Layer 3	77.99	304
Layer 4	89.24	2216
Layer 5	91.11	5943
Multi-layer	90	650
Quad. Prog.	92.01	2132
Linear Prog.	92.08	16761

graphs, comparing structures with smaller size take less time. Using layered tree matching, we first match the query object with the rest of the images at layer 1. This step is performed in short time since the trees consist of a single node at this layer. Then, we discard the HSTs with large matching score and carry on matching at consecutive levels of the HSTs with the remaining images. Thus, progressively we eliminate the images that do not resemble the query image via rough HST matchings at coarser layers and perform rigorous matches with the few remaining images.

To demonstrate our approach on image matching problem, we performed experiments on COIL-20 [16] dataset. The testbed consists of 1440 gray scale images of 20 objects where each object has 72 views obtained by consecutive rotations of 5 degrees around z axis. After obtaining the HST representation of each image, the matching experiment is carried out as follows: an image is removed from the test set, used as query object for the rest of the set, and pairwise similarity scores between this view and the rest are recorded in a similarity matrix. Matching is considered to be successful if the image that is deemed to be most similar to the query object is a neighboring view.

We run four set of experiments on the dataset. First experiment consists of matching images at fixed layers for the entire dataset. This experiment is repeated for each layer, starting from the first layer and ending with the fifth layer. Second experiment is the multi-layer matching where we discarded 1340 of the least similar images according to matching results of the third layer, and further discarded 70 of the remaining 100 images according to matching results of the fourth layer. The number of images to discard are determined empirically. Finally, third and fourth experiment comprise using the QP and LP formulation of metric labeling on the dataset for using as the baseline. First, we compare the results of the QP and LP formulations. As indicated in Fig. 3, success rate for both formulations is identical except the 0.07% difference for 1-NN case, whereas QP formulation is 8 times faster than the LP formulation. Success rate being identical is noteworthy since it shows that the $O(\log n)$ distortion which is introduced by HST embedding in LP formulation does not effect the result of matching. As shown in Table 1, success rate of layered matching increases as the matching is carried out at finer layers. Ultimately, layered matching achieves 91.11% success for matching at the fifth layer while the baseline success rate is 92.08%. Average running time of layered matching at fifth layer is almost three times more than the QP formulation although success rate is lower. On the other

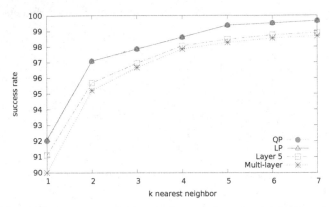

Fig. 3. Success rates for increasing number of neighboring views

hand, multi-layer matching achieves a slightly lower success rate while attaining remarkably better average running time which is 3 and 25 times faster than the QP and LP formulation methods, respectively. Fig. 3 illustrates the increase in success rate if we define successful match as finding an image which is within k view distance of the query image in the dataset. Success rate goes above 95% for all of the methods as $k \geq 2$ and reaches 98% for $k = 4$.

5   Conclusion and Future Work

In this paper, we have presented a novel many-to-many matching method. Our method utilizes the spatial distribution of nodes as well as the node features for the matching process. We use the HST representations of graphs to achieve many-to-many matching where the internal nodes of the HSTs capture the segmentation of features. Utilizing hierarchical structure of HSTs, our method allows to do matching at different levels of detail. In a series of experiments in the domain of image matching, we demonstrated that our method performs very well both in terms of accuracy and running time.

Although we applied our method to only image matching problem, our technique is general and can be applied to any domain that defines a metric. We are aiming to extend this preliminary work in several directions: 1) to come up with a primal-dual algorithm to improve the running time; 2) to introduce a distributed algorithm for labeling; 3) to extend the formulation to cover the case where some of the labels can be discarded from matching. This formulation also has a promising application area in user association problem where idle base stations should be detected and deactivated to optimize overall service quality of the network.

References

1. Almohamad, H.A., Duffuaa, S.O.: A linear programming approach for the weighted graph matching problem. IEEE Transactions on Pattern Analysis and Machine Intelligence 15(5), 522–525 (1993)

2. Ambauen, R., Fischer, S., Bunke, H.: Graph edit distance with node splitting and merging, and its application to diatom identification. In: Hancock, E.R., Vento, M. (eds.) GbRPR 2003. LNCS, vol. 2726, pp. 95–106. Springer, Heidelberg (2003)
3. Bartal, Y.: Probabilistic approximation of metric spaces and its algorithmic applications. In: Proceedings of the 37th Annual Symposium on Foundations of Computer Science, FOCS 1996, p. 184. IEEE Computer Society, Washington, DC (1996)
4. Bartal, Y.: On approximating arbitrary metrices by tree metrics. In: Proceedings of the 30th Annual ACM Symposium on Theory of Computing, STOC 1998, pp. 161–168. ACM, New York (1998)
5. Berretti, S., Del Bimbo, A., Pala, P.: A graph edit distance based on node merging. In: Enser, P.G.B., Kompatsiaris, Y., O'Connor, N.E., Smeaton, A.F., Smeulders, A.W.M. (eds.) CIVR 2004. LNCS, vol. 3115, pp. 464–472. Springer, Heidelberg (2004)
6. Bunke, H.: Error correcting graph matching: On the influence of the underlying cost function. IEEE Transactions on Pattern Analysis and Machine Intelligence 21(9), 917–922 (1999)
7. Caelli, T., Kosinov, S.: An eigenspace projection clustering method for inexact graph matching. IEEE Transactions on Pattern Analysis and Machine Intelligence 26(4), 515–519 (2004)
8. Charikar, M., Chekuri, C., Goel, A., Guha, S., Plotkin, S.: Approximating a finite metric by a small number of tree metrics. In: Proceedings of the 39th Annual Symposium on Foundations of Computer Science, FOCS 1998, pp. 379–388. IEEE Computer Society, Washington, DC (1998)
9. Cook, S.A.: The complexity of theorem-proving procedures. In: Proceedings of the Third Annual ACM Symposium on Theory of Computing, pp. 151–158. ACM (1971)
10. Fatih Demirci, M., Osmanlioglu, Y., Shokoufandeh, A., Dickinson, S.: Efficient many-to-many feature matching under the l1 norm. Comput. Vis. Image Underst. 115(7), 976–983 (2011)
11. Fakcharoenphol, J., Rao, S., Talwar, K.: A tight bound on approximating arbitrary metrics by tree metrics. In: Proceedings of the 35th Annual ACM Symposium on Theory of Computing, STOC 2003, pp. 448–455. ACM, New York (2003)
12. Richard, M.: Karp. A 2k-competitive algorithm for the circle (August 5, 1989) (manuscript)
13. Kleinberg, J., Tardos, É.: Approximation algorithms for classification problems with pairwise relationships: Metric labeling and markov random fields. J. ACM 49(5), 616–639 (2002)
14. Konjevod, G., Ravi, R., Sibel Salman, F.: On approximating planar metrics by tree metrics. Information Processing Letters 80(4), 213–219 (2001)
15. Lowe, D.G.: Distinctive image features from scale-invariant keypoints. International Journal of Computer Vision 60(2), 91–110 (2004)
16. Nene, S.A., Nayar, S.K., Murase, H., et al.: Columbia object image library (coil-20). Technical report, Technical Report CUCS-005-96 (1996)
17. Shapiro, L.G., Haralick, R.M.: Structural descriptions and inexact matching. IEEE Transactions on Pattern Analysis and Machine Intelligence (5), 504–519 (1981)
18. Wildman, J., Osmanlioglu, Y., Weber, S., Shokoufandeh, A.: Delay minimizing user association in cellular networks via hierarchically well-separated trees. arXiv preprint arXiv:1501.02419 (2015)
19. Williams, M.L., Wilson, R.C., Hancock, E.R.: Deterministic search for relational graph matching. Pattern Recognition 32(7), 1255–1271 (1999)
20. Zaslavskiy, M., Bach, F., Vert, J.-P.: Many-to-many graph matching: A continuous relaxation approach. In: Balcázar, J.L., Bonchi, F., Gionis, A., Sebag, M. (eds.) ECML PKDD 2010, Part III. LNCS, vol. 6323, pp. 515–530. Springer, Heidelberg (2010)

Large-Scale Graph Indexing Using Binary Embeddings of Node Contexts

Pau Riba[✉], Josep Lladós, Alicia Fornés, and Anjan Dutta

Computer Vision Center - Computer Science Department
Universitat Autònoma de Barcelona, Barcelona, Spain
{priba,josep,afornes}@cvc.uab.es
http://www.cvc.uab.es

Abstract. Graph-based representations are experiencing a growing usage in visual recognition and retrieval due to their representational power in front of classical appearance-based representations in terms of feature vectors. Retrieving a query graph from a large dataset of graphs has the drawback of the high computational complexity required to compare the query and the target graphs. The most important property for a large-scale retrieval is the search time complexity to be sub-linear in the number of database examples. In this paper we propose a fast indexation formalism for graph retrieval. A binary embedding is defined as hashing keys for graph nodes. Given a database of labeled graphs, graph nodes are complemented with vectors of attributes representing their local context. Hence, each attribute counts the length of a walk of order k originated in a vertex with label l. Each attribute vector is converted to a binary code applying a binary-valued hash function. Therefore, graph retrieval is formulated in terms of finding target graphs in the database whose nodes have a small Hamming distance from the query nodes, easily computed with bitwise logical operators. As an application example, we validate the performance of the proposed methods in a handwritten word spotting scenario in images of historical documents.

Keywords: Graph matching · Graph indexing · Application in document analysis · Word spotting · Binary embedding

1 Introduction

The practical success of machine learning methods applied to simple image representations faded away other schemes representationally richer but practically unfeasible. However, to tackle with complex recognition problems, methods not exclusively based on appearance but enriched with more abstract visual information, such as visual structure of objects, are required. Although the first attempts of part-based descriptors suggesting graph representations were presented long ago [10], it has been in the last decade when a resurgence of structural models has been perceived in computer vision. Graph representations are implicitly or explicitly drivers of more powerful approaches for visual recognition and retrieval.

© Springer International Publishing Switzerland 2015
C.-L. Liu et al. (Eds.): GbRPR 2015, LNCS 9069, pp. 208–217, 2015.
DOI: 10.1007/978-3-319-18224-7_21

Graphs are robust representations offering a representation paradigm able to deal with many-to-many relationships among visual features and their parts. The use of graph matching is an effective solution to deal with visual recognition. Graph matching is among the most important challenges of graph processing. Roughly speaking, the problem consists in finding the best correspondence between the sets of vertices of two graphs preserving the underlying structures and the corresponding labels and attributes. Graph matching plays an important role in many applications of computer vision and pattern recognition, and several algorithms have been proposed in the literature [4]. One of the most popular error-tolerant graph matching methods is based on graph edit distance [19]. However the error-tolerant nature involves an inexact (sub)graph isomorphism computation which is a known NP-Complete problem. Consequently, methods based on graph edit distance are only applicable to graphs of small size. Approximate or suboptimal variations of graph edit distance have been proposed to overcome this difficulty [16]. In the last years new approaches based on graph embeddings and graph kernels have emerged rapidly [6,15]. These methods are based on finding an explicit or implicit transformation of the graph to a n-dimensional space so the problem of graph similarity is elegantly reduced to a machine learning problem using classical classification schemes (e.g. SVM). Other solutions to reduce the complexity of graph matching are based on graph serialization [7,17] consisting in transforming the graph to a sequence so the problem can be solved by an alignment algorithm in quadratic time. More recently, Zhou et al. proposed an efficient approach based on graph factorization [23] applied to deformable object recognition and alignment.

Although the existence of many suboptimal methods for graph matching, the scalability to large scale scenarios is still a challenge. The huge increase of user-generated contents (e.g. image repositories and videos in social networks) has resulted in a need for services including algorithms for searching by content in large databases. As stated before, structural information can play an important role in developing such tools for content-based image retrieval (CBIR). In terms of the complexity of graph matching, it can not be solved by comparing a query graph with thousand or million graphs of the database in a sequential way. Graph indexing approaches must be introduced. This is the motivation of the work presented in this paper.

The problem of graph-based indexing or hashing has been addressed in the literature, especially from the application point of view. In general, it is solved by graph factorization techniques where the database of graphs is decomposed in smaller ones that represent a codebook of compounding ones. The indexation is therefore stated in terms of indexing the constituent graphs organized in a lookup table structure. Messmer in [13] proposed an approach where the constituent graphs are organized in a decision tree. At run time, subgraph isomorphisms are detected by means of decision tree traversal. The complexity for indexing is polynomial in the number of input graph vertices, but the decision tree is of exponential size. A similar approach based on the construction of a graph lattice was proposed in [20]. The performance for large scale retrieval is achieved by matching many overlapping and redundant subgraphs. In [3] an indexing technique for graph databases is proposed. It is based on constructing a nested inverted-index, called FG-index, based on the set of frequent subgraphs. Some works have explored the use of local substructures for indexing. Hence, Yan et al. [22] proposed a graph-based index in terms of frequent subgraphs.

Binary codes are compact descriptors that capture the local context of an image keypoint, according to a local neighborhood pattern, and represent it with a vector of bits. One of the most promising local descriptors is the efficient BRIEF descriptor [2]. BRIEF is a binary descriptor that aims at quickly comparing local features while requiring few amounts of memory. The BRIEF descriptor outputs a set of bits obtained by comparing intensities of pairs of pixels within the local key-region. The good property of binary codes is that, since they are represented as vectors of bits, the comparison between two of them can be quickly computed with basic logical operations (usually XOR) using directly the features of the CPU.

In this paper we propose a graph hashing approach inspired by the ideas of binary encoding for CBIR. We propose to extend the attributes associated to graph nodes by an embedding function describing the local context of the node. By local context we mean the structure of a subgraph centered at the node of radius k (the radius means the length of the path to the farthest node). This vector of attributes is converted to a binary code applying a binary-valued hash function. Therefore, graph retrieval is formulated in terms of finding target (sub)graphs in the database whose nodes have a small Hamming distance from the query nodes. This indexation based on binary codes can be easily computed with bitwise logical operators (XOR) taking advantage of the hardware benefits.

The rest of this paper is organized as follows: in Section 2 we describe the scientific contribution of our work. Section 3 presents an application example where our proposed graph indexing approach is applied to the problem of handwritten word spotting in historical documents. Finally Section 4 draws the conclusion.

2   Binary Embedding Formulation

In this section we describe the main contribution of this work consisting in the encoding of the local topological context of graph nodes by binary vectors. It allows to construct a fast indexing scheme in terms of the Hamming distance.

2.1   Binary Topological Node Features

An *attributed graph* G is defined as a 4-tuple. $G = (V, E, L_V, L_E)$ where V is the set of nodes; $E \subseteq V \times V$ is the set of edges; L_V and L_E are two labeling functions defined as $L_V : V \rightarrow \Sigma_V \times A_V^k$ and $L_E : E \rightarrow \Sigma_E \times A_E^l$, where Σ_V and Σ_E are two sets of symbolic labels for vertices and edges, respectively, A_V and A_E are two sets of attributes for vertices and edges, respectively, and $k, l \in \mathbf{N}$. We will denote the number of vertices in a graph by $|V|$ and the number of edges by $|E|$.

An embedding function $\phi : \mathcal{G} \rightarrow \mathbb{R}^n$ transforms a graph $G \in \mathcal{G}$ to an n-dimensional feature vector. Hence, the distance between two graphs can be computed by a distance in a metric space, and the problem of graph classification can be solved by a statistical learning approach.

The *Morgan index* M is a node feature, originally used to characterize chemical structures [14], that computes the node context in terms of its local context. This index is iteratively computed for each node $v \in V$ as follows:

$$M(v,k) = \begin{cases} 1 & \text{if } k = 0; \\ \sum_u M(u, k-1) & \text{otherwise.} \end{cases}$$

where u is a vertex adjacent to v. The Morgan index of order k associated to a given node v $M(v,k)$ counts the number of paths of length k incident to node v and starting somewhere in the graph. The Morgan index can be computed by the values of the exponentiation of the adjacency matrix. An interesting property of the adjacency matrix A of any graph G is that the (i,j)th entry of A^n denotes the number of walks of length n from the node v_j to the node v_i. Therefore, the Morgan index of order k a node v_i is equivalent to the sum of the cells of the i-th row of the matrix A^k, formally $M(v_i, k) = \sum_j A^k(i,j)$, $j = 1 \ldots |V|$.

Inspired by the topological node features proposed by Dahm et al. [5] we define the *context* of a node v as a node embedding function computed in terms of the topological information of a subgraph centered at v. This context is described in terms of the Morgan index, but it is enriched taking into account the labels of the neighboring nodes. Hence, let us define a variation of the Morgan index concept as follows. Let us denote as $M_l(v,k)$ the Morgan index of node v, order k and label l which counts the number of paths of length k incident at node v and starting at nodes labeled as l. According to this, the *context* of a node v is formally defined as:

$$\nu(v) = [M_{l_1}(v,1), \ldots, M_{l_1}(v,K), M_{l_2}(v,1), \ldots, M_{l_2}(v,K), \ldots, M_{l_{|\Sigma_V|}}(v,K)],$$

where K is the maximum length of the paths incident in v that is considered. The value of K is dependent of each experimental setup. In the application case described in Section 3.3 we have set $K = 3$. Thus, every graph node is attributed by a $K \cdot |\Sigma_V|$ feature vector characterizing the number of paths incident at v of lengths up to K and starting at nodes for all the possible labels in Σ_V.

The context vector $\nu(v)$ is converted to a binary code $\hat{\nu}(v) = \{0,1\}^{K \cdot |\Sigma_V|}$ in terms of a list of corresponding threshold values T_i. These values are application dependent, and in the use case described in Section 3.3 are set to the mean of each dimension.

Figure 1 illustrates the computation of the binary codes. In this example, the codes associated to nodes have length 6 ($K = 3$ and $|\Sigma_V| = 2$). The threshold value is set to the mean of each $M_l(v,k)$.

2.2   Indexing

Given a query graph G_q and a database of graphs $\{G_1, \ldots, G_T\}$, a *focused graph retrieval* problem is defined as finding the subgraphs of G_i similar to G_q. Thus, it consists in finding inexact subgraph matchings between the query and the target

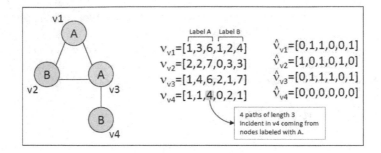

Fig. 1. Example of the binary code computation from a graph

graphs. The graph indexing scheme proposed in this paper follows the paradigm of focused retrieval. In terms of a visual retrieval application, this process can be understood as not only retrieving the images of a database where a query object is likely to appear, but finding the position in each retrieved image. Our proposed graph indexing approach follows this objective.

An inverted file indexing architecture in terms of node contexts is constructed. It stores a mapping from the binary topological features to the nodes of the target graphs in the database. This inverted file is therefore formulated as a lookup table $H : \{0,1\}^b \to \{v_i\}_{v_i \in V}$ that indexes a b-bit vector and returns a list of nodes whose context (binary code) is similar to the input code.

The last step is the actual subgraph matching process. With the indexing table H we only retrieve individual nodes, so it is necessary to implement a node consistency verification. With this purpose, we define a *partition* P of a graph G as a decomposition of it in n small subgraphs, $P(G) = \{g_1, \ldots, g_n\}$, where $g_i \subseteq G$. Hence, the lookup table H is reformulated as a hashing function that instead of returning nodes similar to the input binary code, it returns subgraphs where these target nodes appear. Formally, given a query graph G_q and a database of graphs $\{G_1, \ldots, G_T\}$, for each node of the query graph $v \in V_q$, the indexation function H returns the subgraphs of the database, after a partition has been previously defined, containing this vertex $H(v) = \{g_i\}$, where g_i is a subgraph of one of the target graphs $\{G_1, \ldots, G_T\}$. The definition of the partition under which the database of graphs is decomposed in small graphs is application dependent. The subgraphs g_i can be seen as voting bins, according to a Hough-based principle. Thus, the final result consists of the subgraphs receiving a high number of votes.

Concerning the practical implementation of H that computes the Hamming distance between binary codes, the most straightforward solution is a brute-force linear scan, i.e. to compute the Hamming distance between the query vector and each vector in the database. Computing the Hamming distance between two vectors consists in computing the XOR and counting the number of 1's in the resulting vector. This computation can be computed very fast on modern CPU's, with logic operations being part of the instruction set. A fast hashing process like Locality Sensitive Hashing (LSH) [11] can be added to speed up the indexation.

3   Application Example

This section experimentally illustrates the graph indexing approach with a practical application consisting in word spotting in historical manuscripts.

3.1   Handwritten Word Spotting

The preservation of historical handwritten document collections is of key importance for archives, museums and libraries. Their goal is not only the digitization of paper documents, but also the extraction of the information that these documents contain towards the creation of digital libraries. Since the cost of the manual transcription by human experts is prohibitive, the challenge is to enable the automatic extraction of information through document image analysis techniques.

Since handwriting recognition techniques require large amounts of annotated images to train the recognizer, word spotting is a viable solution to make historical manuscripts amenable to searching and browsing, especially when training data is difficult to obtain. Word spotting is defined as the task of retrieving all the instances of a given query word. In this scenario, the user selects one by looking at the documents, and the system retrieves all words with a similar shape. The first advantage is that word spotting can be performed on-the-fly: the user can crop a word in a new document collection, and the system searches for similar words without any training step. The second advantage is that, since the query word is treated as a shape, these approaches are also able to retrieve graphical elements, such as stamps, symbols, or seals.

Most existing word spotting techniques use statistical representations (e.g. SIFT, HOG) of the word images [1,18]. However, there also exists few approaches using structural representations. The main motivation is that the nature of handwriting suggests than the structure is more stable than the pure appearance of the handwritten strokes. This is specially important when dealing with the elastic deformations of different handwriting styles.

As stated in the comparison of statistical versus structural representations for handwritten word spotting reported in [12], the main disadvantages of structural approaches are the time complexity and scalability to large document collections. Although some methods [9] only use the graph nodes (avoiding the edges), and other approaches [21] propose an embedding using a bag of graphlets (codebook of small graphs, with order 2 or 3), these approaches are still far away of being able to cope with large databases in an efficient way.

However, we believe that the graph indexing using binary embeddings proposed in this paper can be the key to make the graph-based word spotting approaches comparable to statistical ones in terms of time and memory requirements for large document collections.

3.2   Experimental Setup

For the experiments, we have used some pages from the Barcelona Historical Handwritten Marriages Database (BH2M) [8]. It consists of 174 images of manuscripts from the 17th century. The images are part of a collection of marriage records from the archive of the Barcelona Cathedral. The use of word spotting in this collection allows to search names, places, occupations, etc. To illustrate the performance of the method proposed in this paper, 11 pages and 5 instances of 8 query words have been selected. Images are represented by attributed graphs where nodes correspond to basic primitives (graphemes) like loops, vertical lines, arcs, etc. and nodes represent adjacency relations between primitives. This results in 40 query graphs and 11 graphs corresponding to the database images. It has to be noticed that the graphs corresponding to pages of documents contain several connected graphs (between 200-300 in average) corresponding to words or parts of them. In total, the 11 pages contain 3,609 words. In terms of size, query graphs have an average of 25 nodes, and a graph representing a page of a document has an average of 4,500 nodes. If the whole database is considered as a unique large graph, it consists of 50,556 nodes. The partition of the database to define the voting bins is roughly associated to possible words in the images in terms of bounding boxes of connected components. The generation of page graphs takes long time, but it is computed off-line. The extraction of graphemes and the construction of the corresponding graph has a complexity of $O(n^2)$. Concerning the setup of the method, the node contexts have been computed with the paths up to order $K = 3$, and the number of possible node labels is $|\Sigma_V| = 10$. Thus, the length of binary vectors is 30.

3.3   Results

To visually assess the performance of the method, the result of a query word graph is shown in Fig. 2. Figure 2(a) shows a query word and the corresponding graph. Figure 2(b) illustrates the locations where query nodes are detected. It can be appreciated that in the locations where a true positive exists there is a higher density of votes.

In Table 1 we quantitatively report the performance metrics. For each query word, we show the figures of averaging the 5 query instances. For each query, we show the precision, recall and F1-score measures. We can observe that we are obtaining a quite high recall values, i.e. most of the true positives are retrieved, however the precision is quite low, so there is a high number of false positives. It must be noticed that these values depend on the acceptance threshold that is set to consider a retrieval as correct. The graph indexation presented in this paper must not be seen as a final graph matching but a coarse step to quickly locate subgraphs of the database likely to match to the query graph. However, a more accurate matching should be done afterwards with the retrieved subgraphs.

In terms of computational cost, although the implementation is not optimized, the elapsed time for indexing a graph corresponding to a page is 0,02 seconds (target graph of 4,500 nodes). The elapsed time of a standard implementation of a bipartite graph-matching method is 1,02 seconds. Hence, the time is drastically reduced.

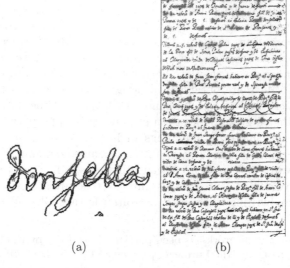

(a)   (b)

Fig. 2. Qualitative results: (a) A query word and its corresponding graph; (b) A full page and the locations where query nodes are detected

Table 1. Quantitative results of word spotting based on graph indexing

Query	Transcription	Precision	Recall	F1-score
Eularia	Eularia	0.0080	0.8462	0.0158
Hieronyma	Hieronyma	0.0118	0.7875	0.0232
Jua$	Jua$	0.0149	0.5389	0.0291
defunct	defunct	0.0271	0.7886	0.0524
donsella	donsella	0.0420	0.8215	0.0796
pages	pages	0.0590	0.9352	0.1107
rebere$	rebere$	0.0645	0.7676	0.1187
viudo	viudo	0.0133	0.6455	0.0261
	Total	**0.0301**	**0.7664**	**0.0569**

4   Conclusions

In this paper we have presented an approach for computing fast inexact sub-graph matching for large scale retrieval purposes. The main contribution of the proposed approach is the definition of a binary embedding for graph nodes based on the called local context. The node context has been defined as the topology of the paths of order k incident in the node and coming from nodes of a given label. A hashing architecture has been designed using binary codes as indexation keys. An application consisting in word spotting into historical handwritten document images has been used as experimental scenario. Although the results are in a preliminary stage, they are encouraging. In terms of a retrieval problem, high recall values are obtained, although the precision is low. The time complexity is linear in terms of the number of nodes of the database. It leads us to conclude that a graph indexation scheme as it is proposed is very useful to compute inexact subgraph matchings in large-scale scenarios as a filtering step aiming to prune the database, so a more accurate matching method can be computed afterwards only in the retrieved subgraphs. Finally, in terms of the application, we have demonstrated that compact structural descriptors are useful signatures for handwriting recognition, despite the variability of handwriting.

Acknowledgement. This work has been partially supported by the Spanish project TIN2012-37475-C02-02 and the European project ERC-2010-AdG-20100407-269796.

References

1. Almazán, J., Gordo, A., Fornés, A., Valveny, E.: Word spotting and recognition with embedded attributes. IEEE Transactions on Pattern Analysis and Machine Intelligence 36(12), 2552–2566 (2014)
2. Calonder, M., Lepetit, V., Strecha, C., Fua, P.: BRIEF: Binary robust independent elementary features. In: Daniilidis, K., Maragos, P., Paragios, N. (eds.) ECCV 2010, Part IV. LNCS, vol. 6314, pp. 778–792. Springer, Heidelberg (2010)
3. Cheng, J., Ke, Y., Ng, W., Lu, A.: Fg-index: Towards verification-free query processing on graph databases. In: International Conference on Management of Data, SIGMOD 2007, pp. 857–872. ACM, New York (2007)
4. Conte, D., Foggia, P., Sansone, C., Vento, M.: Thirty years of graph matching in pattern recognition. International Journal of Pattern Recognition and Artificial Intelligence 18(03), 265–298 (2004)
5. Dahm, N., Bunke, H., Caelli, T., Gao, Y.: A unified framework for strengthening topological node features and its application to subgraph isomorphism detection. In: Kropatsch, W.G., Artner, N.M., Haxhimusa, Y., Jiang, X. (eds.) GbRPR 2013. LNCS, vol. 7877, pp. 11–20. Springer, Heidelberg (2013)
6. Dupé, F.X., Brun, L.: Edition within a graph kernel framework for shape recognition. In: Torsello, A., Escolano, F., Brun, L. (eds.) GbRPR 2009. LNCS, vol. 5534, pp. 11–20. Springer, Heidelberg (2009)
7. Dutta, A., Llados, J., Pal, U.: A symbol spotting approach in graphical documents by hashing serialized graphs. Pattern Recognition 46(3), 752–768 (2013)
8. Fernández-Mota, D., Almazán, J., Cirera, N., Fornés, A., Lladós, J.: Bh2m: The barcelona historical, handwritten marriages database. In: 22nd International Conference on Pattern Recognition, pp. 256–261. IEEE (2014)
9. Fischer, A., Suen, C.Y., Frinken, V., Riesen, K., Bunke, H.: A fast matching algorithm for graph-based handwriting recognition. In: Kropatsch, W.G., Artner, N.M., Haxhimusa, Y., Jiang, X. (eds.) GbRPR 2013. LNCS, vol. 7877, pp. 194–203. Springer, Heidelberg (2013)

10. Fischler, M.A., Elschlager, R.A.: The representation and matching of pictorial structures. IEEE Transactions on Computers C-22(1), 67–92 (1973)
11. Indyk, P., Motwani, R.: Approximate nearest neighbors: Towards removing the curse of dimensionality. In: Proceedings of the Thirtieth Annual ACM Symposium on Theory of Computing, STOC 1998, pp. 604–613. ACM (1998)
12. Llados, J., Rusinol, M., Fornes, A., Fernandez, D., Dutta, A.: On the influence of word representations for handwritten word spotting in historical documents. International Journal of Pattern Recognition and Artificial Intelligence 26(05) (2012)
13. Messmer, B.T., Bunke, H.: A decision tree approach to graph and subgraph isomorphism detection. Pattern Recognition 32(12), 1979–1998 (1999)
14. Morgan, H.L.: The generation of a unique machine description for chemical structures-a technique developed at chemical abstracts service. Journal of Chemical Documentation 5(2), 107–113 (1965)
15. Neuhaus, M., Bunke, H.: Bridging the Gap Between Graph Edit Distance and Kernel Machines. World Scientific Publishing Co., Inc., River Edge (2007)
16. Riesen, K., Bunke, H.: Approximate graph edit distance computation by means of bipartite graph matching. Image and Vision Computing (2009)
17. Robles-Kelly, A., Hancock, E.R.: Graph edit distance from spectral seriation. IEEE Transactions on Pattern Analysis and Machine Intelligence 27(3), 365–378 (2005)
18. Rusiñol, M., Aldavert, D., Toledo, R., Lladós, J.: Efficient segmentation-free keyword spotting in historical document collections. Pattern Recognition 48(2), 545–555 (2015)
19. Sanfeliu, A., Fu, K.S.: A distance measure between attributed relational graphs for pattern recognition. IEEE Transactions on Systems, Man and Cybernetics SMC-13(3), 353–362 (1983)
20. Saund, E.: A graph lattice approach to maintaining and learning dense collections of subgraphs as image features. IEEE Transactions on Pattern Analysis and Machine Intelligence 35(10), 2323–2339 (2013)
21. Wang, P., Eglin, V., Garcia, C., Largeron, C., Lladós, J., Fornés, A.: A coarse-to-fine word spotting approach for historical handwritten documents based on graph embedding and graph edit distance. In: 22nd International Conference on Pattern Recognition, pp. 3074–3079. IEEE (2014)
22. Yan, X., Yu, P.S., Han, J.: Graph indexing: a frequent structure-based approach. In: Proceedings of the 2004 ACM SIGMOD International Conference on Management of Data, pp. 335–346. ACM (2004)
23. Zhou, F., De la Torre, F.: Factorized graph matching. In: IEEE Conference on Computer Vision and Pattern Recognition, pp. 127–134 (June 2012)

Attributed Relational Graph Matching with Sparse Relaxation and Bistochastic Normalization

Bo Jiang, Jin Tang, and Bin Luo$^{(\boxtimes)}$

School of Computer Science and Technology,
Anhui University, Hefei 230039, Anhui, China
zeyiabc@163.com, ahhftang@gmail.com, luobin@ahu.edu.cn

Abstract. Attributed relational graph (ARG) matching problem can usually be formulated as an Integer Quadratic Programming (IQP) problem. Since it is NP-hard, relaxation methods are required. In this paper, we propose a new relaxation method, called Bistochastic Preserving Sparse Relaxation Matching (BPSRM), for ARG matching problem. The main benefit of BPSRM is that the mapping constraints involving both discrete and bistochastic constraint can be well incorporated in BPSRM optimization. Thus, it can generate an approximate binary solution with one-to-one mapping constraint for ARG matching problem. Experimental results show the effectiveness of the proposed method.

Keywords: Attributed relational graph · Graph matching · Bistochastic normalization · Sparse model

1 Introduction

Attributed relational graph (ARG) matching is a fundamental problem in computer vision and pattern recognition. In general, an ARG consists of nodes with unary attributes and edges with binary relationships between nodes. Generally, the goal of ARG matching is to find an one-to-one mapping between two node sets that preserves both unary attributes of nodes and binary relationships of edges as much as possible [1–3].

Recent works have formulated ARG matching problem as an Integer Quadratic Programming (IQP) problem with bistochastic constraint encoding one-to-one mapping constraint [3–7]. Since IQP problem is known to be NP-hard, it is usually solved by relaxation methods. One popular way is to develop a continuous relaxation problem by ignoring the discrete (integer) constraint [3, 4, 6, 8]. Leordeanu and Hebert [3] proposed a spectral relaxation method (spectral matching, SM) to IQP matching problem by ignoring both discrete and bistochastic mapping constraints. Cour et al. [4] extended SM to spectral matching with affine constraint (SMAC) by incorporating the affine mapping constraint. Van Wyk et al. [8] proposed a method which iteratively projected the matching solution onto the bistochastic domain. Cho et al. [6] provided a probabilistic matching method which imposed the bistochastic constraint via a re-weighted random walk model.

© Springer International Publishing Switzerland 2015
C.-L. Liu et al. (Eds.): GbRPR 2015, LNCS 9069, pp. 218–227, 2015.
DOI: 10.1007/978-3-319-18224-7_22

The above relaxation methods generally first optimize the IQP matching problem in a continuous domain and then need a discretization step to compute the final discrete solution. One drawback is that the required discretization step usually leads to weak optima for the problem [5, 7]. Recently, Albarelli et al [9] proposed a relaxation method by adding a ℓ_1-norm constraint on the related solution. This model has been widely used in many matching tasks [10–12]. One important feature of this method is that it can generate a sparse solution for the problem due to the ℓ_1-norm constraint [9, 13]. Rodolà et al [14] further provided a relaxation model with elastic net constraint and thus combined SM [3] and game-theoretic matching [9] simultaneously. Jiang et al [15] proposed a sparse relaxation matching method by exploring ℓ_p-norm constraint. It can be regarded as a balanced model between SM [3] and game-theoretic matching model [9]. The benefit of these sparse matching methods is that they can generate a sparse solution for the IQP matching problem and thus impose the discrete constraint approximately in optimization process. However, one main limitation is that the one-to-one mapping constraint has been ignored generally in these methods.

Following these works, in this paper, we propose a new sparse relaxation method, called bistochastic preserving sparse relaxation matching (BPSRM), for ARG matching problem. The main benefit of BPSRM is that it integrates both sparse (approximate discrete) and bistochastic mapping constraint simultaneously in its optimization process, and thus can generate an approximate binary solution for the IQP matching problem. Experimental results on both synthetic and real-world image matching tasks show the effectiveness of the method.

2   Problem Formulation

Let $G^D = (V^D, E^D, A^D, R^D)$ and $G^M = (V^M, E^M, A^M, R^M)$ be two ARGs to be matched, where V denotes node set, E, edges, A, unary attributes, and R, binary relationships. In general, each node $v_i^D \in V^D$ or edge $e_{ij}^D \in E^D$ has an attribute vector $\mathbf{u}_i^D \in A^D$ or $\mathbf{r}_{ij}^D \in R^D$ [3, 5, 6]. The aim of ARG matching problem is to determine the optimal correspondences $(v_i^D, v_{i'}^M)$ between G^D and G^M. For each match $a_i = (v_i^D, v_{i'}^M)$, there is an unary affinity $f_u(\mathbf{u}_i^D, \mathbf{u}_{i'}^M)$ that measures how well the node $v_i^D \in V^D$ matches the node $v_{i'}^M \in V^M$. For each match pair (a_i, a_j) there is a binary affinity $f_r(\mathbf{r}_{ij}^D, \mathbf{r}_{i'j'}^M)$ that measures the compatibility between node pair (v_i^D, v_j^D) in G^D and $(v_{i'}^M, v_{j'}^M)$ in G^M. Thus, we can use an affinity matrix $\mathbf{W}$ whose non-diagonal elements denote $f_r(\mathbf{r}_{ij}^D, \mathbf{r}_{i'j'}^M)$ and diagonal elements represent $f_u(\mathbf{u}_i^D, \mathbf{u}_{i'}^M)$. The optimal correspondences can also be represented by a matching matrix $\mathbf{Z}$ where $\mathbf{Z}_{ii'} = 1$ implies that the node v_i^D in G^D corresponds to the node $v_{i'}^M$ in G^M, and $\mathbf{Z}_{ii'} = 0$ otherwise. In this paper, we denote $\mathbf{z} \in \{0, 1\}^{mn}$ $(m = |V^D|, n = |V^M|)$ as a row-wise vectorized replica of $\mathbf{Z}$, i.e., $\mathbf{z} = (\mathbf{z}_{11}...\mathbf{z}_{1n}, ..., \mathbf{z}_{m1}...\mathbf{z}_{mn})^{\mathrm{T}} \in \mathbb{R}^{mn \times 1}$ and $\mathbf{z}_{ij} = \mathbf{Z}_{ij}$. We call matrix $\mathbf{Z}$ as the matrix form of vector $\mathbf{z}$. The ARG matching problem can be formulated as an Integer Quadratic Programming (IQP) problem [4–6], i.e.,

$$\max_{\mathbf{z}} \quad J_{gm} = \sum_{i=1}^{m}\sum_{i'=1}^{n}\sum_{j=1}^{m}\sum_{j'=1}^{n} \mathbf{W}_{ii',jj'}\mathbf{z}_{ii'}\mathbf{z}_{jj'} = \mathbf{z}^{\mathrm{T}}\mathbf{W}\mathbf{z} \qquad (1)$$

$$s.t. \quad \forall i \ \sum_{i'=1}^{n}\mathbf{z}_{ii'} = 1, \forall i' \ \sum_{i=1}^{m}\mathbf{z}_{ii'} \leq 1, \mathbf{z}_{ii'} \in \{0,1\}. \qquad (2)$$

The constraints (Eq.(2)) ensure one-to-one mapping between two graphs.

Note that, the above binary constraint $\mathbf{z} \in \{0,1\}^{mn}$ can be replaced by a nonnegative sparse constraints, i.e., $\|\mathbf{z}\|_0 = m, \mathbf{z}_{ii'} \geq 0$, where $\|\mathbf{z}\|_0$ denotes the number of the non-zeros in vector $\mathbf{z}$. Thus, Eqs.(1,2) is equivalent to

$$\max_{\mathbf{z}} \quad J_{gm} = \mathbf{z}^{\mathrm{T}}\mathbf{W}\mathbf{z} \qquad (3)$$

$$s.t. \quad \forall i \ \sum_{i'=1}^{n}\mathbf{z}_{ii'} = 1, \forall i' \ \sum_{i=1}^{m}\mathbf{z}_{ii'} \leq 1, \|\mathbf{z}\|_0 = m, \mathbf{z}_{ii'} \geq 0. \qquad (4)$$

Note that the above problem Eqs.(3,4) is also NP-hard. Our aim in this paper is to propose an approximate method to solve it.

3   Bistochastic Preserving Sparse Relaxation Matching

Indeed, the constraint(Eq.(4)) consists of two parts, i.e., sparse constraint $C_s = \{\mathbf{z}| \ \|\mathbf{z}\|_0 = m\}$ and bistochastic constraint $C_b = \{\mathbf{z}| \sum_{i'=1}^{n}\mathbf{z}_{ii'} = 1, \sum_{i=1}^{m}\mathbf{z}_{ii'} \leq 1, \mathbf{z}_{ii'} \geq 0\}$. Many methods generally aim to optimize the problem under constraint C_b by ignoring sparse constraint C_s. These methods can only return a continuous solution for the problem, and should further use a discretization step to obtain the final discrete solution [4, 6, 8]. Recently, some works also focus on sparse constraint C_s and try to find a sparse solution for the problem [9, 10, 15]. However, the bistochastic constraint C_b encoding one-to-one mapping is generally dropped in these methods.

Our aim in this paper is to propose a new kind of relaxation method, called bistochastic preserving sparse relaxation matching (BPSRM), for ARG matching problem. The key point of BPSRM method is that both sparse constraint C_s and bistochastic constraint C_b can be well incorporated in its optimization process. Thus, BPSRM can generate a sparse (or approximate discrete) solution with one-to-one mapping constraint for the problem. In the following, we first present a kind of sparse relaxation matching (SRM) method [15]. Then, we introduce the bistochastic normalization of Sinkhorn scheme [16]. At last, we propose our BPSRM method based on SRM and bistochastic normalization.

3.1   Sparse Relaxation Matching

The original l_0-norm sparse constraint in Eq.(4) is combinatorial, one popular way is to use the continuous l_p-norm ($p \to 1$ in general) instead of l_0-norm to generate a sparse solution for the problem [15], i.e.,

$$\max_{\mathbf{z}} \quad J_{gm} = \mathbf{z}^{\mathrm{T}}\mathbf{W}\mathbf{z}, \quad s.t. \ \|\mathbf{z}\|_p = 1, \mathbf{z}_{ii'} \geq 0, \qquad (5)$$

where $\|\mathbf{z}\|_p = \left[\sum_{k=1}^{m}\sum_{k'=1}^{n}|\mathbf{z}_{kk'}|^p\right]^{1/p}$ is the l_p-norm of $\mathbf{z}$ and $p \in [1,2]$. In this paper, we call it as sparse relaxation matching (SRM), because it can generate a sparse solution for the problem and the sparsity of the solution can also be controlled by parameter p [15]. Note that SRM is a general matching model. When $p = 1$, SRM degenerates to the game-theoretical matching (GameM) method which has been widely used [9–12]. When $p = 2$, it is equivalent to the popular spectral matching (SM) method [3]. Thus, SRM can be regarded as a balanced model between GameM and SM.

An effective update algorithm has been derived to solve SRM problem. It iteratively updates the current solution $\mathbf{z}^t$ as follows,

$$\mathbf{z}_{kk'}^{t+1} = \left(\mathbf{z}_{kk'}^t \frac{(\mathbf{W}\mathbf{z}^t)_{kk'}}{[\mathbf{z}^t]^{\mathrm{T}}\mathbf{W}\mathbf{z}^t}\right)^{1/p}, \tag{6}$$

where $k = 1, 2 \cdots m, k' = 1, 2 \cdots n$. The correctness and convergence of the algorithm has been proposed in the work [17]. As the iteration increases, $\mathbf{z}^t$ becomes more and more sparse.

Algorithm 1. Bistochastic normalization $\mathbf{X}^* = N_b(\mathbf{Y})$

Input: Any square positive matrix $\mathbf{Y}$
Output: Bistochastic matrix $\mathbf{X}^*$
 1. Initialize $\mathbf{X}^0 = \mathbf{Y}, t = 0$
 2. **while** not convergence **do**
 3.   Normalize across rows by $\tilde{\mathbf{X}}_{hk}^t = \mathbf{X}_{hk}^t / \sum_{h=1}^{n} \mathbf{X}_{hk}^t$, where $h, k = 1...n$
 4.   Normalize across columns by $\mathbf{X}_{hk}^{t+1} = \tilde{\mathbf{X}}_{hk}^t / \sum_{k=1}^{n} \tilde{\mathbf{X}}_{hk}^t$, where $h, k = 1...n$
 5. $t = t + 1$
 6. **end while**
 7. $\mathbf{X}^* = \mathbf{X}^{(t+1)}$

3.2 Bistochastic Normalization

As a continuous analog of the permutation matrix, the bistochastic matrix $\mathbf{X} \in \mathcal{R}^{n \times n}$ is a matrix whose elements are all positive and whose rows and columns all add up to one [6, 16], i.e, $C_b = \{\mathbf{X}|\forall i \ \sum_j \mathbf{X}_{ij} = 1, \forall j \ \sum_i \mathbf{X}_{ij} = 1, \mathbf{X}_{ij} \geq 0\}$. For any square matrix $\mathbf{Y}, \mathbf{Y}_{ij} \geq 0$, the bistochastic normalization proposed by Sinkhorn [16] is to alternatively normalize the rows and columns of the matrix $\mathbf{Y}$ until convergence. This kind of normalization has been used in graph matching optimization to enforce the one-to-one mapping constraint [2, 6].

3.3 Bistochastic Preserving Sparse Relaxation Matching

As discussed before, one benefit of the SRM (Eq.(5)) is that it can generate a sparse solution for the problem, i.e., the solution satisfies the discrete constraint C_s strongly. However, the bistochastic constraint C_b which encodes one-to-one

---

Algorithm 2. Bistochastic preserving sparse relaxation matching (BPSRM)

Input: Affinity matrix $\mathbf{W}$, maximum iteration T, error ϵ, parameter p
Output: Final discrete binary matching solution $\tilde{\mathbf{z}}^*$
1. Initialize $\mathbf{z}^0$ with SM solution, $t = 0$
2. **while** $t < T$ and $|J_{gm}(\mathbf{z}^{(t+1)}) - J_{gm}(\mathbf{z}^{(t)})|/J_{gm}(\mathbf{z}^{(t)}) > \epsilon$ **do**
3.   Update $\mathbf{z}^t$ using the update rule Eq.(7) as

$$\tilde{z}_{hk}^t = \left(z_{hk}^t(\mathbf{W}\mathbf{z}^t)_{hk}\right)^{1/p}$$

4.   Normalize $\tilde{\mathbf{Z}}^t$ (matrix form of $\tilde{\mathbf{z}}^t$) using Algorithm 1 as

$$\mathbf{Z}^{t+1} = N_b(\tilde{\mathbf{Z}}^t)$$

5. $t = t + 1$
6. **end while**
7. $\mathbf{z}^* = \mathbf{z}^{(t+1)}$
8. Compute the binary solution $\tilde{\mathbf{z}}^*$ from $\mathbf{z}^*$ using a discretization step

---

mapping constraint has been entirely ignored in SRM. In this section, we propose our BPSRM method which is motivated by the following two observations.

(1) It is noted that, the update algorithm Eq.(6) of SRM indeed contains the following two steps:
(A) Update step:

$$\tilde{z}_{kk'}^t = \left(z_{kk'}^t(\mathbf{W}\mathbf{z}^t)_{kk'}\right)^{1/p}. \tag{7}$$

(B) l_p-norm normalization step:

$$z_{kk'}^{t+1} = \frac{\tilde{z}_{kk'}^t}{\sum_{k=1}^m \sum_{k'=1}^n [\tilde{z}_{kk'}^t]^p}, \tag{8}$$

where $k = 1 \cdots m, k' = 1 \cdots n$.

(2) The normalization step (B) is used to guarantee that the solution $\mathbf{z}^{t+1}$ satisfies the constraint $\|\mathbf{z}\|_p = 1$ (note that $z_{kk'} \geq 0$, thus $\|\mathbf{z}\|_p = \left[\sum_k \sum_{k'} z_{kk'}^p\right]^{1/p}$). We will show later that, as the iteration t increases, $\mathbf{z}^{t+1}$ becomes more and more sparse. Indeed, the sparse property of $\mathbf{z}^{t+1}$ is generated by step (A). This is because the normalization step (B) does not change the sparsity of $\tilde{\mathbf{z}}^t$.

The core idea of BPSRM is to explore the above **bistochastic normalization** instead of normalization step (B) to further impose the bistochastic mapping constraint C_b in SRM relaxation. The complete algorithm is summarized in Algorithm 2[1]. Generally, there are two main benefits of BPSRM method: (1) the one-to-one mapping constraint is well incorporated in BPSRM optimization while it is entirely dropped in SRM. (2) The sparse property of SRM can be inherited by BPSRM, because the bistochastic normalization generally does not change the sparsity of the vector $\tilde{\mathbf{z}}^t$.

---

[1] In Algorithm 2, we use both vector $\mathbf{z}$ and its matrix form $\mathbf{Z}$ interchangeably.

4 Sparsity and Desirable Discrete Solution

One important benefit of the proposed BPSRM method is that although it optimizes the ARG matching problem in a continuous domain, it can also generate a sparse solution with bistochastic mapping constraint for the problem. That is, the converged solution $\mathbf{z}^*$ of BPSRM can satisfy both C_s and C_b strongly and thus closes to a binary solution. Figure 1 shows the solution vector $\mathbf{z}^t$ of SRM and BPSRM with parameter $p = 1.05$ across different iterations on matching two ARGs generated from CMU sequence (see Experimental section in detail). Here we can note that (1) regardless of initialization, as the iteration increases, both SRM and BPSRM can generate sparse solutions. (2) Comparing with SRM, the solution vector $\mathbf{z}^t$ of BPSRM becomes more and more regularized and almost converges to a binary solution at convergence. This clearly demonstrates the desired effect of bistochastic normalization step of BPSRM method in searching for the optimal solution for ARG matching problem.

Figure 2 top row shows the objective function J_{gm} of the solution vector $\mathbf{z}^t$ under BPSRM algorithm with different initializations and parameter p values. Note that, regardless of initialization and p, the objective $J_{gm}(\mathbf{z}^t)$ increases and converges after some iterations, demonstrating the convergence of BPSRM algorithm. Figure 2 bottom row shows the sparsity[2] of the solution $\mathbf{z}^t$ under BPSRM algorithm. Note that (1) when $p \to 1.0$, the sparsity of $\mathbf{z}^t$ increases and converges after some iterations, i.e., $\mathbf{z}^t$ becomes more and more sparse and converges to a desired approximate discrete solution. (2) The parameter p controls the sparsity of the converged solution.

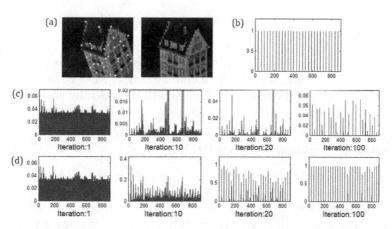

Fig. 1. (a) Image corner points; (b) ground truth binary solution; (c-d) solution matrix $\mathbf{z}^t$ across different iterations under SRM and BPSRM methods, respectively ((c) SRM, (d) BPSRM). Note that, compared with SRM, the solution vector $\mathbf{z}^t$ of BPSRM is more and more regularized and closes to binary solution at convergence.

---

[2] The sparsity of $\mathbf{z}$ is defined as the percentage of zero (or close-to-zero) elements in $\mathbf{z}$.

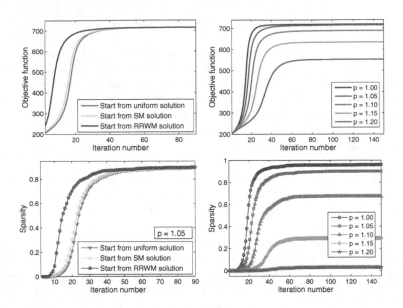

Fig. 2. Top left: objective function J_{gm} of the solution $\mathbf{z}^t$ across the iterations under BPSRM algorithm ($p = 1.05$) with different initializations; **Top right**: objective function of $\mathbf{z}^t$ across the iterations under BPSRM algorithm with different p values. **Bottom left**: sparsity of $\mathbf{z}^t$ across the iterations with different initializations; **Bottom right**: sparsity of $\mathbf{z}^t$ across the iterations with different p values.

5   Experiments

We evaluate our BPSRM method on several matching tasks, and compare it with some recent relaxation methods including SM [3], IPFP [5], SMAC [4], GameM [9], SRM [15] and RRWM [6]. These methods generally have the similar computational complexity and are most related with our method.

5.1   Synthetic Random Graph Matching

Following the experimental setting [6], we first generated two ARGs, G^M and G^D, both of which contain n_{in} nodes, and then added n_{out} outlier nodes in both graphs. The edge is randomly generated between each node pair according to the edge density $\rho \in [0, 1]$. Each edge in G^M was assigned with a random attribute $\mathbf{r}_{ij}^M$ distributed as $\mathbf{r}_{ij}^M \sim U[0, 1]$. The corresponding edge $\mathbf{r}_{i'j'}^D$ in G^D was perturbed by adding a random σ to $\mathbf{r}_{ij}^M$. The affinity matrix $\mathbf{W}$ was computed by $\mathbf{W}_{ii',jj'} = \exp(-\|\mathbf{r}_{ij}^D - \mathbf{r}_{i'j'}^M\|_F^2/0.05)$. We have generated 100 random graph pairs for each noise level and then computed the average performances. Figure 3 summarizes the comparison results. Note that: (1) the sparse matching method SRM generally performs better than SM [3] and GameM [9], indicating the benefit of the SRM method. (2) BPSRM obviously outperforms SRM, which clearly demonstrates the important effect of bistochastic normalization in searching for the optimal solution of ARG matching problem. (3) Comparing with the continuous domain methods SMAC and RRWM, BPSRM further maintains more sparse mapping constraint and thus obtains better performance.

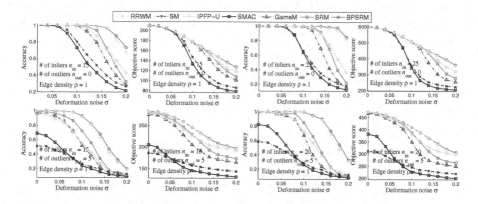

Fig. 3. Comparison results on synthetic graph matching across different noise levels

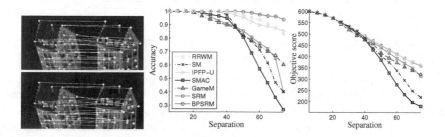

Fig. 4. Comparison results across different separations on CMU image sequences

5.2   Feature Matching across CMU Sequence

For CMU "hotel" sequence, there are 101 images of a toy house [18]. For each image, 25 landmark points were manually marked with known correspondences. We have matched all images spaced by 5, 10 $\cdots$ 70 and 75 frames and computed the average performances. For each image pair, we have generated two ARGs G^M and G^D with the nodes representing the landmark points and the edges denoting the relationship between nodes. The affinity matrix has been computed by $\mathbf{W}_{ii',jj'} = \exp(-\|\mathbf{r}_{ij}^D - \mathbf{r}_{i'j'}^M\|_F^2/0.015)$, where $\mathbf{r}_{ij}^D$ is the Euclidean distance between two points. Figure 4 shows the comparison results. Note that (1) as the separation gap exceeds 40, BPSRM maintains better performance while both GameM and SRM degrades abruptly in general. It demonstrates the robustness of BPSRM on matching two graphs with large variations. (2) BPSRM generally outperforms the continuous method RRWM, which further shows the benefit of the sparse constraint on solving ARG matching problem.

5.3 Real-World Image Matching

We evaluate our method on the image pairs (20 pairs) selected from Zurich Building Image Database (ZuBud) [19]. The candidate correspondences were generated by matching each feature in the first image to the 5 closest features in the second image using SIFT descriptor [19]. The affinity matrix $\mathbf{W}$ was computed as $\mathbf{W}_{ii',jj'} = 1 - \frac{\mathbf{K}_{ij,i'j'}}{\max_{hk,h'k'} \mathbf{K}_{hk,h'k'}}$, where $\mathbf{K}_{ij,i'j'} = |d_{ij} - d_{i'j'}|$ and d_{ij} is the Euclidean distance between the feature i and j. The average performances including true positive, false positive and relative matching score [6] are computed. The results are summarized in Table 1. Note that BPSRM returns higher true positive while maintains lower false positive. It shows the effectiveness of BPSRM on conducting real-world image matching task.

Table 1. Matching results on real-world image datasets

Metric	SM	IPFP	RRWM	GameM	SRM	BPSRM
True Positive	0.9388	0.9425	0.8953	0.9111	0.9223	**0.9479**
False Positive	0.0180	0.0168	0.0308	0.0249	0.0228	**0.0165**
Relative score	0.9885	0.9891	0.9784	0.9745	0.9842	**0.9969**

6 Conclusion

This paper proposes a new sparse relaxation matching method, called bistochastic preserving sparse relaxation matching (BPSRM), for ARG matching problem. Different from many previous works which only focus on either bistochastic or discrete mapping constraint, our BPSRM incorporates these two mapping constraints simultaneously and thus can find an approximate binary solution for ARG matching problem. Experiments show the effectiveness of BPSRM method.

Acknowledgement. This work is supported in part by National High Technology Research and Development Program (863 Program) of China under Grant (2014AA015104); National Nature Science Foundation of China (61472002).

References

1. Conte, D., Foggia, P., Sansone, C., Vento, M.: Thirty years of graph matching in pattern recognition. IJPRAI 18(3), 265–298 (2004)
2. Gold, S., Rangarajan, A.: A graduated assignment algorithm for graph matching. IEEE Trans. on PAMI 18(4), 377–388 (1996)
3. Leordeanu, M., Hebert, M.: A spectral technique for correspondence problem using pairwise constraints. In: ICCV, pp. 1482–1489 (2005)
4. Cour, M., Srinivasan, P., Shi, J.: Balanced graph matching. In: NIPS, pp. 313–320 (2006)

5. Leordeanu, M., Hebert, M., Sukthankar, R.: An integer projected fixed point method for graph macthing and map inference. In: NIPS, pp. 1114–1122 (2009)
6. Cho, M., Lee, J., Lee, K.M.: Reweighted random walks for graph matching. In: Daniilidis, K., Maragos, P., Paragios, N. (eds.) ECCV 2010, Part V. LNCS, vol. 6315, pp. 492–505. Springer, Heidelberg (2010)
7. Zhou, F., la Torre, F.D.: Factorized graph matching. In: CVPR, pp. 127–134 (2012)
8. van Wyk, B.J., van Wyk, M.A.: A pocs-based graph matching algorithm. IEEE Trans. on PAMI 16(11), 1526–1530 (2004)
9. Albarelli, A., Bulo, S.R., Torsello, A., Pelillo, M.: Matching as a non-coorperative game. In: ICCV, pp. 1319–1326 (2009)
10. Albarelli, A., Rodolà, E., Torsello, A.: Imposing semi-local geometric constraints for accurate correspondences selection in structure from motion: a game-theoretic perspective. International Journal of Computer Vision 97(1), 36–53 (2012)
11. Liu, H., Yan, S.: Common visual pattern discovery via spatially coherent correspondences. In: CVPR, pp. 1609–1616 (2010)
12. Rodolà, E., Bronstein, A.M., Albarelli, A., Bergamasco, F., Torsello, A.: A game-theoretic approach to deformable shape matching. In: CVPR, pp. 182–189 (2012)
13. Donoho, D.: Compressed sensing. In: Technical Report, Stanford University (2006)
14. Rodolà, E., Torsello, A., Harada, T., Kuniyoshi, Y., Cremers, D.: Elastic net constraints for shape matching. In: ICCV, pp. 1169–1176 (2013)
15. Jiang, B., Zhao, H.F., Tang, J., Luo, B.: A sparse nonnegative matrix factorization technique for graph matching problem. Pattern Recognition 47(1), 736–747 (2014)
16. Sinkhorn, R.: A relationship between arbitray positive matrices and doubly stochastic matrices. Ann. Mach. Statistics 35(2), 876–879 (1964)
17. Ding, C., Li, T., Jordan, M.I.: Nonnegative matrix factorization for combinatorial optimization: Spectral clustering, graph matching and clique finding. In: ICDM, pp. 183–192 (2008)
18. Caetano, T.S., McAuley, J.J., Cheng, L., Le, Q.V., Smola, A.J.: Learning graph matching. IEEE Trans. on PAMI 31(6), 1048–1058 (2009)
19. Ng, E.S., Kingsbury, N.G.: Matching of interest point groups with pairwise spatial constraints. In: ICIP, pp. 2693–2696 (2010)

Graph Clustering and Classification

On the Influence of Node Centralities
on Graph Edit Distance for Graph Classification

Xavier Cortés, Francesc Serratosa$^{(\boxtimes)}$, and Carlos F. Moreno-García

Universitat Rovira i Virgili, Tarragona, Catalonia, Spain
{francesc.serratosa,xavier.cortes}@urv.cat
carlosfrancisco.moreno@estudiants.urv.cat

Abstract. Classical graph approaches for pattern recognition applications rely on computing distances between graphs in the graph domain. That is, the distance between two graphs is obtained by directly optimizing some objective function which consider node and edge attributes. Bipartite Graph Matching was first published in a journal in 2009 and new versions have appeared to speed up its runtime such as the Fast Bipartite Graph Matching. This algorithm is based on defining a cost matrix between all nodes of both graphs and solving the node correspondence through a linear assignment method. To construct the matrix, several local structures can be defined from the simplest one (only the node) to the most complex (a whole clique or eigenvector structure). In this paper, we propose five different options and we show that the type of local structure and the distance defined between these structures is relevant for graph classification.

Keywords: Graph edit distance · Bipartite graph matching · Fast bipartite graph matching · Levenshtein distance

1   Introduction

Attributed Graphs have been of crucial importance in pattern recognition throughout more than 3 decades [1], [2], [3], [4], [5] and [6]. If elements in pattern recognition are modelled through attributed graphs, error-tolerant graph-matching algorithms are needed that aim to compute a matching between nodes of two attributed graphs that minimizes some kind of objective function. Unfortunately, the time and space complexity to compute the minimum of these objective functions is very high. For this reason, some graph prototyping methods have appeared with the aim of reducing the runtime while querying a graph in a large database [7], [8], [9], [10], [11]. There are two interesting surveys in [2] and [3] about this subject. Recently, Fast Bipartite algorithm (FBP) [12] and Square Fast Bipartite algorithm (SFBP) [13] was presented that solve the graph-matching problem in a similar way. They are variants of Bipartite algorithm (BP) [14] but with a reduced runtime. The three algorithms are based on translating the error-tolerant graph-matching problem into a linear assignation

This research is supported by I+D projects DPI2013-42458-P and TIN2013-47245-C2-2-R.

C.-L. Liu et al. (Eds.): GbRPR 2015, LNCS 9069, pp. 231–241, 2015.
DOI: 10.1007/978-3-319-18224-7_23

problem. They are composed of two steps. In the first one, a cost matrix is constructed with the information of both graphs to be compared. Each cell of the matrix represents the distance between a local sub-structure of one of the graphs and the local sub-structure of the other graph. In the second one, a linear assignation algorithm is applied to this matrix to obtain the best (sub-optimal) isomorphism between nodes of both graphs. Several linear assignation algorithms can be used, such as [15] or [16]. Recently, a new research field has appeared where the human or an expert system can interact on the graph matching algorithm to increase the accuracy of the obtained node correspondence [17] and [18].

In this paper, we compare five local sub-structures. The first three are strictly based on considering the local structure as a sub-graph. The other two are based on the spectral information centred at the involved nodes of both graphs.

Results show that the selected local sub-structure has a great impact on the obtained distance value and also on the runtime although in the past little research have been done. There is a related research in [19] that considers the two eigenvector centralities we propose and it studies their impact on the runtime and the accuracy to obtain the exact distance value. In our paper, we test the five centralities in well-known datasets and we obtain the recognition accuracy and also the runtime.

The outline of the paper is as follows. In the next section, we define Attributed graphs, Graph edit distance and we comment how to compute the Graph edit distance using Square Fast Bipartite algorithm (SFBP) [13]. In section 3, we present five local sub-structures and distances between them. In section 4, we show the experimental validation and finally, we conclude the article in section 5.

2   Graphs and Graph Edit Distance

In this section, we first define Attributed graphs with the concept of a neighbour of a node and Error-tolerant graph matching and then we explain the Graph edit distance.

Attributed Graphs

An Attributed graph is defined as a triplet $G = (\Sigma_v, \Sigma_e, \gamma_v)$, where $\Sigma_v = \{v_a \mid a = 1, \dots, n\}$ is the set of vertices and $\Sigma_e = \{e_{ab} \mid a, b \in 1, \dots, n\}$ is the set of undirected and unattributed edges. Function $\gamma_v \colon \Sigma_v \to \Delta_v$ assigns attribute values in any domain to vertices. The order of graph G is n. We call $E(v_a)$ to the number of neighbours of node v_a. Finally, we define the neighbours of a node v_a, named N_a, on an attributed graph G, as another graph $N_a = (\Sigma_v^{N_a}, \Sigma_e^{N_a}, \gamma_v^{N_a})$ only composed of nodes connected to them by an edge. Formally, $\Sigma_v^{N_a} = \{v_a \mid e_{ab} \in \Sigma_e\}$, $\Sigma_e^{N_a} = \Phi$ and $\gamma_v^{N_a}(v_a) = \gamma_v(v_a)$, $\forall v_a \in \Sigma_v^{N_a}$. Finally, we represent as A the adjacency matrix such that $A[a,b] = 1$ if $e_{ab} \in \Sigma_e$ and $A[a,b] = 0$ otherwise.

Error Correcting Graph Isomorphism

Let $G^p = (\Sigma_v^p, \Sigma_e^p, \gamma_v^p)$ and $G^q = (\Sigma_v^q, \Sigma_e^q, \gamma_v^q)$ be two Attributed graphs of initial order n and m. To allow maximum flexibility in the matching process, graphs are extended with null nodes to be of order $n + m$. We refer to null nodes of G^p and G^q by

$\hat{\Sigma}_v^p \subseteq \Sigma_v^p$ and $\hat{\Sigma}_v^q \subseteq \Sigma_v^q$ respectively. We assume null nodes have indices $a \in [n + 1, \dots, n + m\}$ and $i \in [m + 1, \dots, n + m\}$ for graphs G^p and G^q, respectively. Let T be a set of all possible bijections between two vertex sets Σ_v^p and Σ_v^q. We define the non-existent or null edges as $\hat{\Sigma}_e^p \subseteq \Sigma_e^p$ and $\hat{\Sigma}_e^q \subseteq \Sigma_e^q$. Isomorphism $f^{p,q} \colon \Sigma_v^p \to \Sigma_v^q$, assigns one vertex of G^p to only one vertex of G^q. The isomorphism between edges is defined accordingly to the isomorphism of their terminal nodes.

Graph Edit Distance between Graphs

One of the most widely used methods to evaluate an error-correcting graph isomorphism is the Graph edit distance [1], [6], [20]. The dissimilarity is defined as the minimum amount of required distortion to transform one graph into the other. To this end, a number of distortion or edit operations, consisting of insertion, deletion and substitution of both nodes and edges are defined. Edit cost functions are introduced to quantitatively evaluate the edit operations. The basic idea is to assign a penalty cost to each edit operation according to the amount of distortion that it introduces in the transformation. Deletion and insertion operations are transformed to assignations of a non-null node of the first or second graph to a null node of the second or first graph. Substitutions simply indicate node-to-node assignments. Using this transformation, given two graphs G^p and G^q, and a bijection between their nodes, $f^{p,q}$, the graph edit cost is given by:

$$EditCost_{K_v,K_e}(G^p, G^q, f^{p,q}) =$$

$$\sum_{\substack{v_a^p \in \Sigma_v^p - \hat{\Sigma}_v^p \\ v_i^q \in \Sigma_v^q - \hat{\Sigma}_v^q}} C_{vs}\left(v_a^p, v_i^q\right) + \sum_{\substack{v_a^p \in \Sigma_v^p - \hat{\Sigma}_v^p \\ v_i^q \in \hat{\Sigma}_v^q}} K_v + \sum_{\substack{v_a^p \in \hat{\Sigma}_v^p \\ v_i^q \in \Sigma_v^q - \hat{\Sigma}_v^q}} K_v + \tag{1}$$

$$\sum_{\substack{e_{ab}^p \in \Sigma_e^p - \hat{\Sigma}_e^p \\ e_{ij}^q \in \hat{\Sigma}_e^q}} K_e + \sum_{\substack{e_{ab}^p \in \hat{\Sigma}_e^p \\ e_{ij}^q \in \Sigma_e^q - \hat{\Sigma}_e^q}} K_e$$

Where $f^{p,q}\left(v_a^p\right) = v_i^q$ and $f_e^{p,q}\left(e_{ai}^p\right) = e_{ij}^q$

where C_{vs} is a function that represents the cost of substituting node v_a^p of G^p by node $f^{p,q}\left(v_a^p\right)$ of G^q. Constant K_v is the cost of deleting node v_a^p of G^p or inserting node v_i^q of G^q. Likewise for the edges, K_e is the cost of assigning edge e_{ab}^p of G^p to a non-existing edge of G^q or assigning edge e_{ab}^q of G^q to a non-existing edge of G^p. Note that we have not considered the cases in which two null nodes or null arcs are mapped, this is because this cost is zero by definition. In the same way, we do not have considered the cost of substituting two edges since they are unattributed and so, its substitution has a null cost. Note the definitions exposed in this paper can be easily generalised by adding attributes on edges.

The Graph edit distance is defined as the minimum cost under any bijection in T:

$$EditDist_{K_v,K_e}(G^p, G^q) = \min_{f^{p,q} \in T}\{EditCost_{K_v,K_e}(G^p, G^q, f^{p,q})\} \tag{2}$$

The assignment problem considers the task of finding an optimal assignment of the elements of a set P to the elements of another set Q, where both sets have the same cardinality $N = |P| = |Q|$. Let us assume there is a $N \times N$ cost matrix C. The matrix elements $C_{i,j}$ correspond to the cost of assigning the i-th element of P to the j-th element of Q. An optimal linear assignment is the one that minimises the sum of the assignment costs and so, the assignment problem can be stated as finding the permutation p that minimises $\sum_{i=1}^{N} C_{i,p(i)}$. There are several algorithms that solve the linear assignation problem [15], [16]. In the worst case, the maximum number of operations needed by these algorithms is $O(N^3)$.

Square Fast Bipartite algorithm (SFBP) [13] is an efficient algorithm to Edit Distance computation for general graphs that in a first step generates a matrix costs and in a second step, applies an optimal linear assignment algorithm on this matrix. The algorithm is similar to Bipartite BP but with a different cost matrix. In fact, SFBP defines two matrices, depending on the order of the graphs (BP only defines one matrix).

If $m \geq n$ then the cost matrix is,

$$
C_{m \geq n}^{SFBP} = m \left\{ \begin{bmatrix} C_{1,1} & C_{1,2} & \cdots & C_{1,m} \\ C_{2,1} & C_{2,2} & & C_{2,m} \\ \vdots & \vdots & & \vdots \\ C_{n,1} & C_{n,2} & \cdots & C_{n,m} \\ C_{\varepsilon,1} & C_{\varepsilon,2} & \cdots & C_{\varepsilon,m} \\ C_{\varepsilon,1} & C_{\varepsilon,2} & & C_{\varepsilon,m} \\ \vdots & \vdots & & \vdots \\ C_{\varepsilon,1} & C_{\varepsilon,2} & \cdots & C_{\varepsilon,m} \end{bmatrix} \right.
$$

If $m \leq n$ then the cost matrix is,

$$
C_{m \leq n}^{SFBP} = n \left\{ \begin{bmatrix} C_{1,1} & C_{1,2} & \cdots & C_{1,m} & C_{1,\varepsilon} & C_{1,\varepsilon} & \cdots & C_{1,\varepsilon} \\ C_{2,1} & C_{2,2} & & C_{2,m} & C_{2,\varepsilon} & C_{2,\varepsilon} & & C_{2,\varepsilon} \\ \vdots & \vdots & & \vdots & \vdots & \vdots & & \vdots \\ C_{n,1} & C_{n,2} & \cdots & C_{n,m} & C_{n,\varepsilon} & C_{n,\varepsilon} & \cdots & C_{n,\varepsilon} \end{bmatrix} \right.
$$

Where $C_{a,i}$ denotes the cost of substituting nodes v_a^p and v_i^q and its local sub-structure. $C_{a,\varepsilon}$ denotes the cost of deleting node v_a^p and its local sub-structure, and $C_{\varepsilon,i}$ denotes the cost of inserting node v_i^q and its local sub-structure. Obviously, as described in [14], this minimum cost is a sub-optimal Edit distance value between the involved graphs since cost matrix rows are related to local sub-structures of graph G^p and columns are related to local sub-structures of G^q. The computational cost of the SFBP is $O((\max(n,m))^3)$. Note that the linear assignment algorithms used in its original form are optimal for solving the assignment problem, but they are suboptimal for finding the Graph edit distance. This is due to the fact that nodes and their local sub-structure are considered individually. If $f_m^{p,q*}$ is the obtained isomorphism through method or algorithm m, then we have that $EditCost_{K_v,K_e}(G^p,G^q,f_m^{p,q*}) \geq EditDist_{K_v,K_e}(G^p,G^q)$. In some cases, it is not possible to compute the real distance value $EditDist_{K_v,K_e}(G^p,G^q)$ for runtime reasons due to it has to be computed through an A^* algorithm. In these cases, if we want to evaluate two suboptimal methods to compute the Graph edit distance, the lower the cost, the better the method.

3   Node Centralities and Distances between Them

In this section, different methods are proposed to obtain values $C_{a,i}$, $C_{a,\varepsilon}$ and $C_{\varepsilon,i}$ in the cost matrices used by SFBP algorithm to compute the isomorphism between nodes. The whole values on the cost matrices depend on two weighted disjoint costs. The first one only depends on the nodes and the second one depends on the rest of the local sub-structure.

When the original nodes $v_a^p \in \Sigma_v^p - \hat{\Sigma}_v^p$ and $v_i^q \in \Sigma_v^q - \hat{\Sigma}_v^q$ are mapped:

$$C_{a,i} = \beta \cdot C_{vs}(v_a^p, v_i^q) + (1-\beta) \cdot C_{cs}(v_a^p, v_i^q) \tag{3}$$

When the original node $v_a^p \in \Sigma_v^p - \hat{\Sigma}_v^p$ in G^p is deleted and so mapped to a null node $v_i^q \in \hat{\Sigma}_v^q$ in G^q:

$$C_{a,\varepsilon} = \beta \cdot k_v + (1-\beta) \cdot C_{cd}(v_a^p, v_i^q) \tag{4}$$

When the original node $v_i^q \in \Sigma_v^q - \hat{\Sigma}_v^q$ in G^q is inserted and so mapped from a null node $v_a^p \in \hat{\Sigma}_v^p$ in G^p:

$$C_{\varepsilon,i} = \beta \cdot k_v + (1-\beta) \cdot C_{ci}(v_a^p, v_i^q) \tag{5}$$

As commented in section 2, C_{vs} is a distance function defined through the node attribute values and k_v gauges the importance of deleting or inserting nodes in the matching process. C_{cs} is the cost to substitute the local sub-structure and C_{cd} and C_{ci} are the costs to delete and insert it, respectively. These costs depend on the used local sub-structures. Finally, the weighting parameter β has to be set in a validation or learning process.

We propose the following five node centralities and distances:

-**The degree**: The local sub-structure is composed of a node and its connected arcs. Although it is easily generalizable, in this paper we have considered edges do not have attributes. Therefore, these costs are based on counting the number of edges,

$$
\begin{aligned}
C_{cs}(v_a^p, v_i^q) &= k_e \cdot |E(v_a^p) - E(v_i^q)| \text{ where } v_a^p \in \Sigma_v^p - \hat{\Sigma}_v^p \text{ and } v_i^q \in \Sigma_v^q - \hat{\Sigma}_v^q. \\
C_{cd}(v_a^p, v_i^q) &= k_e \cdot E(v_a^p) \text{ where } v_a^p \in \Sigma_v^p - \hat{\Sigma}_v^p \text{ and } v_i^q \in \hat{\Sigma}_v^q. \\
C_{ci}(v_a^p, v_i^q) &= k_e \cdot E(v_i^q) \text{ where } v_a^p \in \hat{\Sigma}_v^p \text{ and } v_i^q \in \Sigma_v^q - \hat{\Sigma}_v^q.
\end{aligned}
\tag{6}
$$

-**The clique**: The local sub-structure is composed of a node, its arcs and the connected nodes. Clearly, we could define more levels of complexity but we decided not to consider them. Some excluded examples are the clique plus the edges of the neighbouring nodes, since the combinations of structures exponentially explode. To increase the complexity of the local sub-structure but not the computational complexity, the spectral centralities are proposed. In the clique structure, the costs on the centralities are defined as follows,

$$C_{cs}\left(v_a^p, v_i^q\right) = EditDistance_{k_v+k_e,0}\left(N_a^p, N_i^q\right)$$
$$\text{where } v_a^p \in \Sigma_v^p - \hat{\Sigma}_v^p \text{ and } v_i^q \in \Sigma_v^q - \hat{\Sigma}_v^q. \tag{7}$$

Function $EditDistance$ is computed through SFBP in the same way it is done with graphs. A neighbour node and its connecting edge have to be seen as an indivisible structure. Note, the neighbour structures N_a^p and N_i^q are defined as non-connected graphs with only the neighbouring nodes and without edges. Yet, the centrality cost has to consider the cost of the edges that connect the central node with the neighbouring nodes. To do so, the cost of deleting and inserting nodes in function $EditDistance$ is defined as $k_v + k_e$. Due to there are no edges, any cost of deleting and inserting edges could be set at function $EditDistance$ and we imposed 0.

The deletion and insertion centrality costs depend on the number of neighbours,

$$C_{cd}\left(v_a^p, v_i^q\right) = (k_v + k_e) \cdot E\left(v_a^p\right) \text{ where } v_a^p \in \Sigma_v^p - \hat{\Sigma}_v^p \text{ and } v_i^q \in \hat{\Sigma}_v^q.$$
$$C_{ci}\left(v_a^p, v_i^q\right) = (k_v + k_e) \cdot E\left(v_i^q\right) \text{ where } v_a^p \in \hat{\Sigma}_v^p \text{ and } v_i^q \in \Sigma_v^q - \hat{\Sigma}_v^q. \tag{8}$$

-**The planar**: The local sub-structure is exactly the same as the clique but in this case the relative position is considered. Costs C_{cd} and C_{ci} are defined in the same way as above. The difference resides in C_{cs}. The only allowed isomorphisms $f^{N_a^p, N_i^q}$ are the ones that are generated from cyclic combinations of the neighbours. Therefore, sets N_a^p and N_i^q are seen as strings and the cost is computed through the Cyclic Levenshtein distance [21], [22].

$$C_{cs}\left(v_a^p, v_i^q\right) = CyclicLevensh\left(N_a^p, N_i^q, k_v + k_e\right)$$
$$\text{where } v_a^p \in \Sigma_v^p - \hat{\Sigma}_v^p \ \& \ v_i^q \in \Sigma_v^q - \hat{\Sigma}_v^q. \tag{9}$$

Figure 1 shows an example of the Clique and Planar centralities. Red lines represent the optimal correspondence while considering Clique centrality and green lines while considering Planar centrality. Numbers on the nodes are the attributes $\gamma_v^p\left(v_a^p\right)$ or $\gamma_v^q\left(v_i^q\right)$. Edges do not have attributes. Suppose $C_{cs}\left(v_a^p, v_i^q\right) = \left|\gamma_v^p\left(v_a^p\right) - \gamma_v^q\left(v_a^q\right)\right|$. In the Clique case (red lines), $C_{cs}\left(v_a^p, v_i^q\right) = 1 + 0 + 1 + 0 = 2$. In the Planar case (green lines), $C_{cs}\left(v_a^p, v_i^q\right) = 1 + 17 + 1 + 17 = 36$. Note the restriction to be the labelling cyclic makes the distance value to be larger.

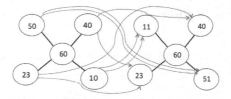

Fig. 1. In red: optimal correspondence in Clique centrality. In green: optimal correspondence in Planar centrality

-The Eigenvector: The local sub-structure takes into consideration the degree of the neighbouring nodes considering the eigenvector that has the largest eigenvalue. The eigenvector centralities given a specific node in both graphs are defined by,

$$c_a^p = \frac{1}{\lambda_1^p} \cdot \sum_{v_{a'}^p \in N_a^p} |V_{a'}^p| \quad \text{and} \quad c_i^q = \frac{1}{\lambda_1^q} \cdot \sum_{v_{i'}^q \in N_i^q} |V_{i'}^q| \tag{10}$$

where $V_{a'}^p$ and $V_{i'}^q$ are the values of the a'-th and i'-th positions of the eigenvectors with the largest eigenvalues obtained through adjacency matrices A^p and A^q. Besides, λ_1^p and λ_1^q are the largest eigenvalues of these adjacency matrices. Thus, the substitution cost is simply computed as,

$$C_{cs}\left(v_a^p, v_i^q\right) = \left|c_a^p - c_i^q\right|$$
$$\text{where } v_a^p \in \Sigma_v^p - \hat{\Sigma}_v^p \ \& \ v_i^q \in \Sigma_v^q - \hat{\Sigma}_v^q. \tag{11}$$

The deletion and insertion costs are computed assuming that the centrality of a null node is 0.

$$C_{cd}\left(v_a^p, v_i^q\right) = c_a^p \text{ where } v_a^p \in \Sigma_v^p - \hat{\Sigma}_v^p \text{ and } v_i^q \in \hat{\Sigma}_v^q.$$
$$C_{ci}\left(v_a^p, v_i^q\right) = c_i^q \text{ where } v_a^p \in \hat{\Sigma}_v^p \text{ and } v_i^q \in \Sigma_v^q - \hat{\Sigma}_v^q. \tag{12}$$

-The PageRank: This centrality is a variation of the Eigenvector centrality. The difference consists in each eigenvector element is normalised by the number of neighbours of the node it represents.

$$c_a^p = \frac{1}{\lambda_1^p} \cdot \sum_{v_{a'}^p \in N_a^p} \frac{|V_{a'}^p|}{\max\left(1, E(V_{a'}^p)\right)} \quad \text{and} \quad C_i^q = \frac{1}{\lambda_1^q} \cdot \sum_{v_{i'}^q \in N_i^q} \frac{|V_{i'}^q|}{\max\left(1, E(V_{i'}^q)\right)} \tag{13}$$

This centrality is normalised by $\max\left(1, E(V_{a'}^p)\right)$ instead of $E(V_{a'}^p)$ to avoid dividing by 0. Then, the substitution, insertion and deletion costs are computed in a similar why than the Eigenvector centrality but using the PageRank centrality.

4   Experimental Validation

Table 1. Recognition ratio of the five proposed centralities given the 5 datasets and different combinations of parameters

	Kv	Ke	β	Degree	Clique	Planar		Kv	β	Eigen Vector	Pagerank
LETTER LOW	1	1	0.5	0.95	0.98	0.98		1	0.9	0.94	0.94
	1	0.1	0.5	0.96	**0.99**	0.98		1	0.5	0.96	0.96
	0.1	1	0.5	0.96	0.98	0.98		1	0.1	0.93	0.93
	0.1	0.1	0.5	0.98	**0.99**	**0.99**		1	0.01	0.92	0.92
LETTER MED	1	1	0.5	0.87	**0.94**	**0.94**		1	0.9	0.66	0.65
	1	0.1	0.5	0.73	**0.94**	**0.94**		1	0.5	0.85	0.84
	0.1	1	0.5	0.86	0.89	0.89		1	0.1	0.84	0.84
	0.1	0.1	0.5	0.18	0.09	0.09		1	0.01	0.83	0.83
LETTER HIGH	1	1	0.5	0.79	0.88	**0.90**		1	0.9	0.65	0.65
	1	0.1	0.5	0.74	0.88	**0.90**		1	0.5	0.72	0.71
	0.1	1	0.5	0.81	0.80	0.80		1	0.1	0.71	0.74
	0.1	0.1	0.5	0.38	0.07	0.07		1	0.01	0.59	0.61
GREC	10	10	0.5	0.96	0.84	0.83		10	0.9	0.71	0.71
	10	5	0.5	0.92	0.65	0.63		10	0.5	0.93	0.93
	5	10	0.5	0.95	0.39	0.37		10	0.1	**0.99**	**0.99**
	5	5	0.5	0.87	0.14	0.14		10	0.01	0.97	0.96
AIDS	1	1	0.5	0.96	0.84	0.84		1	0.9	0.83	0.83
	1	0.1	0.5	0.83	0.82	0.82		1	0.5	0.91	0.92
	0.1	1	0.5	0.90	0.82	0.82		1	0.1	**0.99**	**0.99**
	0.1	0.1	0.5	0.82	0.81	0.81		1	0.01	**0.99**	**0.99**

Table 1 and table 2 show the recognition ratio and mean runtime of the five centralities described in section 3 for graph classification using the five graph databases LETTER LOW, LETTER MEDIUM, LETTER HIGH, GREC and AIDS from the IAM repository [23]. Each database consists on a set of different graph instances divided in different classes where each class is composed of a reference set and a test set. We used the K-NN classifier where K = 3 and the graph-matching algorithm SFBP summarised in section 2 with Jonker-Volgenant solver [16]. Note the computation of the edit operations on local structures C_{cs}, C_{cd} and C_{ci} depends on the following parameters: In case of Degree, Clique and Planar, parameters are k_v, k_e and β. In case of Eigenvector and PageRank, parameters are k_v and β. Best results for each database are marked in bold and underlined.

Table 2. Mean runtime spent for a graph classification of the 5 proposed centralities given the 5 datasets and different parameters (I7 3.07 Ghz, Windows, Matlab 2014a)

	Kv	Ke	β	Degree	Clique	Planar		Kv	β	Eigen Vector	Pagerank
	1	1	0.5	0.61	3.14	0.95		1	0.9	0.58	0.59
LETTER	1	0.1	0.5	0.56	3.15	0.90		1	0.5	0.62	0.62
LOW	0.1	1	0.5	**0.55**	3.09	0.85		1	0.1	0.76	0.79
	0.1	0.1	0.5	0.48	3.15	0.80		1	0.01	2.04	2.30
	1	1	0.5	0.61	3.13	0.94		1	0.9	0.57	0.58
LETTER	1	0.1	0.5	0.56	3.16	0.91		1	0.5	0.63	0.64
MED	0.1	1	0.5	**0.55**	3.09	0.86		1	0.1	0.77	0.81
	0.1	0.1	0.5	0.46	3.16	0.81		1	0.01	2.04	2.38
	1	1	0.5	0.61	3.68	1.24		1	0.9	0.58	0.59
LETTER	1	0.1	0.5	0.60	3.60	1.26		1	0.5	0.65	0.69
HIGH	0.1	1	0.5	**0.57**	3.52	1.17		1	0.1	0.84	0.88
	0.1	0.1	0.5	0.47	3.51	1.10		1	0.01	2.40	2.71
	10	10	0.5	0.45	7.34	1.99		10	0.9	0.49	0.48
GREC	10	5	0.5	**0.42**	7.48	2.01		10	0.5	0.53	0.52
	5	10	0.5	0.42	7.35	1.99		10	0.1	0.99	0.97
	5	5	0.5	0.40	7.34	1.97		10	0.01	4.10	4.35
	1	1	0.5	0.64	6.29	1.78		1	0.9	0.41	0.41
AIDS	1	0.1	0.5	0.37	6.27	1.76		1	0.5	0.55	0.59
	0.1	1	0.5	0.47	5.94	1.60		1	0.1	3.04	39.23
	0.1	0.1	0.5	**0.27**	6.12	1.65		1	0.01	23.73	21.41

Table 3 summarises the achievements that can be deducted from tables 1 and 2. We realise that the structure that obtains higher recognition ratio clearly depends on the database. We know from [23] that graphs in GREC and AIDS have more variability than graphs in the three LETTER databases. From the runtime point of view, it is clear that Degree is the fastest structure. Considering edit costs, lower are values of k_v and k_e, lower is the runtime. Contrarily, lower is β, higher is the runtime.

Table 3. Summary of the achievements of tables 1 and 2

	Highest recognition ratio	**Lowest runtime**
LETTER LOW	Clique, Planar	Degree
LETTER MEDIUM	Clique, Planar	Degree
LETTER HIGH	Planar	Degree
GREC	Eigenvector/PageRank	Degree
AIDS	Eigenvector/PageRank	Degree

5   Conclusions

In this paper, we have shown the relevance of the node centrality on graph edit distance for graph classification. We have presented five different centralities, viz. Clique, Degree, Planar, Eigenvector and PageRank. We have defined them and we have shown their recognition ratio and runtime in 5 different databases and different configurations of Edit costs. With respect to the runtime, it is clear that the Degree is the best. But it is not so clear which one is the best with respect to the recognition ratio. It seems that the method that presents a best balance between recognition ratio and runtime is the Degree centrality but clearly, it depends on the application or database. For this reason, the decision to use a particular centrality can require an accurate application-dependent analysis to maximize the specific goals.

References

1. Sanfeliu, A., Alquézar, R., Andrade, J., Climent, J., Serratosa, F., Vergés, J.: Graph-based Representations and Techniques for Image Processing and Image Analysis. Pattern Recognition 35(3), 639–650 (2002)
2. Conte, D., Foggia, P., Sansone, C., Vento, M.: Thirty Years Of Graph Matching In Pattern Recognition. IJPRAI 18(3), 265–298 (2004)
3. Vento, M.: A One Hour Trip in the World of Graphs, Looking at the Papers of the Last Ten Years. In: Kropatsch, W.G., Artner, N.M., Haxhimusa, Y., Jiang, X. (eds.) GbRPR 2013. LNCS, vol. 7877, pp. 1–10. Springer, Heidelberg (2013)
4. Hancock, E.R., Wilson, R.C.: Pattern analysis with graphs: Parallel work at Bern and York. Pattern Recognition Letters 33(7), 833–841 (2012)
5. Serratosa, F., Cortés, X., Solé-Ribalta, A.: Component Retrieval based on a Database of Graphs for Hand-Written Electronic-Scheme Digitalisation. Expert Systems With Applications, ESWA 40, 2493–2502 (2013)
6. Solé, A., Serratosa, F., Sanfeliu, A.: On the Graph Edit Distance cost: Properties and Applications. International Journal of Pattern Recognition and Artificial Intelligence 26(5) (2012)
7. Serratosa, F., Alquézar, R., Sanfeliu, A.: Estimating the Joint Probability Distribution of Random Vertices and Arcs by Means of Second-Order Random Graphs. In: Caelli, T.M., Amin, A., Duin, R.P.W., Kamel, M.S., de Ridder, D. (eds.) SPR 2002 and SSPR 2002. LNCS, vol. 2396, pp. 252–262. Springer, Heidelberg (2002)
8. Ferrer, M., Valveny, E., Serratosa, F.: Median graphs: A genetic approach based on new theoretical properties. Pattern Recognition 42(9), 2003–2012 (2009)
9. Ferrer, M., Valveny, E., Serratosa, F.: Median graph: A new exact algorithm using a distance based on the maximum common subgraph. Pattern Recognition Letters 30(5), 579–588 (2009)
10. Serratosa, F., Alquézar, R., Sanfeliu, A.: Estimating the Joint Probability Distribution of Random Vertices and Arcs by Means of Second-Order Random Graphs. In: Caelli, T.M., Amin, A., Duin, R.P.W., Kamel, M.S., de Ridder, D. (eds.) SPR 2002 and SSPR 2002. LNCS, vol. 2396, pp. 252–262. Springer, Heidelberg (2002)
11. Serratosa, F., Sanfeliu, A.: Function-Described Graphs applied to 3D object recognition. In: Del Bimbo, A. (ed.) ICIAP 1997. LNCS, vol. 1310, pp. 701–708. Springer, Heidelberg (1997)

12. Serratosa, F.: Fast Computation of Bipartite Graph Matching. Pattern Recognition Letters 45, 244–250 (2014)
13. Serratosa, F.: Speeding up Fast Bipartite Graph Matching trough a new cost matrix, International Journal of Pattern Recognition and Artificial Intelligence 29(2) (2015)
14. Riesen, K., Bunke, H.: Approximate graph edit distance computation by means of bipartite graph matching. Image Vision Comput. 27(7), 950–959 (2009)
15. Munkres, J.: Algorithms for the assignment and transportation problems. Journal of the Society for Industrial & Applied Mathematics 5, 32–38 (1957)
16. Jonker, R., Volgenant, T.: A shortest augmenting path algorithm for dense and sparse linear assignment problems. Computing 38, 325–340 (1987)
17. Cortés, X., Serratosa, F.: An Interactive Method for the Image Alignment problem based on Partially Supervised Correspondence. Expert Systems With Applications 42(1), 179–192 (2015)
18. Serratosa, F., Cortés, X.: Interactive Graph-Matching using Active Query Strategies. Pattern Recognition 48, 1360–1369 (2015)
19. Riesen, K., Bunke, H., Fischer, A.: Improving Graph Edit Distance Approximation by Centrality Measures. In: International Congress on Pattern Recognition (2014)
20. Cortés, X., Serratosa, F.: Learning Graph-Matching Edit-Costs based on the Optimality of the Oracle's Node Correspondences. Pattern Recognition Letters (2015)
21. Levenshtein, V.I.: Binary codes capable of correcting deletions, insertions and reversals. Soviet Physics Doklady, Cybernetics and Control Theory 10, 707–710 (1966)
22. Peris, G., Marzal, A.: Fast Cyclic Edit Distance Computation with Weighted Edit Costs in Classification. In: ICPR 2002, vol. 4, pp. 184–187 (2002)
23. Riesen, K., Bunke, H.: IAM graph database repository for graph based pattern recognition and machine learning. In: da Vitoria Lobo, N., Kasparis, T., Roli, F., Kwok, J.T., Georgiopoulos, M., Anagnostopoulos, G.C., Loog, M. (eds.) S+SSPR 2008. LNCS, vol. 5342, pp. 287–297. Springer, Heidelberg (2008)

A Mixed Weisfeiler-Lehman Graph Kernel

Lixiang Xu[1,2,4], Jin Xie[2,4], Xiaofeng Wang[3], and Bin Luo[1(✉)]

[1] School of Computer Science and Technology
Anhui University, Hefei 230601, Anhui, People's Republic of China
{xulixianghf,luobinahu}@163.com
[2] Department of Mathmatics & Physics
Hefei University, Hefei 230601, Anhui, People's Republic of China
hfuuxiejin@126.com
[3] Department of Computer Science and Technology
Hefei University, Hefei 230601, Anhui, People's Republic of China
xfwang@iim.ac.cn
[4] Institute of Scientific Computing
Hefei University, Hefei 230601, Anhui, People's Republic of China
hfuuxiejin@126.com

Abstract. Using concepts from the Weisfeiler-Lehman (WL) test of isomorphism, we propose a mixed WL graph kernel (MWLGK) framework based on a family of efficient WL graph kernels for constructing mixed graph kernel. This family of kernels can be defined based on the WL sequence of graphs. We apply the MWLGK framework on WL graph sequence taking into account the structural information which was overlooked. Our MWLGK is competitive with or outperforms the corresponding single WL graph kernel on several classification benchmark data sets.

Keywords: graph kernel · Graph classification · Weisfeiler-Lehman algorithm · Mixed graph kernel

1 Introduction

Machine learning in domains such as bioinformatics, drug discovery, and web data mining involves the study of relationships between objects. Graphs are natural data structures to model such relations, with nodes representing objects and edges the relationships between them. Simple ways of comparing graphs are based on pairwise comparison of nodes or edges, and are possible in quadratic time. There exist many other graph similarity measures based on graph isomorphism or related concepts such as subgraph isomorphism or the largest common subgraph. Kernel methods [14] offer a natural framework to study graph comparison. The graph kernels were originally proposed by Gärtner et al. [9] [11], where the structural information of graphs were taken into account. They work by counting the number of common random walks between two graphs. Later, many different graph kernels have been defined, which focus on different types of substructures in graphs, such as random walks [9] [11], shortest path kernel [4],

© Springer International Publishing Switzerland 2015
C.-L. Liu et al. (Eds.): GbRPR 2015, LNCS 9069, pp. 242–251, 2015.
DOI: 10.1007/978-3-319-18224-7_24

subtrees kernel [13], edge kernel [17], graphlet kernels [15], Weisfeiler-Lehman graph kernels [16], and quantum Jensen-Shannon graph kernel [3].

Several studies have recently shown that these graph kernels can achieve results competitive with the state of the art on benchmark data sets from bioinformatics and chemistry. For example, Borgwardt et al. [5] presented a graph model for proteins and defined a protein graph kernel that measures similarity between these graphs, and successfully tested the performance of this classifier on two function prediction tasks. The next year, they [6] extended common concepts from linear algebra to Reproducing Kernel Hilbert Spaces, and used these extensions to define a unifying framework for random walk kernels. Moreover, Vishwanathan et al. [17] presented a unified framework to study graph kernels, special cases of which included the random walk and marginalized graph kernels [10] [12]. Through reduction to a Sylvester equation they improved the time complexity of kernel computation between unlabeled graphs with n vertices. However, aforementioned state-of-the-art graph kernels did not scale to large graphs with hundreds of nodes and thousands of edges. So Shervashidze et al. [15] proposed efficient graph kernels based on counting or sampling limited size subgraphs in a graph. To consider higher order relations in the neighborhood to iteratively compute a kernel matrix, Camps-Valls et al. [12] presented a graph kernel for spatio-spectral remote sensing image classification with support vector machines. Another algebraic approach to graph kernels was appeared in [8], it showed how to represent scenes as graphs that encode models and their semantic relationships. To deal with high dimensional data and measure the mutual information between pairs of graphs, Bai et al. [2] showed how to construct Jensen Shannon kernels for graph data sets using the von-Neumann entropy and Shannon entropy.

However, aforementioned graph kernel methods rarely evolved to mixture of graph kernels. The mixed graph kernel can enhance classification accuracy as compared to single graph kernel that takes into account one aspect or part of the structural information only. Furthermore, the mixed graph kernel may flexibly capture structure information among the subtree, edge and shortest path in graph data sets in the classification.

The main contributions of this paper are as follows. Firstly, using concepts from the WL test of isomorphism, we propose a MWLGK framework based on a family of efficient WL graph kernels. This family of kernels is defined based on the WL sequence of graphs. Secondly, we apply the MWLGK framework on WL graph sequence taking into account the structural information which was overlooked. Thirdly, key to our approach is the WL test of isomorphism and mixed graph kernel, which allow us to compute a sequence of graphs which capture the topological and label information of the original graph. It may flexibly capture structure information among the subtree, edge and shortest path in graph data sets in the classification tasks.

The remainder of this article is structured as follows. In Section 2, we describe the related works, including the WL graph sequence and the general graph kernel framework based on them. In Section 3, we first describe three instances of WL graph kernels. We then propose a mixed graph kernel framework based on a family of three instances of efficient WL graph kernels for graphs. In Section 4, we compare MWLGK to these single WL graph kernels. We report results on kernel computation efficiency and classification accuracy on graph benchmark data sets. Section 5 summarizes conclusion and future work.

2   Related Works

In order to know the principle of mixed WL graph kernel, we must make clear the concepts below. In this section, we define the WL graph sequence and the general graph kernel framework based on them. The content of this section refers mainly to the work in [16]. Our graph kernels use concepts from the WL test of isomorphism [18]. Assume we are given two graphs G and G' and we would like to test whether they are isomorphic. The 1-dimensional WL test is an iterative algorithm described in [16]. The key idea of the algorithm is to augment the node labels by the sorted set of node labels of neighbouring nodes, and compress these augmented labels into new, short labels. These steps are then repeated until the node label sets of G and G' differ, or the number of iterations reaches n. In each iteration i of the WL algorithm (see Algorithm 1 [16]), we get a new labeling for all nodes v. Recall that this labeling is concordant in G and G' meaning that if nodes in G and G' have identical multi-set labels, and only in this case, they will get identical new labels. Therefore, we can imagine that one iteration of WL relabeling as a function $r(V, E, l_i) = (V, E, l_{i+1})$ that transforms all graphs in the same manner. Note that r depends on the set of graphs that we consider. The 1-dimensional WL algorithm has been shown to be a valid isomorphism test for almost all graphs [1].

Definition 1. Define the WL graph at height i of the graph $G = (V, E, l) = G = (V, E, l_0)$ as the graph $G_i = (V, E, l_i)$. We call the sequence of WL graphs

$$\{G_0, G_1, ..., G_h\} = \{(V, E, l_0), (V, E, l_1), ..., (V, E, l_h)\}, \tag{1}$$

where $G_0 = G$ and $l_0 = l$, the WL sequence up to height h of G. G_0 is the original graph, $G_1 = r(G_0)$ is the graph resulting from the first relabeling, and so on. Note that neither V, nor E ever change in this sequence, but we define it as a sequence of graphs rather than a sequence of labeling functions for the sake of clarity of definitions that follow.

Definition 2. Let k be any kernel for graphs, that we will call the base kernel. Then the WL graph kernel with h iterations with the base kernel k is defined as

$$k_{WL}^{(h)}(G, G') = k(G_0, G_0') + k(G_1, G_1') + ... + k(G_h, G_h'), \tag{2}$$

where h is the number of WL iterations and $\{G_0, G_1, ..., G_h\}$ and $\{G_0', G_1', ..., G_h'\}$ are the WL sequences of G and G', respectively.

3   The Mixed Graph Kernels Framework

In this section, we first present three instances of WL graph kernels, the WL Subtree Kernel (WLSK), the WL Edge Kernel (WLEK) and the WL Shortest Path Kernel (WLSPK).We then propose a mixed graph kernels framework based on a family of three instances of efficient WL graph kernels for graphs.

3.1   The WL Subtree Kernel

In this section we present the WL subtree kernel, which is a natural instance of Definition 2.

Definition 3. Let G and G' be graphs. Define $\sum_i \subseteq \sum$ as the set of letters that occur as node labels at least once in G or G' at the end of the i-th iteration of the WL algorithm. Let $\sum_0$ be the set of original node labels of G and G'. Assume all $\sum_i$ are pairwise disjoint. Without loss of generality, assume that every $\sum_i = \{\sigma_{i1}, \sigma_{i2}, ..., \sigma_{i|\sum_i|}\}$ is ordered. Define a map $c_i : \{G, G'\} \times \sum_i \to N$ such that $c_i(G, \sigma_{ij})$ is the number of occurrences of the letter σ_{ij} in the graph G.

The WL subtree kernel on two graphs G and G' with h iterations is defined as:

$$k^{(h)}_{WLsubtree}(G, G') = \langle \phi^{(h)}_{WLsubtree}(G), \phi^{(h)}_{WLsubtree}(G') \rangle, \tag{3}$$

where

$$\phi^{(h)}_{WLsubtree}(G) = (c_0(G, \sigma_{01}), ..., c_0(G, \sigma_{0|\sum_0|}), ..., c_h(G, \sigma_{h1}), ..., c_h(G, \sigma_{h|\sum_h|})), \tag{4}$$

and

$$\phi^{(h)}_{WLsubtree}(G') = (c_0(G', \sigma_{01}), ..., c_0(G', \sigma_{0|\sum_0|}), ..., c_h(G', \sigma_{h1}), ..., c_h(G', \sigma_{h|\sum_h|})). \tag{5}$$

That is, the WL subtree kernel counts common original and compressed labels in two graphs.

The following Lemma 1 shows that (3) is indeed a special case of the general WL graph kernel (2).

Lemma 1. Let the base kernel k be a function counting pairs of matching node labels in two graphs:

$$k(G, G') = \sum_{v \in V} \sum_{v' \in V'} \delta(l(v), l(v')), \tag{6}$$

where δ is the Dirac kernel, that is, it is 1 when its arguments are equal and 0 otherwise. Then $k^{(h)}_{WL}(G, G') = k^{(h)}_{WLsubtree}(G, G')$ for all G, G'.

3.2   The WL Edge Kernel

The WL edge kernel is another instance of the WL graph kernel framework. In the case of graphs with unweighted edges, we consider the base kernel that counts with identically labeled endpoints in two graphs. In other words, the base kernel is defined as

$$k_E = \langle \phi_E(G), \phi_E(G') \rangle, \tag{7}$$

where $\phi_E(G)$ is a vector of numbers of occurrences of pairs $(a, b), a, b \in \sum$ which represent ordered labels of endpoints of an edge in G. Denoting (a, b) and (a', b') the ordered labels of endpoints of edges e and e' respectively, and σ the Dirac kernel, k_E can equivalently be expressed as $\sum_{e \in E} \sum_{e' \in E'} \delta(a, a')\delta(b, b')$. If the edges are weighted by a function w that assigns weights, the base kernel k_E can be defined as $k(G, G') = \sum_{e \in E} \sum_{e' \in E'} \delta(a, a')\delta(b, b')k_w(w(e), w(e'))$, where k_w is a kernel comparing edge weights. Following (2), we have

$$k_{WLedge}^{(h)}(G) = k_E(G_0, G_0') + k_E(G_1, G_1') + \ldots + k_E(G_h, G_h'). \tag{8}$$

3.3   The WL Shortest Path Kernel

Another example of the general WL graph kernels that we consider is the WL shortest path kernel. Here we use a node-labeled shortest path kernel as the base kernel [4]. In the particular case of graphs with unweighted edges, we consider the base kernel k_{sp} of the form

$$k_{SP}(G, G') = \langle \phi_{SP}(G), \phi_{SP}(G') \rangle, \tag{9}$$

where $\phi_{SP}(G)$ (resp. $\phi_{SP}(G')$) is a vector whose components are numbers of occurrences of triplets of the form (a, b, p) in G (resp. G'), where $a, b \in \sum$ are ordered endpoint labels of a shortest path and $p \in N_0$ is the shortest path length. According to (2), we have

$$k_{WLshortestpath}^{(h)}(G) = k_{SP}(G_0, G_0') + k_{SP}(G_1, G_1') + \ldots + k_{SP}(G_h, G_h'). \tag{10}$$

3.4   The Mixed WL Graph Kernel

In the following, we introduce the mixed WL graph kernel.

Mixed WL Graph Kernel. A graph G consists of a set of nodes (or vertices) V and edges E. In this article, n denotes the number of nodes in a graph and m the number of edges in a graph. Key to our exposition is the notion of a complement graph which we define below. Because each WL graph kernel can capture a different features. However, existing single WL graph kernels cannot take into account all features of graph data sets. Based on three instances of efficient WL graph kernels for graphs, we propose to modify them appropriately. In the following, using the concept of a complement graph, we present a unifying framework which includes the above mentioned three instances of efficient WL graph kernels as special cases. We now define a novel graph kernel called the mixed graph kernel.

$$k_{comp}^{(h)}(G, G') = \alpha \cdot k_{WLsubtree}^{(h)}(G_h, G_h') + \beta \cdot k_{WLedge}^{(h)}(G_h, G_h') \qquad (11)$$
$$+\gamma \cdot k_{WLshortestpath}^{(h)}(G_h, G_h'),$$

where α, β, γ is weight, $\alpha + \beta + \gamma = 1$. Although this kernel seems simple minded at first, it is in fact rather useful. This simple kernel improves substantial gains in performance in our experiments comparing corresponding single WL graph kernel.

4   Experiments

Here we compare the performance of the WLSK, the WLEK and the WLSPK to the proposed Mixed WL Graph Kernel in terms of classification accuracy and MSE on graph benchmark data sets. These data sets include MUTAG, EN-ZYMES, PTC, NCI1 and NCI109. Some statistics concerning the data sets are given in Table 1. Our WL graph kernel matrix is computed from the description of Shervashidze [16].

MUTAG. The MUTAG benchmark is based on graphs representing 188 chemical compounds, and aims to predict whether each compound possesses mutagenicity. The maximum and average number of vertices are 28 and 17.93 respectively. As the vertices and edges of each compound are labeled with a real number, we transform these graphs into unweighted graphs.

ENZYMES. The ENZYMES data set is a data set based on graphs representing protein tertiary structures consisting of 600 enzymes from the BRENDA enzyme database. In this case the task is to correctly assign each enzyme to one of the 6 EC top-level classes. The maximum and average number of vertices are 126 and 32.63 respectively.

PTC. The Predictive Toxicology Challenge data set reports the carcinogenicity of several hundred chemical compounds for Male Mice (MM), Female Mice (FM), Male Rats (MR) and Female Rats (FR).

NCI1 and NCI109. The NCI1 and NCI109 data sets consist of graphs representing two balanced subsets of data sets of chemical compounds screened for activity against non-small cell lung cancer and ovarian cancer cell lines respectively. There are 4110 and 4127 graph based structures in NCI1 and NCI109 respectively. The maximum, minimum and average number of vertices in NCI1 and NCI109 are 111, 3 and 29.87, and 111, 4 and 29.68 respectively.

In the experiments, we perform 10-fold cross-validation of C-Support Vector Machine Classification using LIBSVM [7], using 9 folds for training and 1 for testing. All parameters of the SVM were optimised on the training data sets only. To exclude random effects of fold assignments, we repeat the whole experiment 20 times. We report average prediction accuracies and standard deviations in Tables 2.

Table 1. Information of the Graph based Data sets

Data Set	Size	Classes	Avg Nodes	Avg Edges
MATUG	188	2	17.9	39.5
ENZYME	600	6	32.6	124.2
PTC	344	2	25.5	51.9
NCI1	4110	2	29.8	64.6
NCI109	4127	2	29.6	62.2

Due to the large number of graph kernels in the literatures, we could not compare every graph kernel. But the representative instances of the major families of graph kernels were compared in [16]. On these data sets, we only compared our MWLGK to the corresponding WL graph kernels.

We chose h for our all the WL graph kernels and mixed WL graph kernel by cross-validation on the training data sets for $h = 0, 1, 2, 5, 10, 20$ in MUTAG and ENZYMES data sets, which means that we compute 6 different WL graph kernel matrices in each experiment. We report the total classification accuracy of these computations. It is worth mentioning that small values of h, such as 2 or 5, systematically can give the best results for all data sets used, and the accuracy obtains relatively stable value. When h equals 5, 10 or more larger, this kernel is more computationally expensive. However, While h equals 0 or 1, classification accuracy happen phenomenon of tottering and instability (see Fig. 1). So for the rest of the experimental data sets, we only need to choose $h = 2$.

In each experiment, Our MWLGK use two of the three WL graph kernels for reducing the complexity of the calculation in this paper (i.e., $k_{comp} = (1-\alpha)K_1 + \alpha K_2, (0 \leq \alpha \leq 1)$, where α is the mixed coefficient. K_1, K_2 denotes the WLSK or WLEK or WLSPK, respectively. If some instances of the MWLGK can't obtain better performance, they are not showed and discussed in this paper. In Fig. 2(a) we show the efficiency of mixing a WLSK and a WLEK on PTC, where K_1 denotes WLSK, K_2 denotes WLEK. The classification accuracy of the WLEK reaches higher accuracy than the WLSK. Fig. 2(a) shows that the mixed kernel

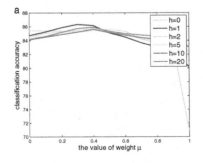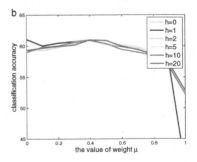

Fig. 1. a. The mixture of WLSPK and WLSK on graph data set MUTAG, b. The mixture of WLSPK and WLSK on graph data set ENZYMES.

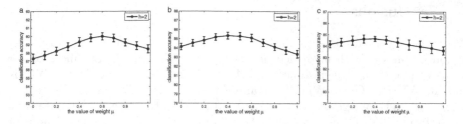

Fig. 2. a. The mixture performance of WLSK and WLEK on PTC, b. The mixture of WLEK and WLSPK on NCI1, c. The mixture of WLEK and WLSPK on NCI109. Error bars correspond to standard errors.

not only has a WLSK efficiency, but also a WLEK effect. By increasing the mixing coefficient, the influence of the WLEK is also increased. Furthermore, the results have the highest accuracy using the mixed weight α in the range [0.5, 0.7].

In Fig. 2(b) and Fig. 2(c) we show the efficiency of mixing a WLEK and a WLSPK on NCI1 and NCI109, where K_1 denotes WLEK, K_2 denotes WLSPK. The classification accuracy of the WLEK reaches higher accuracy than the WLSPK on two data sets. Fig. 2(b) and Fig. 2(c) show that the results have the highest accuracy using the mixed weight α in the range [0.3, 0.5].

Furthermore, from the Fig. 2, we obtain the influence of WLEK is stronger in WL graph kernels. At the same time, the highest accuracy of MWLGK is higher than the corresponding single WL graph kernel.

Our MWLGK is competitive with or outperform corresponding single WL graph kernel on several classification benchmark data sets. Even the MWLGK can reach the highest accuracy level on all data sets. In our experimental evaluation, our mixed WL graph kernel framework is an attempt to applications of the mixture of graph kernels in various disciplines such as computational biology and social network analysis. To summarize, the MWLGK turns out to be competitive in terms of classification accuracy on all data sets, and its accuracy levels are highest on all data sets in experiments.

Table 2. Prediction accuracy (± standard deviation) on graph classification benchmark data sets (- : the bad performance of the WL graph kernels)

Kernel	MUTAG	ENZYMES	PTC	NCI1	NCI109
WLSK	82.34±0.57	52.40±1.23	57.32±0.56	-	-
WLEK	-	-	58.47±0.48	84.13±0.35	84.16±0.29
WLSPK	84.53±0.49	59.26±1.17	-	83.36±0.41	83.59±0.37
MWLGK	85.76±0.47	60.93±1.09	60.00±0.43	85.34±0.37	84.64±0.34

5   Conclusions

We define a general mixed graph kernel framework for constructing graph kernel on graph sets with unlabeled or discretely labeled nodes. Instances of our framework base on three instances of WL graph kernels, the WLSK, the WLEK and the WLSPK. Our MWLGK is competitive in terms of accuracy with corresponding single WL graph kernel on several classification benchmark data sets. In the future, we will be to consider the MWLGK on graphs with continuous or high-dimensional node labels and their efficient computation.

Acknowledgement. This work was supported by National High Technology Research and Development Program (863 Program) of China under Grant (2014AA015104), National Nature Science Foundation of China (61472002), Key Project of Scientific Research, Education Department of Anhui Province (KJ2014ZD30), Key Construction Disciplines of Applied Mathematics of Hefei University (2014XK08), Anhui Provincial Natural Science Foundation (1308085MF84), Training Object Project for Academic Leader of Hefei University (2014dtr08), Natural Science Research Project of Colleges of Anhui Province (KJ2013B234), MOE Youth Project of Humanities and Social Sciences (14YJCZH169).

References

1. Babai, L., Kucera, L.: Canonical labelling of graphs in linear average time. In: 20th Annual Symposium on Foundations of Computer Science, pp. 39–46. IEEE (1979)
2. Bai, L., Hancock, E.R.: Graph clustering using the jensen-shannon kernel. In: Real, P., Diaz-Pernil, D., Molina-Abril, H., Berciano, A., Kropatsch, W. (eds.) CAIP 2011, Part I. LNCS, vol. 6854, pp. 394–401. Springer, Heidelberg (2011)
3. Bai, L., Hancock, E.R., Torsello, A., Rossi, L.: A quantum jensen-shannon graph kernel using the continuous-time quantum walk. In: Kropatsch, W.G., Artner, N.M., Haxhimusa, Y., Jiang, X. (eds.) GbRPR 2013. LNCS, vol. 7877, pp. 121–131. Springer, Heidelberg (2013)
4. Borgwardt, K.M., Kriegel, H.-P.: Shortest-path kernels on graphs. In: Fifth IEEE International Conference on Data Mining, 8 p. IEEE (2005)
5. Borgwardt, K.M., Ong, C.S., Schönauer, S., Vishwanathan, S.V.N., Smola, A.J., Kriegel, H.-P.: Protein function prediction via graph kernels. Bioinformatics 21(suppl. 1), i47–i56 (2005)
6. Borgwardt, K.M., Schraudolph, N.N., Vishwanathan, S.V.N.: Fast computation of graph kernels. In: Advances in Neural Information Processing Systems, pp. 1449–1456 (2006)
7. Chang, C.-C., Lin, C.-J.: Libsvm: a library for support vector machines. ACM Transactions on Intelligent Systems and Technology (TIST) 2(3), 27 (2011)
8. Fisher, M., Savva, M., Hanrahan, P.: Characterizing structural relationships in scenes using graph kernels. ACM Transactions on Graphics (TOG) 30, 34 (2011)
9. Gärtner, T., Flach, P., Wrobel, S.: On Graph Kernels: Hardness Results and Efficient Alternatives. In: Schölkopf, B., Warmuth, M.K. (eds.) COLT/Kernel 2003. LNCS (LNAI), vol. 2777, pp. 129–143. Springer, Heidelberg (2003)
10. Kashima, H., Tsuda, K., Inokuchi, A.: Marginalized kernels between labeled graphs. In: ICML, vol. 3, pp. 321–328 (2003)

11. Kashima, H., Tsuda, K., Inokuchi, A.: Kernels for graphs. Kernel Methods in Computational Biology 39(1), 101–113 (2004)
12. Mahé, P., Ueda, N., Akutsu, T., Perret, J.-L., Vert, J.-P.: Extensions of marginalized graph kernels. In: Proceedings of the Twenty-First International Conference on Machine Learning, p. 70. ACM (2004)
13. Ramon, J., Gärtner, T.: Expressivity versus efficiency of graph kernels. In: First International Workshop on Mining Graphs, Trees and Sequences, pp. 65–74. Citeseer (2003)
14. Schölkopf, B., Smola, A.J.: Learning with kernels: support vector machines, regularization, optimization, and beyond. MIT Press (2002)
15. Shervashidze, N., Petri, T., Mehlhorn, K., Borgwardt, K.M., Vishwanathan, S.V.N.: Efficient graphlet kernels for large graph comparison. In: International Conference on Artificial Intelligence and Statistics, pp. 488–495 (2009)
16. Shervashidze, N., Schweitzer, P., Leeuwen, E.J.V., Mehlhorn, K., Borgwardt, K.M.: Weisfeiler-lehman graph kernels. The Journal of Machine Learning Research 12, 2539–2561 (2011)
17. Vichy, S., Vishwanathan, N., Schraudolph, N.N., Kondor, R., Borgwardt, K.M.: Graph kernels. The Journal of Machine Learning Research 11, 1201–1242 (2010)
18. Ju Weisfeiler, B., Leman, A.A.: Reduction of a graph to a canonical form and an algebra which appears in the process. NTI, Ser. 2(9), 12–16 (1968)

A Quantum Jensen-Shannon Graph Kernel Using Discrete-Time Quantum Walks

Lu Bai[1,2], Luca Rossi[3], Peng Ren[4], Zhihong Zhang[5(✉)],
and Edwin R. Hancock[2]

[1] School of Information, Central University of Finance and Economics, Beijing, China
[2] Department of Computer Science, University of York, York, UK
[3] School of Computer Science, University of Birmingham, Birmingham, UK
[4] College of Information and Control Engineering, China University of Petroleum
(Huadong), Shandong Province, Qingdao, P.R. China
[5] Software school, Xiamen University, Fujian, Xiamen, China
zhihong@xmu.edu.cn

Abstract. In this paper, we develop a new graph kernel by using the quantum Jensen-Shannon divergence and the discrete-time quantum walk. To this end, we commence by performing a discrete-time quantum walk to compute a density matrix over each graph being compared. For a pair of graphs, we compare the mixed quantum states represented by their density matrices using the quantum Jensen-Shannon divergence. With the density matrices for a pair of graphs to hand, the quantum graph kernel between the pair of graphs is defined by exponentiating the negative quantum Jensen-Shannon divergence between the graph density matrices. We evaluate the performance of our kernel on several standard graph datasets, and demonstrate the effectiveness of the new kernel.

1 Introduction

Graph based representations are powerful tools for structural analysis in pattern recognition. One challenge of classifying graphs is how to convert the discrete graph structures into numeric features. One way is to use graph kernels. The main advantage of using graph kernels is that they characterize graph features in a high dimensional space and thus better preserve graph structures.

Generally speaking, a graph kernel is a similarity measure between a pair of graphs [1]. To extend the large spectrum of kernel methods from the general machine learning domain to the graph domain, Haussler [2] has proposed a generic way, namely the R-convolution, to define a graph kernel. For a pair of graphs, an R-convolution kernel is computed by decomposing each graph into smaller subgraphs and counting the number of isomorphic subgraph pairs between the two original graphs. Thus, a new type of decomposition of a graph usually results in a new graph kernel. Following this scenario, Kashima et al. [3] introduced the random walk kernel, which is based on the enumeration of common random walks between two graphs. Borgwardt et al. [4], on the other hand, proposed a shortest path kernel by counting the numbers of matching shortest paths over the graphs.

© Springer International Publishing Switzerland 2015
C.-L. Liu et al. (Eds.): GbRPR 2015, LNCS 9069, pp. 252–261, 2015.
DOI: 10.1007/978-3-319-18224-7_25

Aziz et al. [5] introduced a backtrackless kernel using the cycles identified by the Ihara zeta function [6] in a pair of graphs. Shervashidze et al. [7] developed a fast subtree kernel by comparing pairs of subtrees identified by the Weisfeiler-Lehman (WL) algorithm. Some other alternative R-convolution kernels include a) the segmentation graph kernel developed by Harchaoui and Bach [8], b) the point cloud kernel developed by Bach [9], and c) the (hyper)graph kernel based on directed subtree isomorphism tests [10].

Recently, a number of alternative graph kernel measures have been introduced in the literature. These are based on the computation of the mutual information between two graphs in terms of the classical Jensen-Shannon divergence. In information theory, the classical Jensen-Shannon divergence is a dissimilarity measure between probability distributions. In [11–13], we have used the classical Jensen-Shannon divergence to define a Jensen-Shannon graph kernel. Unlike the R-convolution kernels that count the number of isomorphic substructure pairs, the Jensen-Shannon graph kernel is defined in terms of the entropy difference between a pair of graphs and their composite graph (e.g., the disjoint union graph or the product graph formed by the pair of graphs). Here, the entropy of a graph can be either the von Neumann entropy (associated with the graph spectrum information) or the Shannon entropy (associated with the steady state random walk or the information functional). Both the von Neumann entropy and the Shannon entropy of a graph can be directly computed without the need to decompose the graph into substructures. As a result, the Jensen-Shannon graph kernel avoids the computational burden of comparing all pairs of substructures for a pair of graphs. To develop this work further, in [14–16] we have introduced a new quantum Jensen-Shannon graph kernel using the quantum Jensen-Shannon divergence [17, 18] and the continuous-time quantum walk [19]. Here the basic idea is to associate with each graph a mixed quantum state representing the time evolution of a quantum walk. The kernel between a pair of graphs is then defined as the quantum Jensen-Shannon divergence between their corresponding density matrices [14].

The aim of this paper is to develop our previous quantum Jensen-Shannon kernels [14–16] one step further. We propose a new quantum Jensen-Shannon kernel for graphs using the discrete-time quantum walk. The discrete-time quantum walk is the quantum analogue of the discrete-time classical random walk [19]. Remarkably, the discrete-time quantum walk possesses a number of interesting properties that are not exhibited by its classical counterpart. In fact, the behaviour of the discrete-time quantum walk is governed by a unitary matrix rather than a stochastic matrix, as in the case of the classical random walk. As a consequence, its evolution is reversible and non-ergodic. However, unlike the continuous-time quantum walk, where the state space is the graph vertex set, the state space of the discrete-time quantum walk is the set of arcs residing on the graph edges. More specifically, given an undirected graph $G(V, E)$, each edge $\{u, v\} \in E$ is replaced by a pair of directed arcs (u, v) and (v, u), and the set of arcs is denoted by E_d. Then, the state space for the discrete-time quantum walk is the set of arcs E_d. We are interested in developing a new quantum Jensen-Shannon kernel where the graph structure is probed by means of a discrete-time quantum walk. To commence, we perform a discrete-time quantum walk on each graph and we compute a mixed quantum state represented by a density matrix. With the density matrices for a pair of graphs to hand, the quantum graph kernel between the pair of graphs is defined by exponentiating the negative quantum Jensen-Shannon divergence between the graph density matrices. We evaluate the performance of our new kernel on several standard graph datasets from both bioinformatics and computer vision. The experimental results demonstrate the effectiveness of the proposed graph kernel. Our new kernel is shown to be competitive to state-of-the-art graph kernels.

2   Quantum Mechanical Background

In this section, we introduce the quantum mechanical formalism that will be used in this work. We commence by reviewing the concept of discrete-time quantum walk on a graph. Furthermore, we describe how to associate with each graph a density matrix describing the quantum walk evolution. Then, we show how to compute the von Neumann entropy of a graph through its density matrix. Finally, we discuss the relationship between the Perron-Frobenius operator [20] and the transition matrix of the discrete-time quantum walk, and thus explain the advantage of the discrete-time quantum walk over its continuous-time version.

2.1   Discrete-Time Quantum Walks

The discrete-time quantum walk is the quantum counterpart of the discrete-time classical random walk [21]. To simulate the evolution of a discrete-time quantum walk on a graph $G(V, E)$, we first replace each edge $e(u, v) \in E$ with a pair of directed arcs $e_d(u, v)$ and $e_d(v, u)$. This in turn ensures the reversibility of the quantum process. Let us denote the new set of arcs as E_d. Then, the state space for the discrete-time quantum walk is E_d, and we denote the state corresponding to the walker being on the arc $e_d(u, v)$ as $|uv\rangle$. A general state of the walk is

$$|\psi\rangle = \sum_{e_d(u,v) \in E_d} \alpha_{uv} |uv\rangle, \tag{1}$$

where the quantum amplitudes α_{uv} are complex, i.e., $\alpha_{uv} \in \mathbb{C}$. The probability that the walk is in the state $|uv\rangle$ is given by $\Pr(|uv\rangle) = \alpha_{uv} \alpha_{uv}^*$, where α_{uv}^* is the complex conjugate of α_{uv}.

The evolution of the state vector between the steps t and $t+1$ is determined by the transition matrix U. The entries of U determine the transition probabilities between states, i.e., $|\psi_{t+1}\rangle = U |\psi_t\rangle$. Since the evolution of the walk is linear and conserves probability, the matrix U must be unitary, i.e., $U^{-1} = U^\dagger$, where $U^\dagger$ denotes the Hermitian transpose of U.

It is usual to adopt the Grover diffusion matrix [22] as the transition matrix. Using the Grover diffusion matrices, the transition matrix U has entries

$$U_{(u,v),(w,x)} = \begin{cases} \frac{2}{d_x} - \delta_{ux}, & v = w; \\ 0, & \text{otherwise,} \end{cases} \tag{2}$$

where $U_{(u,v),(w,x)}$ gives the quantum amplitude for the transition $e_d(u,v) \rightarrow e_d(w,x)$ and δ_{ux} is the Kronecker delta, i.e., $\delta_{ux} = 1$ if $u = x$ and 0 otherwise. Given a state $|u_1 v\rangle$, the Grover matrix assigns the same amplitudes to all transitions $|u_1 v\rangle \rightarrow |vu_i\rangle$, and a different amplitude to the transition $|u_1 v\rangle \rightarrow |vu_1\rangle$, where u_i denotes a neighbour of v. Finally, note that although the entries of U are real, they can be negative as well as positive. It is important to stress that, as a consequence of this, negative quantum amplitudes can arise during the evolution of the walk. In other words, the definition in Eq.(2) allows *destructive interference* to take place.

2.2   A Density Matrix from the Mixed State

In quantum mechanics, a pure state can be described as a single ket vector. A quantum system, however, can also be in a mixed state, i.e., a statistical ensemble of pure quantum states $|\psi_i\rangle$, each with probability p_i. The density matrix (or density operator) of such a system is defined as

$$\rho = \sum_i p_i |\psi_i\rangle \langle \psi_i| \tag{3}$$

Assume a sample graph $G(V,E)$. Let $|\psi_t\rangle$ denote the state corresponding to a discrete-time quantum walk that has evolved from the step $t = 0$ to the step $t = T$. We define the time-averaged density matrix ρ_G^T for $G(V,E)$ as

$$\rho_G^T = \frac{1}{T+1} \sum_{t=0}^{T} |\psi_t\rangle \langle \psi_t|. \tag{4}$$

Since $|\psi_t\rangle = U^t |\psi_0\rangle$, where U is the transition matrix of the discrete-time quantum walk, Eq.(4) can be re-written in terms of the initial state $|\psi_0\rangle$ as

$$\rho_G^T = \frac{1}{T+1} \sum_{t=0}^{T} (U)^t |\psi_0\rangle \langle \psi_0| (U^\top)^t. \tag{5}$$

As a result, the density matrix ρ_G^T describes a quantum system that has an equal probability of being in each of the pure states defined by the evolution of the discrete-time quantum walk from step $t = 0$ to step $t = T$.

2.3   The von Neumann Entropy of a Graph

In quantum mechanics, the *von Neumann entropy* [23] H_N of a density matrix ρ is defined as $H_N = -\mathrm{tr}(\rho \log \rho) = -\sum_i \xi_i \ln \xi_i$, where $\xi_1, \ldots, \xi_n$ denote the eigenvalues of ρ. Note that if the quantum system is in a pure state $|\psi_i\rangle$ with probability $p_i = 1$, then the Von Neumann entropy $H_N(\rho) = -\mathrm{tr}(\rho \log \rho)$ is zero. On the other hand, a mixed state generally has a non-zero Von Neumann entropy associated with its density matrix. Here we propose to compute the von

Neumann entropy for each graph using the density matrix defined in Eq.(5). Consider a graph $G(V, E)$, the von Neumann entropy of $G(V, E)$ is defined as

$$H_N(\rho_G) = -\text{tr}(\rho_G^T \log \rho_G^T) = -\sum_j^{|V|} \lambda_j^G \log \lambda_j^G, \tag{6}$$

where $\lambda_1^G, \ldots, \lambda_j^G, \ldots, \lambda_{|V|}^G$ are the eigenvalues of ρ_G^T.

2.4   Relation to the Perron-Frobenius Operator

In [20], Ren et al. have demonstrated that the Perron-Frobenius operator can be represented in terms of the transition matrix of discrete-time quantum walks. To show this connection, we first introduce the definitions of the directed line graph and the positive support of a matrix.

Definition 1. For a sample graph $G(V, E)$, the directed line graph $OLG(V_L, E_{dL})$ is a dual representation of $G(V, E)$. To obtain $OLG(V_L, E_{dL})$, we first construct the associated symmetric digraph $SDG(V, E_d)$ of $G(V, E)$, by replacing every edge $e(u, w) \in E(G)$ by a pair of reverse arcs, i.e., directed edges $e_d(u, w) \in E_d(G)$ and $e_d(w, u) \in E_d(G)$ for $u, w \in V$. The directed line graph $OLG(V_L, E_{dL})$ is the directed graph with vertex set V_L and arc set E_{dL} defined as follows

$$\begin{aligned} V_L &= E_d(SDG), \\ E_{dL} &= \{(e_d(u, v), e_d(v, w)) \in E_d(SDG) \times E_d(SDG) \mid u, v, w \in V, u \neq w\}. \end{aligned} \tag{7}$$

The Perron-Frobenius operator $\boldsymbol{T} = [T_{i,j}]_{|V_L| \times |V_L|}$ of $G(V, E)$ is the adjacency matrix of the associated directed line graph $OLG(V_L, E_{dL})$.   □

Definition 2. The positive support $\text{S}^+(\boldsymbol{M}) = [s_{i,j}]_{m \times n}$ of the matrix $\boldsymbol{M} = [M_{i,j}]_{m \times n}$ is defined to be a matrix with entries

$$s_{i,j} = \begin{cases} 1, & M_{i,j} > 0, \\ 0, & \text{otherwise}, \end{cases} \tag{8}$$

where $1 \leq i \leq m, 1 \leq j \leq n$.   □

Based on the definition in [20], we can re-define the Perron-Frobenius operator in terms of the unitary matrix of the discrete-time quantum walk. Let $G(V, E)$ be a sample graph and $\boldsymbol{U}$ be the unitary matrix associated with the discrete-time quantum walk on $G(V, E)$. The Perron-Frobenius operator $\boldsymbol{U}$ of $G(V, E)$ is

$$\boldsymbol{T} = \boldsymbol{S}^+(\boldsymbol{U}^\top). \tag{9}$$

Def.1, Def.2 and Eq.(9) show us how the discrete-time quantum walk and the Perron-Frobenius operator (i.e., the directed line graph) are co-related. For a graph $G(V, E)$ and its directed line graph $OLG(V_L, E_{dL})$, V_L is just the state space of the discrete-time quantum walk on $G(V, E)$, i.e., each vertex in $OLG(V_L, E_{dL})$ corresponds to a unique directed arc residing on the corresponding edge in $G(V, E)$. Moreover, if there is a directed edge from a vertex $v_L \in V_L$

to a vertex $u_L \in V_L$, the transition of the quantum walk on $G(V, E)$ is allowed from the arc corresponding to v_L to the arc corresponding to u_L, and vice versa. As a result, the discrete-time quantum walk on a graph can also be seen as a walk performed on its directed line graph. The state space of the walk is the vertex set of the line graph, and the transition of the walk relies on the connections between pairs of vertices in the line graph.

Furthermore, in [10, 20], we have observed that the directed line graph of a graph possesses some interesting properties that are not available on the original graph. For instance, compared to the original graph the line graph spans a higher dimensional feature space and thus exposes richer graph characteristics. This is because the cardinality of the vertex set for the line graph is much greater than, or at least equal to, that of the original graph. This property suggests that the discrete-time quantum walk may reflect richer graph characteristics than the continuous-time quantum walk on the original graph.

Finally, since the discrete-time quantum walk can be seen as a walk on the line graph and the state space of the walk is the vertex set of the line graph, we propose to use the rooting of the in-degree distribution of the line graph as the initial state of the discrete-time quantum walk.

3   A Quantum Jensen-Shannon Graph Kernel

In this section, we develop a new quantum Jensen-Shannon kernel for graphs by using the quantum Jensen-Shannon divergence and the discrete-time quantum walk. We commence by reviewing the concept of the classical and quantum Jensen-Shannon divergence. Finally, we give the definition of the new kernel.

3.1   Classical and Quantum Jensen-Shannon Divergence

The classical Jensen-Shannon divergence is a non-extensive mutual information measure defined between probability distributions. Consider two (discrete) probability distributions $\mathcal{P} = (p_1, \ldots, p_a, \ldots, p_A)$ and $\mathcal{Q} = (q_1, \ldots, q_b, \ldots, q_B)$, then the classical Jensen-Shannon divergence between $\mathcal{P}$ and $\mathcal{Q}$ is defined as

$$D_{JS}(\mathcal{P}, \mathcal{Q}) = H_S\left(\frac{\mathcal{P} + \mathcal{Q}}{2}\right) - \frac{1}{2}H_S(\mathcal{P}) - \frac{1}{2}H_S(\mathcal{Q}), \tag{10}$$

where $H_S(\mathcal{P}) = \sum_{a=1}^{A} p_a \log p_a$ is the Shannon entropy of distribution $\mathcal{P}$. D_{JS} is always well defined, symmetric, negative definite and bounded, i.e., $0 \leq D_{JS} \leq 1$.

The quantum Jensen-Shannon divergence has recently been developed as a generalization of the classical Jensen-Shannon divergence to quantum states by Lamberti et al. [17]. Given two density operators ρ and σ, the quantum Jensen-Shannon divergence between them is defined as

$$D_{QJS}(\rho, \sigma) = H_N\left(\frac{\rho + \sigma}{2}\right) - \frac{1}{2}H_N(\rho) - \frac{1}{2}H_N(\sigma). \tag{11}$$

D_{QJS} is always well defined, symmetric, negative definite and bounded, i.e., $0 \leq D_{QJS} \leq 1$ [17].

3.2 A Quantum Kernel Using the Discrete-Time Quantum Walk

We propose a novel quantum Jensen-Shannon kernel for graphs using the quantum Jensen-Shannon divergence associated with the discrete-time quantum walk. Given a set of graphs $\{G_1, \ldots, G_a, \ldots, G_b, \ldots, G_N\}$, we simulate a discrete-time quantum walks on each $G_a(V_a, E_a)$ and $G_b(V_b, E_b)$ for $t = 0, 1, \ldots, T$). Then, the density matrices $\rho_{G;a}^S$ and $\sigma_{G;b}^T$ of $G_a(V_a, E_a)$ and $G_b(V_b, E_b)$ can be computed using Eq.(5). With the density matrices $\rho_{G;a}^T$ and $\sigma_{G;b}^T$ to hand, the quantum Jensen-Shannon divergence $D_{QJS}(\rho_{G;a}, \sigma_{G;b})$ is computed as in Eq.(11). Finally, the quantum Jensen-Shannon kernel $k_{QJS}(G_a, G_b)$ between the pair of graphs $G_a(V_a, E_a)$ and $G_b(V_b, E_b)$ is defined as

$$
\begin{aligned}
k_{QJS}(G_a, G_b) &= \exp(-\alpha D_{QJS}(\rho_{G;a}^T, \sigma_{G;b}^T)) \\
&= \exp\left\{-\alpha H_N\left(\frac{\rho_{G;a}^T + \sigma_{G;b}^T}{2}\right) + \alpha\frac{1}{2}H_N(\rho_{G;a}^T) + \alpha\frac{1}{2}H_N(\sigma_{G;b}^T)\right\}.
\end{aligned}
\tag{12}
$$

where $\frac{\rho_{G;a}^T + \sigma_{G;b}^T}{2}$ is a mixed state, α is a decay factor satisfying $0 \leq \alpha \leq 1$, and $H_N(\cdot)$ is the von Neumann entropy defined in Eq.(6). For simplification, in this work we set α as 1.

Lemma. *The quantum Jensen-Shannon kernel k_{QJS} is positive definite* **pd**.

Proof. This follows the definitions in [17, 18]. The quantum Jensen-Shannon divergence between a pair of density operators $\rho_{G;a}^T$ and $\sigma_{G;b}^T$ is symmetric and is a dissimilarity measure. Thus, the proposed quantum kernel k_{QJS} that is computed by exponentiating the negative divergence measure is **pd**. ∎

For a pair of graphs, each of which has n vertices and m edges, the quantum kernel k_{QJS} requires time complexity $O(m^3)$. This is because the state space of the discrete-time quantum walk for a graph corresponds to the vertex set of its line graph. The number of the vertex in the line graph is double in the number of the edges of the original graph, i.e., the size of the unitary matrix or the density matrix for a graph is $m \times m$. The von Neumann entropy relies on the eigen decomposition of the density matrix, and thus requires time complexity $O(m^3)$. As a result, the whole time complexity of the kernel k_{QJS} is $O(m^3)$.

4 Experimental Evaluations

4.1 Graph Datasets

We explore our new kernel on five standard graph datasets from bioinformatics and computer vision. These datasets include: MUTAG, PPIs, PTC(MR), COIL5 and Shock. Some statistic concerning the datasets are given in Table 1.

MUTAG: The MUTAG dataset consists of graphs representing 188 chemical compounds, and aims to predict whether each compound possesses mutagenicity.

PPIs: The PPIs dataset consists of protein-protein interaction networks (PPIs). The graphs describe the interaction relationships between histidine kinase in

Table 1. Information of the Graph based Datasets

Datasets	MUTAG	PPIs	PTC	COIL5	Shock
Max # vertices	28	232	109	241	33
Min # vertices	10	3	2	72	4
Ave # vertices	17.93	109.60	25.60	144.90	13.16
# graphs	188	86	344	360	150
# classes	2	2	2	5	10

different species of bacteria. There are 219 PPIs in this dataset and they are collected from 5 different kinds of bacteria. Here we select two kinds of bacteria, i.e., Proteobacteria40 PPIs and Acidobacteria46 PPIs, as the testing graphs.

PTC: The PTC (The Predictive Toxicology Challenge) dataset records the carcinogenicity of several hundred chemical compounds for male rats (MR), female rats (FR), male mice (MM) and female mice (FM). These graphs are very small (i.e., $20 - 30$ vertices), and sparse (i.e., $25 - 40$ edges. We select the graphs of male rats (MR) for evaluation. There are 344 test graphs in the MR class.

COIL5: We create a dataset referred to as COIL5 from the COIL image database. The COIL database consists of images of 100 3D objects. In our experiments, we use the images for the first five objects. For each of these objects we employ 72 images captured from different viewpoints. For each image we first extract corner points using the Harris detector, and then establish Delaunay graphs based on the corner points as vertices. Each vertex is used as the seed of a Voronoi region, which expands radially with a constant speed. The linear collision fronts of the regions delineate the image plane into polygons, and the Delaunay graph is the region adjacency graph for the Voronoi polygons.

Shock: The Shock dataset consists of graphs from the Shock 2D shape database. Each graph is a skeletal-based representation of the differential structure of the boundary of a 2D shape. There are 150 graphs divided into 10 classes.

4.2   Experiments on Standard Graph Datasets from Bioinformatics

Experimental Setup: We compare the performance of our new quantum Jensen-Shannon kernel (QJSD) with that of several alternative state-of-the-art graph kernels. These kernels include 1) the unaligned quantum Jensen-Shannon kernel (UQJS) associated with the continuous-time quantum walk [14], 2) the Weisfeiler-Lehman subtree kernel (WL) [7], 3) the shortest path graph kernel (SPGK) [4], 4) the Jensen-Shannon graph kernel associated with the steady state random walk (JSGK) [11], 5) the backtrackless random walk kernel using the Ihara zeta function based cycles (BRWK) [5], and 6) the random-walk graph kernel [3]. For our QJSD kernel, we let $T = 40$. In fact, as we let $T \geqslant 30$ we observe that the von Neumann entropy of the density matrices reaches an asymptote. While the optimal procedure would be that of selecting the value of T through cross-validation, the computational complexity of the kernel makes it unfeasible to do so. Moreover, previous work has shown that letting $T \to \infty$ allows us to achieve a good trade-off in terms of accuracy and computational effort [16]. For the Weisfeiler-Lehman subtree kernel, we set the dimension of the Weisfeiler-Lehman

Table 2. Accuracy Comparisons (In % ± Standard Errors) on Graph Datasets

Datasets	MUTAG	PPIs	PTC(MR)	COIL5	Shock
QJSD	83.16 ± .86	70.57 ± 1.20	**58.23** ± .80	69.78 ± .37	**44.86** ± .64
QJSU	82.72 ± .44	69.50 ± 1.20	56.70 ± .49	**70.11** ± .61	40.60 ± .92
WL	82.05 ± .57	**78.50** ± 1.40	56.05 ± .51	33.16 ± 1.01	36.40 ± 1.00
SPGK	**83.38** ± .81	61.12 ± 1.09	56.55 ± .53	69.66 ± .52	37.88 ± .93
JSGK	83.11 ± .80	57.87 ± 1.36	57.29 ± .41	69.13 ± .79	21.73 ± .76
BRWK	77.50 ± .75	53.50 ± 1.47	53.97 ± .31	14.63 ± .21	0.33 ± .37
RWGK	80.77 ± .72	55.00 ± .88	55.91 ± .37	20.80 ± .47	2.26 ± 1.01

isomorphism as 10. Based on the definition in [7], this means that we compute 10 different Weisfeiler-Lehman subtree kernel matrices (i.e., $k(1), k(2), \ldots, k(10)$) with different subtree heights $h(h = 1, 2, \ldots, 10)$, respectively. Note that, the WL and SPGK kernels are able to accommodate attributed graphs. In our experiments, we use the vertex degree as a vertex label for the WL and SPGK kernels.

Given these datasets and kernels, we perform a 10-fold cross-validation using a C-Support Vector Machine (C-SVM) to evaluate the classification accuracies of the different kernels. More specifically, we use the C-SVM implementation of LIBSVM. For each class, we use 90% of the samples for training and the remaining 10% for testing. The parameters of the C-SVMs are optimized separately for each dataset. We report the average classification accuracies (± standard error) of each kernel in Table 2. **Results:** Overall, in terms of classification accuracies our QJSK kernel outperforms or is competitive with the state-of-the-art kernels. In particular, the classification accuracies of our quantum kernel are significantly better than those of the graph kernels based on the classical random walk and the backtrackless random walk, over all the datasets. This suggests that our kernel using discrete-time quantum walks has better ability of capturing the graph characteristics. With respect to the quantum Jensen-Shannon kernel based on the continuous-time quantum walk, we observe a significant improvement on the PTC and Shock datasets. This is because the discrete-time quantum walk can be seen as a walk on line graphs, and reflects richer graph characteristics than the continuous-time quantum walk on the original graphs.

5   Conclusion

In this paper, we develop a new quantum Jensen-Shannon kernel for graphs by using the quantum Jensen-Shannon divergence and the discrete-time quantum walk. Our new quantum kernel can reflect richer graph characteristics than our previous quantum Jensen-Shannon kernel using the continuous-time quantum walk. Experiments demonstrate the effectiveness of the new quantum kernel.

Acknowledgments. Lu Bai and Zhihong Zhang are supported by National Natural Science Foundation of China (Grant No.61402389). Edwin R. Hancock is supported by a Royal Society Wolfson Research Merit Award.

References

1. Schölkopf, B., Smola, A.: Learning with Kernels. MIT Press (2002)
2. Haussler, D.: Convolution kernels on discrete structures. In: Technical Report UCS-CRL-99-10, Santa Cruz, CA, USA (1999)

3. Kashima, H., Tsuda, K., Inokuchi, A.: Marginalized kernels between labeled graphs. In: Proceedings of ICML, pp. 321–328 (2003)
4. Borgwardt, K.M., Kriegel, H.-P.: Shortest-path kernels on graphs. In: Proceedings of the IEEE International Conference on Data Mining, pp. 74–81 (2005)
5. Aziz, F., Wilson, R.C., Hancock, E.R.: Backtrackless walks on a graph. IEEE Transactions on Neural Networks and Learning Systems 24, 977–989 (2013)
6. Ren, P., Wilson, R.C., Hancock, E.R.: Graph characterization via ihara coefficients. IEEE Transactions on Neural Networks 22, 233–245 (2011)
7. Shervashidze, N., Schweitzer, P., van Leeuwen, E.J., Mehlhorn, K., Borgwardt, K.M.: Weisfeiler-lehman graph kernels. Journal of Machine Learning Research 1, 1–48 (2010)
8. Harchaoui, Z., Bach, F.: Image classification with segmentation graph kernels. In: Proceedings of CVPR (2007)
9. Bach, F.R.: Graph kernels between point clouds. In: Proceedings of ICML, pp. 25–32 (2008)
10. Bai, L., Ren, P., Hancock, E.R.: A hypergraph kernel from isomorphism tests. In: Proceedings of ICPR, pp. 3880–3885 (2014)
11. Bai, L., Hancock, E.R.: Graph kernels from the jensen-shannon divergence. Journal of Mathematical Imaging and Vision 47, 60–69 (2013)
12. Bai, L., Hancock, E.R.: Graph clustering using the jensen-shannon kernel. In: Real, P., Diaz-Pernil, D., Molina-Abril, H., Berciano, A., Kropatsch, W. (eds.) CAIP 2011, Part I. LNCS, vol. 6854, pp. 394–401. Springer, Heidelberg (2011)
13. Bai, L., Hancock, E.R., Ren, P.: Jensen-shannon graph kernel using information functionals. In: Proceedings of ICPR, pp. 2877–2880 (2012)
14. Bai, L., Rossi, L., Torsello, A., Hancock, E.R.: A quantum jensen-shannon graph kernel for unattributed graphs. Pattern Recognition 48(2), 344–355 (2015)
15. Bai, L., Hancock, E.R., Torsello, A., Rossi, L.: A quantum jensen-shannon graph kernel using the continuous-time quantum walk. In: Kropatsch, W.G., Artner, N.M., Haxhimusa, Y., Jiang, X. (eds.) GbRPR 2013. LNCS, vol. 7877, pp. 121–131. Springer, Heidelberg (2013)
16. Rossi, L., Torsello, A., Hancock, E.R.: A continuous-time quantum walk kernel for unattributed graphs. In: Kropatsch, W.G., Artner, N.M., Haxhimusa, Y., Jiang, X. (eds.) GbRPR 2013. LNCS, vol. 7877, pp. 101–110. Springer, Heidelberg (2013)
17. Lamberti, P., Majtey, A., Borras, A., Casas, M., Plastino, A.: Metric character of the quantum jensen-shannon divergence. Physical Review A 77, 052311 (2008)
18. Majtey, A., Lamberti, P., Prato, D.: Jensen-shannon divergence as a measure of distinguishability between mixed quantum states. Physical Review A 72, 052310 (2005)
19. Farhi, E., Gutmann, S.: Quantum computation and decision trees. Physical Review A 58, 915 (1998)
20. Ren, P., Aleksic, T., Emms, D., Wilson, R.C., Hancock, E.R.: Quantum walks, ihara zeta functions and cospectrality in regular graphs. Quantum Information Process 10, 405–417 (2011)
21. Emms, D., Severini, S., Wilson, R.C., Hancock, E.R.: Coined quantum walks lift the cospectrality of graphs and trees. Pattern Recognition 42, 1988–2002 (2009)
22. L.G.: A fast quantum mechanical algorithm for database search. In: Proceedings of ACM Symposium on the Theory of Computation, pp. 212–219 (1996)
23. Nielsen, M., Chuang, I.: Quantum computation and quantum information. Cambridge university press (2010)

Density Based Cluster Extension and Dominant Sets Clustering

Jian Hou[1(✉)], Chunshi Sha[2], Xu E[1], Qi Xia[3], and Naiming Qi[3]

[1] School of Information Science and Technology, Bohai University,
Jinzhou, China, 121013
dr.houjian@gmail.com
[2] School of Engineering, Bohai University, Jinzhou, 121013, China
[3] School of Astronautics, Harbin Institute of Technology, Harbin, 150001, China

Abstract. With the pairwise data similarity matrix as input, dominant sets clustering has been shown to be a promising clustering approach with some nice properties. However, its clustering results are found to be influenced by the similarity parameter used in building the similarity matrix. While histogram equalization transformation of the similarity matrices removes this influence effectively, this transformation causes over-segmentation in the clustering results. In this paper we present a density based cluster extension algorithm to solve the over-segmentation problem. Specifically, we determine the density threshold based on the minimum possible density inside the dominant sets and then add new members into clusters if the density requirement is satisfied. Our algorithm is shown to perform better than the original dominant sets algorithm and also some other state-of-the-art clustering algorithms in data clustering and image segmentation experiments.

Keywords: Dominant set · Clustering · Over-segmentation · Cluster extension

1 Introduction

In various clustering approaches, graph-based clustering has received much interest and achieved impressive success in recent decades. This is partly due to the fact that graph-based requires as the input the pairwise similarity matrix of the data, which captures rich information of the data distribution and structure. As a typical graph-based clustering approach, spectral clustering relies on the eigenvalues of the similarity matrix to perform dimension reduction and data partitioning. The well-known normalized cuts (NCuts) algorithm [17] is one example of spectral clustering, and some other works in this field include [7,21,14]. Affinity propagation [1] is another popular clustering approach with the similarity matrix as the input. By passing affinity messages among data, the affinity propagation algorithm gradually identifies cluster centers and cluster members. Some extensions of the affinity propagation algorithm include [5,10,9].

© Springer International Publishing Switzerland 2015
C.-L. Liu et al. (Eds.): GbRPR 2015, LNCS 9069, pp. 262–271, 2015.
DOI: 10.1007/978-3-319-18224-7_26

However, all the above mentioned graph-based clustering algorithms, including normalized cuts and affinity propagation, requires additional parameters input besides the similarity matrix. With the normalized cuts algorithm, we need to specify the appropriate number of clusters, which is not easy to determine in many cases. The affinity propagation algorithm does not explicitly require the number of clusters to be known beforehand. Instead, it involves the so-called preference values of all data, which influence the number of clusters and clustering results. The appropriate selection of these preference values is not a trivial task either. These additional parameters input requirement implies that in order to obtain satisfactory clustering results, we need a careful parameter tuning process. Obviously this is not convenient in practice.

In contrast to normalized cuts and affinity propagation, dominant sets (DSets) clustering [15,16] is another graph-based clustering approach which does not require any parameter as input explicitly. The dominant set concept is an extension of the clique in graph to edge-weighted graph, and this enable a dominant set to be regarded as a cluster. With the pairwise similarity matrix as input, DSets clustering extracts all the clusters sequentially and determine the number of clusters automatically. DSets clustering works with asymmetric and even negative similarity matrices and can be used for soft clustering [19]. With these nice properties, DSets clustering has been applied to various domains [22,13,11] successfully.

However, we have found in previous work [12] that DSets clustering requires parameter input implicitly. In building the pairwise similarity matrix, one common approach is to measure the similarity by $s(x,y) = exp(-d(x,y)/\sigma)$, where $d(x,y)$ is the distance between two data items x and y, and σ is the regulation parameter. Evidently different σ's lead to different similarity matrices, and this is shown to result in different clustering results in [12]. In order to obtain satisfactory clustering results, one must select an appropriate σ to build the input similarity matrix. In this sense, we say that DSets clustering requires the regulation parameter as input.

Similar as in the case of normalized cuts and affinity propagation, it is not a trivial task to determine the appropriate σ for DSets clustering. In [12] the authors propose to solve this problem from another perspective. Specifically, we use histogram equalization to transform the similarity matrices before they are used in DSets clustering. This transformation is shown to remove the dependence of DSets clustering results on the regulation parameter σ's, and enable different σ's to generate identical clustering results. In other words, it is not necessary to select σ any more. However, we also found that the transformation usually leads to over-segmentation in clustering results. While [12] presents a cluster extension method to overcome this problem, the method involves some user-selected parameters, which make it not very flexible in practical application.

In this paper we present a new cluster extension method by making use of the available information inside the dominant sets. We firstly analyze why DSets clustering tends to cause over-segmentation after histogram equalization transformation. As a result, we propose to use local density constraint to add new members into dominant sets and form clusters. Our method is shown to over-come the over-segmentation problem quite effectively and improves the clustering quality significantly. In data clustering and image segmentation experiments our method perform also comparable to some other clustering methods.

The remaining of this paper is organized as follows. In Section 2 we briefly introduce the dominant sets clustering method and its properties. Then we discuss the over-segmentation problem of DSets clustering after histogram equalization and present our solution in Section 3. The effectiveness of our method is validated in data clustering and image segmentation experiments in Section 4. Finally, Section 5 concludes this paper.

2   Dominant Sets Clustering

2.1   The Concept

As a graph-theoretic concept of a cluster, dominant set is in fact an extension of the clique concept in graph to edge-weighted graph. Informally, a dominant set can be regarded a subset of data with high internal similarity and low similarity with those data outside the subset. A formal definition of dominant sets is presented briefly in the following, and the details can be found in [15,16,18].

With n data to be clustered and the corresponding pairwise similarity matrix $A = (a_{ij})$, we represent these data with an undirected edge-weighted graph $G = (V, E, w)$ with no self-loops. Here V means the vertex (data) set, E is the edge set and w denotes the weight function. Since there is no self-loops in the graph, we have $a_{kk} = 0$ for all $k \in V$. With a non-empty subset $D \subseteq V$ and $i \in D, j \notin D$, we define the following

$$\phi_D(i,j) = a_{ij} - \frac{1}{n} \sum_{k \in D} a_{ik}, \tag{1}$$

$$w_D(i) = \begin{cases} 1, & \text{if } |D| = 1, \\ \sum_{k \in D \setminus \{i\}} \phi_{D \setminus \{i\}}(k,i) w_{D \setminus \{i\}}(k), & \text{otherwise.} \end{cases} \tag{2}$$

From the above definitions, we see that $w_D(i)$ reflects the relationship between two values, i.e., the average similarity between i and $D \setminus \{i\}$, and the overall average similarity in $D \setminus \{i\}$. This relationship can be understood as follows intuitively. A positive $w_D(i)$ indicates that i is very similar to $D \setminus \{i\}$ and D has larger internal coherency than $D \setminus \{i\}$. In other words, D is more suitable to regarded as a cluster than $D \setminus \{i\}$. In contrast, a negative $w_D(i)$ means that adding i into the set $D \setminus \{i\}$ will reduce the internal coherency inside $D \setminus \{i\}$, and therefore i is not suitable to be placed in the same cluster as $D \setminus \{i\}$.

Defining the total weight of D as $W(D) = \sum_{i \in D} w_D(i)$, we present the formal definition of dominant sets in the following. A non-empty subset $D \subseteq V$ such that $W(T) > 0$ for any non-empty $T \subseteq D$ is called a dominant set if the following conditions are satisfied:

1. $w_D(i) > 0$, for all $i \in D$.
2. $w_{D \bigcup \{i\}}(i) < 0$, for all $i \notin D$.

In this definition the first condition guarantees that a dominant set is internally coherent, i.e., the data inside a dominant set are highly similar to each other. The second condition, on the other hand, indicates that the internal coherency inside the dominant set will be destroyed if any more data from outside are added into the dominant set. These two conditions together define a dominant set as a subset of data with high internal similarity inside the subset, and low similarity with data outside the subset. This is exactly what is required of a cluster, and enable a dominant set to be regarded as a cluster.

After we define a dominant set as a graph-theoretic concept of a cluster, the remaining problem is how to extract a dominant set (cluster) and accomplish the DSets clustering process. In [15,16] the authors show that a dominant set can be extracted with a game dynamics, e.g., the Replicator Dynamics or the Infection and Immunization Dynamics [2]. After we extract a dominant set, we remove the included data from the whole data set, and extract another dominant set from the remaining data. Repeating this process until all data are grouped into clusters or some criterion is met, we are able to accomplish the clustering process and determine the number of cluster automatically.

2.2 Existing Problem

DSets clustering uses only the pairwise similarity matrix of data as input, and does not require any parameter to be specified beforehand explicitly. However, we notice that the similarity between two data items is often in the form of $s(x, y) = exp(-d(x, y)/\sigma)$, and different σ's result in different similarity matrices. While we expect DSets clustering results to be invariant to σ's, experiments indicate that this is not the case. For illustration, we report in Figure 1 the DSets clustering results (measured by F-measure) with different σ's on eight datasets, i.e., Aggregation [8], Pathbased [3], D31 [20], R15 [20], and four UCI datasets Thyroid, Wine, Glass and Breast. Note that in Figure 1(a) the values of σ's are the product of the horizonal coordinates and $\bar{d}$, which is the average of all pairwise distances.

From Figure 1(a) we see that the σ has a significant influence on DSets clustering results. This implies that in order to obtain satisfactory clustering results, we must select an appropriate σ in building the similarity matrix. However, it is not a trivial task to select the appropriate σ. The reason lies in that σ is only a regulation parameter, and it is hard to figure out the correction between σ and the DSets clustering results.

Instead of insisting on the selection of σ, [12] proposes to use histogram equalization to remove the influence of σ's on DSets clustering result. Specifically, [12] finds that the difference of similarity matrices from different σ's lies mainly in the magnitudes of similarity contrast. Since this contrast difference is a little similar to that of images with different intensity contrast, [12] proposes to use the histogram equalization technique in image enhancement domain to transform

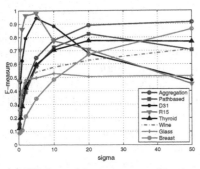

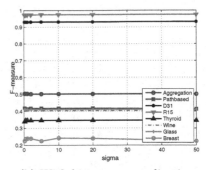

(a) Without histogram equalization   (b) With histogram equalization

Fig. 1. The clustering results of 8 datasets with different σ. (a) With the original similarity matrices. (b) With the similarity matrices transformed by histogram equalization.

Table 1. The comparison of numbers of clusters from DSets and ground truth

	Aggregation	Pathbased	D31	R15	Thyroid	Wine	Glass	Breast
Ground truth	7	6	3	31	15	3	6	2
DSets	27	24	19	51	20	23	25	39

similarity matrices before they are used in DSets clustering. This transformation is shown to make different σ's generate nearly identical similarity matrices and then nearly identical DSets clustering results. This means that we can randomly select one σ in building the similarity matrix, without worrying about its performance in clustering. However, this transformation also causes severe over-segmentation and then harms the clustering results. The DSets clustering results of the eight datasets after histogram equalization are reported in Figure 1(b). Besides, we also report the resulted numbers of clusters and compare with the ground-truth numbers of clusters in Table 1.

In Figure 1(b) and Table 1 the over-segmentation is quite evident. In Figure 1(b) the clustering qualities with different σ's are almost the same, and much smaller than the best performance observed in Figure 1(a) in most cases. In order to solve the over-segmentation, [12] presents a similarity based cluster extension method. However, this method is quite intuitive and the performance depends heavily on user-specified parameters.

3   Solution

In this section we firstly analyze why DSets clustering tends to generate over-segmented clustering results after histogram equalization. Based on the analysis, we then present our cluster extension method to solve the over-segmentation problem.

3.1   Why over-Segmentation

In order to understand why DSets clustering generates over-segmentation after the histogram equalization transformation, we firstly review the definition of the dominant set. From Eq. (2) we have known that a positive $w_D(i)$ means that the average similarity between one data item i and $D\backslash\{i\}$ is greater than the overall similarity within $D\backslash\{i\}$. Then in the formal definition of dominant set we see that each data item in a dominant set corresponds to a positive $w_D(i)$. This means that for each i in a dominant set D, its average similarity with $D\backslash\{i\}$ is greater than the overall similarity within $D\backslash\{i\}$. For ease of expression, we define the average similarity between one data item i and a subset S, and the overall similarity inside S as

$$\delta(i, S) = \frac{1}{\|S\|} \sum_{k \in S} a_{ik}, \tag{3}$$

$$\delta(S) = \delta(S, S) = \frac{1}{\|S\|(\|S\| - 1)} \sum_{j \in S, k \in S} a_{jk}. \tag{4}$$

Based on this definition, the condition of D being a dominant set becomes

$$\delta(i, D\backslash\{i\}) > \delta(D\backslash\{i\}), \forall i \in D. \tag{5}$$

This is a very strict condition as it is applied to *all* the data inside the dominant set, and the average similarity in the left of the inequality is calculated between one data i and *all* the other data in the dominant set. In order for this condition to be satisfied, all the pairwise similarity values of data inside dominant set D must be very close to each other. To some extent, this strict condition is adverse to the generation of large dominant sets inherently.

Now it is quite easy to see why after histogram equalization transformation, DSets clustering tends to generate over-segmented clustering results. By definition, the pairwise similarity values of the data in a dominant set must be very close to each other. However, histogram equalization increases the global contrast inside the similarity matrix. After histogram equalization transformation, the similarity values inside the similarity matrix will be distributed in the range [0,1] more uniformly. This reduces the possibility of many similarity values gathering together, and then reduces the possibility of generating large dominant sets. As a result, an over-large number of clusters of over-small sizes are generated by DSets clustering, and we have an over-segmented clustering result.

3.2   Dominant Set Extension

In the original definition of dominant set D, the average similarity between each data item i and all the other data in D must be greater than the overall similarity within $D\backslash\{i\}$. This global density constraint is very strong and is adverse to the generation of large dominant sets. In this sense, DSets clustering is likely to cause over-segmentation in itself, even if the histogram equalization transformation is not applied. In order to solve the over-segmentation problem, we decide to relax the somewhat over-strict, global density constraint imposed by the dominant set definition. Based on the original dominant set definition, one data item i outside a dominant set D must satisfy $\delta(i, D) > \delta(D)$ in order to be admitted into D. It is usually difficult to satisfy this condition as one data i must be compared with *all* the data in D, including both the nearest and the farthest ones.

Starting from this observation, we propose to replace the global density constraint by a local one. Specifically, for each data outside a dominant set, we evaluate the density of the subset consisting of this data and its k nearest neighbors in the dominant set. If the density of this subset is above a threshold, we add the data into the extended dominant set. Instead of choosing a density threshold manually, here we determine the threshold adaptively by making use of available information in the dominant set. The idea is to calculate the density of each such subset and use the minimum one as the threshold. The threshold selected this way reflects the minimum possible density allowed in the dominant set, and therefore suitable to be used in our algorithm. In implementation we use the similarity with the k-th nearest neighbor to measure the density, where k is selected to be 3 as this value generates the best overall performance. Although k is selected empirically, $k = 3$ is the best or near-best option for most datasets in our experiments. In contrast, the best σ's of different datasets vary widely and it is hard to find a fixed *sigma* applicable to all datasets. Besides, the sizes of dominant sets are usually small and our method uses only a subset of each dominant set. As a result, the value of k is restricted to a quite small range and the candidates of k's are actually rather limited. This also make the selection of k much easier than that of *sigma*.

Similar to the original DSets clustering, our algorithm extracts the clusters sequentially. We extract a dominant set, extend it to be a cluster, and remove the cluster from the data set. Repeating this process until all data are clustered, we accomplish the clustering task. The dominant set extension method is one major contribution of this paper, and the detailed steps of extending a dominant set to be a cluster are as follows

1. For each data j in the dominant set D, find its similarity with the third nearest neighbor in the dominant set and denote it by s_j.
2. Use $th = \min_{j \in D} s_j$ as the local density threshold.
3. For each data i outside the dominant set D, find its similarity with the third nearest neighbor in the dominant set, denote it by s_i. If $s_i > th$, add data i into the dominant set D.
4. Treat the extended dominant set as a cluster.

4   Experiments

4.1   Data Clustering

The data clustering experiments are conducted on the eight datasets used in previous sections. We compare our algorithm with the original DSets clustering,

Table 2. The comparison of different clustering algorithms

	Aggregation	Pathbased	D31	R15	Thyroid	Wine	Glass	Breast	average
DSets	0.92	0.83	0.67	0.66	0.77	**0.69**	0.52	0.75	0.73
Ours	**0.93**	**0.88**	0.73	**0.98**	0.75	**0.69**	0.52	0.93	**0.80**
k-means	0.85	0.70	0.81	0.82	**0.84**	0.64	0.54	**0.96**	0.77
DBSCAN	0.84	0.71	0.75	0.73	0.68	0.51	**0.55**	0.87	0.71
NCuts	0.76	0.71	**0.86**	0.92	0.79	0.60	0.54	0.94	0.77

Fig. 2. Segmentation results. From left to right, the results are obtained by DSets, our algorithm and NCuts respectively. Best viewed in color version.

k-means, DBSCAN [6] and NCuts. With the original DSets clustering, we select $\sigma = 25\overline{d}$ as this value generates the best overall clustering results in testing values from $0.2\overline{d}$ to $50\overline{d}$. For k-means and NCuts we set the numbers of clusters as the ground truth. With DBSCAN we use the implementation presented in [4] and set $MinPts$ as 3, which is selected from testing values 2, 3, $\cdots$, 7, and Eps is determined from $MinPts$. The comparison of these five algorithms on the eight datasets is reported in Table 2. Our algorithm generates the best clustering results on 4 out of the 8 datasets, and outperforms the other four algorithms in average clustering quality. It should be noted that all the other four algorithms benefit from carefully selected parameters.

4.2   Image Segmentation

We also apply our clustering algorithm to image segmentation and compare with other clustering algorithms. The testing images are selected from the Berkeley Segmentation Dataset. The segmentation results with our algorithm, the original DSets algorithm and NCuts algorithm are reported in Figure 2. Although the σ's used in the original DSets and the numbers of segments used in NCuts are carefully tuned manually, our algorithm generates better segmentation results without any parameter input.

5   Conclusions

Based on a graph-theoretic concept of a cluster, dominant sets clustering has shown evident potential in applications. However, we find that dominant sets clustering results are influenced by the similarity parameters used in building the similarity matrices. While histogram equalization transformation of the similarity matrices removes this influence effectively, it causes over-segmentation in clustering results. Starting from the very definition of the dominant set, we discuss the clustering mechanism of the dominant sets method. As a result, we attribute the over-segmentation to the somewhat over-strict density constraint imposed by dominant set definition. We then propose to relax the original global density constraint to be a local one and use a cluster extension process to solve the over-segmentation problem. Specifically, we use the original dominant sets clustering method to extract dominant sets, and then add new members if the local density is above a threshold. Our algorithm determines the density threshold from the existing dominant sets adaptively, and avoids the possible influences from inappropriate user-specified thresholds. We validate the effectiveness of our method in both data clustering and image segmentation experiments.

Acknowledgments. This work is supported by National Natural Science Foundation of China under Grant No. 61473045 and No. 61171189, and by the Scientific Research Fund of Liaoning Provincial Education Department under Grant No. L2012400 and No. L2012397.

References

1. Brendan, J.F., Delbert, D.: Clustering by passing messages between data points. Science 315, 972–976 (2007)
2. Bulo, S.R., Pelillo, M., Bomze, I.M.: Graph-based quadratic optimization: A fast evolutionary approach. Computer Vision and Image Understanding 115(7), 984–995 (2011)

3. Chang, H., Yeung, D.Y.: Robust path-based spectral clustering. Pattern Recognition 41(1), 191–203 (2008)
4. Daszykowski, M., Walczak, B., Massart, D.L.: Looking for natural patterns in data: Part 1. density-based approach. Chemometrics and Intelligent Laboratory Systems 56(2), 83–92 (2001)
5. Dueck, D., Frey, B.J.: Non-metric affinity propagation for unsupervised image categorization. In: IEEE International Conference on Computer Vision, pp. 1–8 (2007)
6. Ester, M., Kriegel, H.P., Sander, J., Xu, X.W.: A density-based algorithm for discovering clusters in large spatial databases with noise. In: International Conference on Knowledge Discovery and Data Mining, pp. 226–231 (1996)
7. Fowlkes, C., Belongie, S., Fan, C., Malik, J.: Spectral grouping using the nystrom method. IEEE Transactions on Pattern Analysis and Machine Intelligence 26(2), 214–225 (2004)
8. Gionis, A., Mannila, H., Tsaparas, P.: Clustering aggregation. ACM Transactions on Knowledge Discovery from Data 1(1), 1–30 (2007)
9. Givoni, I.E., Chung, C., Frey, B.J.: Hierarchical affinity propagation. In: Conference on Uncertainty in Artificial Intelligence, pp. 238–246 (2009)
10. Givoni, I.E., Frey, B.J.: Semi-supervised affinity propagation with instance-level constraints. In: International Conference on Artificial Intelligence and Statistics, pp. 161–168 (2009)
11. Hamid, R., Maddi, S., Johnson, A.Y., Bobick, A.F., Essa, I.A., Isbell, C.: A novel sequence representation for unsupervised analysis of human activities. Artificial Intelligence 173, 1221–1244 (2009)
12. Hou, J., Xu, E., Chi, L., Xia, Q., Qi, N.M.: Dset++: a robust clustering algorithm. In: International Conference on Image Processing, pp. 3795–3799 (2013)
13. Hou, J., Pelillo, M.: A simple feature combination method based on dominant sets. Pattern Recognition 46(11), 3129–3139 (2013)
14. Niu, D., Dy, J.G., Jordan, M.I.: Dimensionality reduction for spectral clustering. In: International Conference on Artificial Intelligence and Statistics, pp. 552–560 (2011)
15. Pavan, M., Pelillo, M.: A graph-theoretic approach to clustering and segmentation. In: IEEE International Conference on Computer Vision and Pattern Recognition, pp. 145–152 (2003)
16. Pavan, M., Pelillo, M.: Dominant sets and pairwise clustering. IEEE Transactions on Pattern Analysis and Machine Intelligence 29(1), 167–172 (2007)
17. Shi, J., Malik, J.: Normalized cuts and image segmentation. IEEE Transactions on Pattern Analysis and Machine Intelligence 22(8), 167–172 (2000)
18. Torsello, A., Bulo, S.R., Pelillo, M.: Grouping with asymmetric affinities: a game-theoretic perspective. In: IEEE International Conference on Computer Vision and Pattern Recognition, vol. 1, pp. 292–299 (2006)
19. Torsello, A., Bulo, S.R., Pelillo, M.: Beyond partitions: Allowing overlapping groups in pairwise clustering. In: International Conference on Pattern Recognition, pp. 1–4 (2008)
20. Veenman, C.J., Reinders, M., Backer, E.: A maximum variance cluster algorithm. IEEE Transactions on Pattern Analysis and Machine Intelligence 24(9), 1273–1280 (2002)
21. Yan, D., Huang, L., Jordan, M.I.: Fast approximate spectral clustering. In: International Conference on Knowledge Discovery and Data Mining, pp. 907–916 (2009)
22. Yang, X.W., Liu, H.R., Laecki, L.J.: Contour-based object detection as dominant set computation. Pattern Recognition 45, 1927–1936 (2012)

Salient Object Segmentation from Stereoscopic Images

Xingxing Fan, Zhi Liu$^{(\boxtimes)}$, and Linwei Ye

School of Communication and Information Engineering, Shanghai University, Shanghai, China
fanxx9527@gmail.com, {liuzhisjtu,yelinweimail@163.com}

Abstract. In this paper, we propose for stereoscopic images an effective object segmentation approach by incorporating saliency and depth information into graph cut. A saliency model based on color and depth is first used to generate the saliency map. Then the graph cut based on saliency and depth information as well as with the introduction of saliency weighted histogram is proposed to segment salient objects in one cut. Experimental results on a public stereoscopic image dataset with ground truths of salient objects demonstrate that the proposed approach outperforms the state-of-the-art salient object segmentation approaches.

Keywords: Salient object segmentation · Saliency map · Depth information · Graph cut · Saliency weighted histogram

1 Introduction

Salient object segmentation is essential for many applications such as object recognition, image retrieval, object-based image/video adaptation, etc. We can quickly and accurately identify the most visually noticeable salient object in the scene, and adaptively focus our attention on such important regions. The representative and popular image segmentation methods are the graph-based approaches [1] which treat the segmentation problem as a minimum cut or maximum flow problem [2] through a graph partitioning structure. For salient object segmentation, a two-phase graph cut method is proposed in [3], which exploits saliency map based graph cut to obtain an initial segmentation result in the first phase, and uses graph cut for iterative seed adjustment based on the analysis of minimum cut, to refine the object segmentation result in the second phase. As an effective interactive segmentation tool using graph cut, GrabCut [4] simplifies the user interaction to simply dragging a rectangle around the interested object, and iteratively optimizes the energy function for better segmentation results. In [5], an improved iterative version of GrabCut with the initialization of saliency detection, namely SaliencyCut is proposed to achieve better performance on salient object segmentation. In [6], the OneCut method which incorporates an appearance overlap term into the graph cut is proposed for image segmentation to guarantee the global minimum in one cut.

With the recent increase in 3D contents such as 3D movies, it has ignited the rapid development of 3D displays and depth cameras. This has aroused the necessity on image segmentation approaches with the use of depth information. In [7], the depth

© Springer International Publishing Switzerland 2015
C.-L. Liu et al. (Eds.): GbRPR 2015, LNCS 9069, pp. 272–281, 2015.
DOI: 10.1007/978-3-319-18224-7_27

feature is used to identify boundary edges and border ownership, and the point-based segmentation strategy is used to segment the salient object. In [8], depth discontinuity is used for identifying object boundaries, and then depth and color information are combined to segment the salient object. In [9], region-level depth, color and spatial information are combined for saliency detection in stereoscopic image pairs. In [10], a saliency model which incorporates depth feature to suppress background regions through piecewise functions is proposed for salient object segmentation. In [11], a saliency model is proposed for stereoscopic images based on the global contrast on disparity map and domain knowledge in stereoscopic photography, and it is used for the task of salient object segmentation to examine the performance of their stereo saliency analysis.

In this paper, we explore how to combine depth information and saliency map for effective graph cut. Our contributions are two-fold. First, for constructing the graph, saliency map and depth map are combined to reasonably define the regional cost term. Second, the saliency weighted histogram is proposed to better exploit the appearance overlap based cost term on colors of object and background for effective salient object segmentation.

The rest of this paper is organized as follows. Section 2 describes the proposed salient object segmentation approach in detail. Experimental results are shown in Section 3, and conclusions are given in Section 4.

2   Proposed Salient Object Segmentation Approach

2.1   Saliency Map Generation

In natural images, the appearance of salient object is usually distinctive and shows contrast with the surrounding background regions. However, when salient object is not distinctive on its appearance but shows very similar colors with its surrounding regions, it is still difficult for effective saliency detection using only visual features as discussed in [12]. Depth information can be used as an additional and important cue for saliency detection when salient objects and background regions have similar visual appearances. Here we use our previous work [9] on saliency model for stereoscopic images. This model outputs the region-level saliency map. For each region $R_i (i = 1,...,n)$, its saliency value is defined as follows:

$$S(R_i) = DC(R_i) + DWCC(R_i) + SC(R_i) \qquad (1)$$

where the saliency value $S(R_i)$ is the sum of depth contrast measure $DC(R_i)$, depth weighted color contrast measure $DWCC(R_i)$ and spatial compactness measure $SC(R_i)$. The saliency map S is normalized into the range of [0, 1]. Please refer to [9] for more details on each measure.

2.2   Basic Terminologies of Graph Cut

We use the framework of graph cut to segment the salient object. The input image is represented as an undirected graph $g = \{v, \varepsilon\}$, where v is a set of nodes and ε is a set of undirected edges connecting these nodes. Each node in the graph represents each pixel in the image, and there are two additional nodes in the graph, i.e., object terminal $\mathcal{S}$ and background terminal $\mathcal{T}$, and two types of edges, i.e., n-links and t-links. Let Ω denote the set of all pixels in the input image, and N denotes the set of all pairs of neighboring pixels $\{p, q\}$ in Ω. Correspondingly, the two sets v and ε are defined as $v = \Omega \cup \{\mathcal{S}, \mathcal{T}\}$ and $\varepsilon = N \cup \{(p, \mathcal{S}), (p, \mathcal{T})\}_{p \in \Omega}$.

A cut is defined as a subset of edges $C \subset \varepsilon$, and nodes in the graph are separated by this subset of edges. Graph cut seeks to minimize a cost function with the following form to determine the optimal label configuration.

$$E(L) = R(L) + \lambda \cdot B(L) \tag{2}$$

$$R(L) = \sum_{p \in \Omega} R_p(L_p) \tag{3}$$

$$B(L) = \sum_{\{p,q\} \in N} B_{\{p,q\}} \cdot \left| L_p - L_q \right| \tag{4}$$

where $L = \{L_p\}$ is a binary vector, $L_p \in \{0, 1\}$ is a binary value which designates pixels to "bkg" (background) or "obj" (object). $R_p(L_p)$ is a regional cost term based on the label L_p, $B_{\{p,q\}}$ is a boundary cost term for neighboring pixels $\{p, q\}$, and λ is the balancing weight between regional cost term and boundary cost term.

Besides, in the OneCut method [6], an appearance overlap based cost term is also incorporate into the energy function with the following form:

$$E(\theta_{obj}, \theta_{bkg}) = \sum_{k=1}^{K} \min[\theta_{obj}(k), \theta_{bkg}(k)] \tag{5}$$

where $\theta_{obj}(k) / \theta_{bkg}(k)$ denotes the number of object/background pixels that fall into the kth bin of image histogram without normalization. Graph construction is then updated as follows: K auxiliary nodes $A_1, A_2 \ldots \ldots A_K$ are added into the graph, in which the kth auxiliary node is connected to all the nodes for those pixels belonging to the kth bin of image histogram. In this way, each node representing each pixel is connected to its corresponding auxiliary node. In [6], the capacity of each edge that connects auxiliary node with pixel is uniformly set to 1. Therefore, according to Eq. (5), the optimal cut separating object and background will cut off $\min[\theta_{obj}(k), \theta_{bkg}(k)]$ edges that connect those pixels with the auxiliary node A_k, so as to obtain the global minimum where the colors of object and background are best separated.

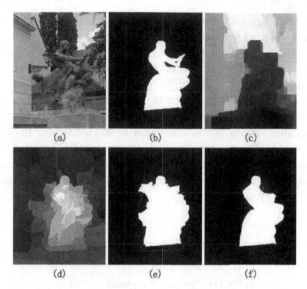

Fig. 1. (a) Left-view image, (b) ground truth, (c) pixel-level depth map, (d) region-level salien-cy map, (e) initial segmentation result, (f) final segmentation result

2.3   Graph Cut Incorporating Saliency and Depth in One Cut

The thresholding operation on the saliency map is first exploited to roughly separate salient objects from background using the scheme in [10], which sets the threshold T as follows:

$$T = b + \alpha \cdot [\max(S) - b] \qquad (6)$$

where α is a constant coefficient and $\max(S)$ is the maximal value of the saliency map S. The parameter b is determined as follows:

$$H_S[a,b] = \eta \cdot H_S[a, \max(S)] \qquad (7)$$

where $H_S[a,b]$ is the sum of histogram values from the bin a to the bin b for S. The bin a is set to correspond to the saliency value of 30/255 for eliminating the interference of background pixels with lower saliency, and both α and η are set to a moderate value, 0.35, as suggested in [10]. Using the above thresholding operation, each pixel p obtains its initial label L_p^i, either as "obj" or "bkg".

For a pair of stereoscopic images, the SIFT flow method [13] is used to estimate the disparity map, which is normalized into the range of [0, 1] and used as a represen-tation of depth map. A pixel closer to the camera is assigned with a lower depth value. For the stereoscopic image in Fig. 1(a), its pixel-level depth map is shown in Fig. 1(c). By thresholding the region-level saliency map in Fig. 1(d), the initial seg-mentation result is shown in Fig. 1(e), which is a reasonable segmentation by refer-ring to the saliency map in Fig. 1(d) but contains some redundant background regions.

Depth information has been used as an additional and important cue for saliency detection when salient objects and background regions have similar visual appearances, and can also be used for better segmentation of salient objects with higher quality under the framework of graph cut. Here we incorporate the depth information into the regional cost term only for pixels that are labeled as object after the initial segmentation, because the depth values of object pixels are usually in a narrower range than those of background pixels. For the example in Fig. 1 (c), the depth values of salient object (the statue) pixels are much lower than most background pixels, which are obviously higher, but the depth values of some background regions such as the stone base are even lower than those of salient object. Therefore, it is reliable to only incorporate the depth information into the cost term of those pixels with the initial label $L_p^i = "obj"$. For clarity, the energy function for graph cut is redefined as follows:

$$E(L) = R(L) + \lambda \cdot B(L) + \beta \cdot E(\theta_{obj}, \theta_{bkg}) \tag{8}$$

Table 1. Definition of regional cost term

L_p^i	$R(L_p^f)$
"obj"	$\dfrac{1-S(p)}{\mu_S} + \dfrac{D(p)}{\mu_D}$, if $L_p^f = "obj"$.
	$\dfrac{S(p)}{\mu_S} + \dfrac{1-D(p)}{\mu_D}$, if $L_p^f = "bkg"$
"bkg"	$\dfrac{1-S(p)}{\mu_S}$, if $L_p^f = "obj"$
	$\dfrac{S(p)}{\mu_S}$, if $L_p^f = "bkg"$

Based on the above analysis of the use of depth information, the definition of regional cost term is shown in Table 1. L_p^i and L_p^f denotes the label of pixel p after initial segmentation and final segmentation, respectively. $S(p)$ and $D(p)$ denote the saliency value and depth value, respectively, of the pixel p, and both are normalized into the range of [0, 1], μ_S and μ_D are the average value of saliency map and depth map, respectively. As shown in Table 1, under the condition of the initial label $L_p^i = "obj"$, if the pixel p is with a higher saliency value and a lower depth value, the final label of the pixel p is likely to be kept as $L_p^f = "obj"$ due that a lower cost will be incurred; on the contrary, if the pixel p is with a lower saliency value and a higher depth value, the final label of the pixel p is likely to be changed as $L_p^f = "bkg"$. Under the condition of the initial label $L_p^i = "bkg"$, only the factor of saliency is considered, and the pixel p with a lower/higher saliency value is likely to obtain the final label $"bkg"/"obj"$.

The boundary cost term, which penalizes the pixels in the 4-connected neighborhood with different labels based on the color contrast, is defined as follows:

$$B(L) = \sum_{\{p,q\} \in N} d(p,q)^{-1} \cdot \left| L_p - L_q \right| \cdot e^{\frac{-\|c_p - c_q\|^2}{2\sigma^2}}. \tag{9}$$

where $d(p,q)$ denotes the Euclidean distance between the spatial position of p and q, $\|c_p - c_q\|$ denotes the Euclidean distance between the color of p and q, and σ^2 is set as the average of $\|c_p - c_q\|^2$ over the set N. The balancing weight for the boundary cost term, λ, is set to 9 in our implementation, for a moderate effect on label smoothness.

We found that the saliency value can be used to adjust the capacity of each edge that connects auxiliary node with pixel, so a weighting-based scheme is exploited to incorporate saliency values into the computation of color histogram. Referring to Eq. (5), here we let $\theta^i_{obj}(k) / \theta^i_{bkg}(k)$ specifically denote the number of object/background pixels that fall into the kth bin of color histogram after the initial segmentation, and let Q_k denote the color range of the kth bin, the saliency weighted color histogram is then defined as follows:

$$H(k) = \begin{cases} \sum_{p \in \Omega} \dfrac{S(p)}{T} \delta(c_p \in Q_k), \text{ if } \theta^i_{obj}(k) > \theta^i_{bkg}(k) \\ \sum_{p \in \Omega} \dfrac{1 - S(p)}{T} \delta(c_p \in Q_k), \text{ otherwise} \end{cases} \tag{10}$$

where the indicator function $\delta(.)$ is set to 1 in case that the predicate is true, otherwise is set to 0. If $\theta^i_{obj}(k) > \theta^i_{bkg}(k)$, a majority of pixels that fall into the kth bin have higher saliency values and they are more likely to belong to salient objects. Then the capacity of each edge that connects A_k with p is set to $S(p)/T$, and thus the optimal cut will cut off the edges with the cost defined as follows:

$$e(k) = \min \left[\sum_{L^f_p = "obj"} \frac{S(p)}{T} \delta(c_p \in Q_k), \ \sum_{L^f_p = "bkg"} \frac{S(p)}{T} \delta(c_p \in Q_k) \right] \tag{11}$$

It can be seen from Eq. (11) that the edges connecting the auxiliary node with background pixels are more likely to be cut off due to the saliency values of background pixels are relatively lower. On the contrary, if $\theta^i_{obj}(k) \le \theta^i_{bkg}(k)$, the capacity of each edge that connects A_k with p is set to $[1 - S(p)]/T$, and the optimal cut will cut off the edges with the cost defined as follows:

$$e(k) = \min\left[\sum_{L_p^f = "obj"} \frac{1-S(p)}{T}\delta(\mathbf{c}_p \in Q_k), \sum_{L_p^f = "bkg"} \frac{1-S(p)}{T}\delta(\mathbf{c}_p \in Q_k)\right] \qquad (12)$$

Then the appearance overlap based cost term is defined as follows:

$$E(\theta_{obj}, \theta_{bkg}) = \sum_{k=1}^{K} e(k). \qquad (13)$$

The balancing weight β for this cost term is defined as follows:

$$\beta = 0.8 \cdot \frac{\sum_{p \in \Omega} \delta(L_p^i = "obj")}{|\Omega|/2 - \sum_{k=1}^{K} \min[\theta_{obj}^i(k), \theta_{bkg}^i(k)]}. \qquad (14)$$

The weight β set by Eq. (14) is used to adjust the contribution of the appearance overlap based term to the energy function, and the coefficient 0.8 is used to moderately control the contribution of this term. If the overlap between θ_{obj}^i and θ_{bkg}^i is higher, i.e., the colors of object and background are not separated well after the initial segmentation, the weight β will be set to a relatively higher value to elevate the contribution of appearance overlap based term. Otherwise, a lower overlap between θ_{obj}^i and θ_{bkg}^i will decrease the contribution of appearance overlap based term accordingly.

The graph is constructed based on the above defined three cost terms, and the minimum cut of the graph is efficiently solved using the max-flow algorithm [2] to obtain the final labels of pixels, which are used to represent the final salient object segmentation result. For the example in Fig. 1(a), the final segmentation result is shown in Fig. 1(f), which segments the salient object with well-defined boundaries more accurately than the initial segmentation result in Fig. 1(e).

3   Experimental Results

In order to evaluate the performance of the proposed salient object segmentation approach, we perform experiments on a public dataset [11], which provides the web links for downloading stereoscopic images and the manually segmented ground truths for salient objects in these stereoscopic images. Similarly as [9], we used a total of 797 pairs of stereoscopic images from all the available web links. We compared the segmentation results generated using our approach with the segmentation results generated using Liu's two-phase graph cut approach [3] and Cheng's SaliencyCut approach [5]. We used the source codes with default parameter settings [3] and the executables [5] provided by the authors to generate saliency maps and segmentation results. However, the segmentation results, executables or source codes of [11], which uses both depth and saliency information, are not available for a comparison.

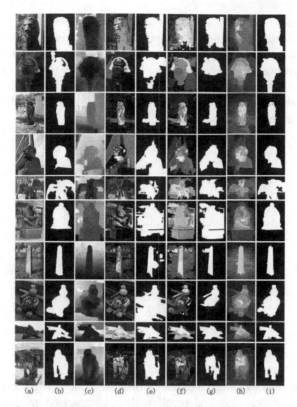

Fig. 2. Examples of salient object segmentation. (a) Left-view images, (b) ground truths, (c) pixel-level depth maps, (d) Liu's saliency maps, (e) Liu's segmentation results, (f) Cheng's saliency maps, (g) Cheng's segmentation results, (h) our saliency maps, (i) our segmentation results.

Fig. 3. Failure cases. (a) Left-view images, (b) ground truths, (c) pixel-level depth maps, (d) Liu's saliency maps, (e) Liu's segmentation results, (f) Cheng's saliency maps, (g) Cheng's segmentation results, (h) our saliency maps, (i) our segmentation results.

Some segmentation results generated using our approach and the other two approaches are shown in Fig. 2 for a subjective comparison. Compared with other approaches, we can observe from Fig. 2 that our approach can segment the complete salient objects with well-defined boundaries. There are two reasons that our approach can achieve better segmentation results. On one hand, our saliency maps are with relatively high quality due to the incorporation of depth information, and they can better suppress background regions. On the other hand, we reasonably use both saliency maps and depth information for graph cut, and effectively integrate the saliency weighted histogram into the appearance overlap based cost term for further better segmentation. However, there are also some failure cases as shown in Fig. 3, in which some background regions are wrongly segmented as salient objects, while the other two approaches also fail for such examples due to the low-quality saliency maps with falsely highlighted background regions.

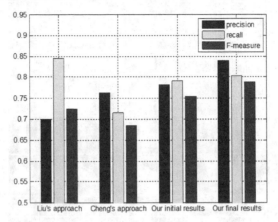

Fig. 4. Objective comparison of salient object segmentation performance

We also objectively evaluated the quality of salient object segmentation results using the measures of precision and recall with a comparison to the binary ground truths. The overall performance measure, F-measure, is then defined as follows:

$$F - measure = \frac{(1 + \gamma) \cdot precision \cdot recall}{\gamma \cdot precision + recall} . \tag{15}$$

where the coefficient γ is set to 1 in our experiments to put the same importance on precision and recall. We can see from Fig. 4 that our approach achieves the highest F-measure as well as the highest precision compared to the other two approaches. In terms of F-measure, which is commonly used to evaluate salient object segmentation performance, our approach outperforms the other two segmentation approaches. Besides, as shown by the values of precision, recall and F-measure achieved using our initial and final segmentation results, we can clearly see the contribution of the proposed graph cut approach which effectively integrates saliency and depth in one cut.

4   Conclusion

In this paper, we have presented a salient object segmentation approach, which effectively utilizes saliency map, depth information and saliency weighted histogram to construct cost terms in the graph cut and segments salient objects with one cut from stereoscopic images. Experimental results on a public stereoscopic image dataset demonstrate the better salient object segmentation performance of our approach.

Acknowledgment. This work was supported by National Natural Science Foundation of China under Grant No. 61171144 and No. 61471230, the Key (Key grant) Project of Chinese Ministry of Education (No. 212053), and the Program for Professor of Special Appointment (Eastern Scholar) at Shanghai Institutions of Higher Learning.

References

1. Boykov, Y., Funka-Lea, G.: Graph cuts and efficient N-D image segmentation. International Journal of Computer Vision 70(2), 109–131 (2006)
2. Boykov, Y., Kolmogorov, V.: An experimental comparison of mincut/max-flow algorithms for energy minimization in vision. IEEE Trans. Pattern Anal. Machine Intell. 26(9), 1124–1137 (2004)
3. Liu, Z., Shi, R., Shen, L., Xue, Y., Ngan, K.N., Zhang, Z.: Unsupervised salient object segmentation based on kernel density estimation and two-phase graph cut. IEEE Trans. Multimedia 14(4), 1275–1289 (2012)
4. Rother, C., Kolmogorov, V., Blake, A.: Grabcut - interactive foreground extraction using iterated graph cuts. ACM Transactions on Graphics 23(3), 309–314 (2004)
5. Cheng, M.M., Mitra, N.J., Huang, X.L., Torr, P.H.S., Hu, S.M.: Global contrast based salient region detection. IEEE Trans. Pattern Anal. Machine Intell. (2014), doi.: 10.1109/TPAMI, 2345401
6. Tang, M., Gorelick, L., Veksler, L., Boykov, Y.: GrabCut in one cut. In: Proc. IEEE ICCV, pp. 1769–1776. IEEE Press, Sydney (2013)
7. Mishra, A.K., Shrivastava, A., Aloimonos, Y.: Segmenting "simple" objects using RGB-D. In: Proc. IEEE ICRA, pp. 4406–4413. IEEE Press, Saint Paul (2012)
8. Liu, H., Philipose, M., Sun, M.T.: Automatic objects segmentation with RGB-D cameras. Journal of Visual Communication and Image Representation 25(4), 709–718 (2014)
9. Fan, X., Liu, Z., Sun, G.: Salient region detection for stereoscopic images. In: Proc. IEEE DSP, pp. 454–458. IEEE Press, Hong Kong (2014)
10. Lei, J., Zhang, H., You, L., Hou, C., Wang, L.: Evaluation and modeling of depth feature incorporated visual attention for salient object segmentation. Neurocomputing 120, 24–33 (2013)
11. Niu, Y., Geng, Y., Li, X., Liu, F.: Leveraging stereopsis for saliency analysis. In: Proc. IEEE CVPR, pp. 454–461. IEEE Press, Providence (2012)
12. Liu, Z., Zou, W., Le Meur, O.: Saliency tree: A novel saliency detection framework. IEEE Trans. Image Process. 23(5), 1937–1952 (2014)
13. Liu, C., Yuen, J., Torralba, A.: Sift flow: Dense correspondence across scenes and its application. IEEE Trans. Pattern Anal. Mach. Intell. 33(5), 978–994 (2011)

Causal Video Segmentation
Using Superseeds and Graph Matching

Vijay N. Gangapure[1], Susmit Nanda[1], Ananda S. Chowdhury[1(✉)],
and Xiaoyi Jiang[2]

[1] Department of Electronics and Telecommunication Engg.
Jadavpur University, Kolkata, India
aschowdhury@etce.jdvu.ac.in
[2] Department of Mathematics and Computer Science,
University of Münster, Münster, Germany

Abstract. The goal of video segmentation is to group pixels into meaningful spatiotemporal regions that exhibit coherence in appearance and motion. Causal video segmentation methods use only past video frames to achieve the final segmentation. The problem of causal video segmentation becomes extremely challenging due to size of the input, camera motion, occlusions, non-rigid object motion, and uneven illumination. In this paper, we propose a novel framework for semantic segmentation of causal video using superseeds and graph matching. We first employ SLIC for the extraction of superpixels in a causal video frame. A set of superseeds is chosen from the superpixels in each frame using color and texture based spatial affinity measure. Temporal coherence is ensured through propagation of labels of the superseeds across each pair of adjacent frames. A graph matching procedure based on comparison of the eigenvalues of graph Laplacians is employed for label propagation. Watershed algorithm is applied finally to label the remaining pixels to achieve final segmentation. Experimental results clearly indicate the advantage of the proposed approach over some recently reported works.

Keywords: Causal video segmentation · Superseeds · Spatial affinity · Graph matching

1 Introduction

Video segmentation [1, 2, 7] aims at grouping pixels into meaningful spatiotemporal regions that exhibit coherence in appearance and motion. The problem of video segmentation [8–10] becomes extremely challenging due to size of the input, camera motion, occlusions, non-rigid object motion, and uneven illumination. Video segmentation techniques can be classified into non-causal (off-line) and causal (on-line) categories. While non-causal segmentation techniques make use of both the past and future video frames, causal segmentation approaches rely only on the past frames. For some recently reported causal video segmentation works, please see [3–6]. Some of these algorithms employ superpixels to

© Springer International Publishing Switzerland 2015
C.-L. Liu et al. (Eds.): GbRPR 2015, LNCS 9069, pp. 282–291, 2015.
DOI: 10.1007/978-3-319-18224-7_28

reduce computational complexity and to achieve powerful within-frame representation [3,6]. The method in [5] does not guarantee temporal consistency. Miksik et al. [4] performs semantic segmentation using optical flow to ensure temporal consistency. But, complexity of pixel-level optical flow computation poses a serious constraint for its use in real-time applications. Couprie et al. [3] proposed an efficient causal graph-based video segmentation method using minimum spanning tree. However, the method uses some heuristics in both the pre and post processing stages. In this paper, we propose a novel framework for semantic segmentation of causal video using superseeds and local graph matching [21]. The major contribution of the work is to propose a novel method of label propagation based on graph matching. Secondly, we have used superseeds for achieving better segmentation. Thirdly, unlike some of the existing approaches [3], we do not use any post-processing steps to achieve superior segmentation performance. Experimental results clearly indicate the advantage of the proposed approach over some of the recently published works [3–5]. The rest of the paper is organized in the following manner: in Section 2, we describe the proposed method. In Section 3, we present the experimental results along with necessary comparisons. We conclude the paper in Section 4 with an outline for directions of future research.

2   Proposed Method

The proposed framework is illustrated in Fig. 1 as shown below. SLIC [11] is applied for the generation of superpixels in each frame of a causal video. As a part of the initialization step, we apply the DBSCAN [13] method with some modifications resulting from our spatial consistency measure to achieve the final segmentation of the first frame. Some representative superpixels are then chosen using the above spatial affinity measure. We deem the centers of such superpixels as superseeds. Labels of these superseeds are propagated to the current frame from the previous frame by using local graph matching. Entries and exits are also handled efficiently to achieve temporal consistency. Watershed is applied

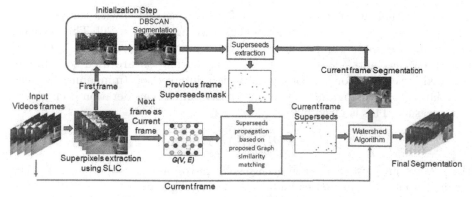

Fig. 1. Schematic of the proposed method

to label the remaining pixels (other than the superseeds) to achieve complete segmentation of the current frame.

2.1 Superpixel Extraction

Superpixel extraction significantly reduces computational complexity in video segmentation algorithms [3,6]. We use the SLIC algorithm [11] for the extraction of superpixels in each frame of a causal video. So, we can write:

$$I_{t,SLIC} = SLIC(I_t, k) \qquad (1)$$

where I_t is the current frame and $I_{t,SLIC}$ is the frame with extracted super-pixels. The inputs to SLIC are the current frame I_t and the desired number of superpixels k. The CIELAB color space is used for clustering color images. In an initialization step, k initial cluster centers $C_i, i = 1, ..., k$, are sampled on a regular grid with spacing S pixels. Hence, we can write:

$$C_i = [l_i, a_i, b_i, x_i, y_i]^T \qquad (2)$$

$$S = \sqrt{\frac{N}{k}} \qquad (3)$$

where N is the number of pixels in the image. The seed centers C_i are moved to locations with lowest gradient position in 3×3 neighborhood. Then, each pixel i is associated with the nearest cluster center. Limiting the size of search region to $2S \times 2S$ around the center significantly reduces the computation compared to the k-means clustering. A new distance measure D which is a combination of color distance (d_c) in CIELAB space and spatial distance (d_s) is used for that purpose. The update step then adjusts each cluster center to be the mean $[l, a, b, x, y]^T$ vector of all the pixels of that cluster. For our work. we find 10 iterations to be sufficient to reach the convergence.

2.2 Spatial Consistency Measure

A hexagonal neighborhood graph $G = (V, E)$ is constructed with the extracted superpixels as the nodes using hexagonal grid as suggested by [20]. This is shown in Fig. 2. The spatial affinity between two superpixels S_i and S_j is captured by the edge weights w_{ij}. Color and texture information are used to compute these edge weights. For the color information, intersection (minimum) between cumulative color histograms of two superpixels under consideration is employed as a measure. This is given by:

$$c_{ij} = N\left[Hist(S_i) \cap Hist(S_j)\right] \qquad (4)$$

Here, $Hist(\cdot)$ represents the cumulative color histogram of a superpixel. N is the normalization constant, set equal to $1/max(c_{ij})$. The larger the value of c_{ij}, the higher is the color affinity between the superpixels S_i and S_j. For the texture information measure, we use a gray-scale local binary pattern (LBP) [12] based measure. The $LBP_{P,R}$ number characterizes the local image structure and can be computed as follows:

$$LBP_{P,R} = \sum_{p=0}^{P-1} s(g_p - g_c)2^p \tag{5}$$

where p denotes a pixel having intensity g_p within a circular neighborhood of radius R centering the pixel c with intensity g_c. We have chosen $P=8$ and $R=1$ for our problem. The function s is given by:

$$s(x) = \begin{cases} 1 & \text{if} \quad x \geq 0 \\ 0 & \text{if} \quad x < 0 \end{cases} \tag{6}$$

In fact, we compute the $\overline{LBP}$ binary vector corresponding to the above LBP number for every pixel in a superpixel. For a superpixel S_i of size n, the texture measure is given by the ordered collection of n such individual vectors:

$$\overline{ST_i} = \{\overline{LBP_{P,R,1}}, \overline{LBP_{P,R,2}}, ..., \overline{LBP_{P,R,n}}\} \tag{7}$$

The normalized texture affinity measure t_{ij} between two superpixels S_i and S_j is given by:

$$t_{ij} = 1 - \frac{W_H(\overline{ST_i} \oplus \overline{ST_j})}{max_{\forall i,j}[W_H(\overline{ST_i} \oplus \overline{ST_j})]} \tag{8}$$

where $\overline{ST_i}$ is truncated to the length of $\overline{ST_j}$ (assuming without loss of generality $|ST_j| < |ST_i|$), $\oplus$ denotes the bitwise XOR operation, and W_H is the Hamming weight function on binary vectors. Larger value of t_{ij} indicates higher texture affinity. Finally, we present the proposed spatial affinity measure between the superpixels S_i and S_j as:

$$\omega_{ij} = c_{ij} \times t_{ij} \tag{9}$$

Note that $\omega_{ij} \in [0\ 1]$.

2.3   Label Propagation Using Graph Similarity

We now mention the various steps linked with propagation of labels from the previous frame to the current frame. These steps are discussed below:

Selection of Superseeds. In the initialization step, only the first frame is segmented by the modified DBSCAN [13] using the above spatial affinity measure. Each segment consists of multiple superpixels and we discard those segments which have less than two superpixels. The geometric centers of the remaining segments are extracted and treated as superseeds.

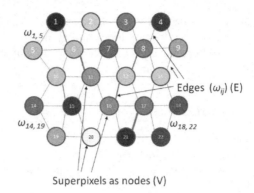

Fig. 2. Superpixel neighborhood graph

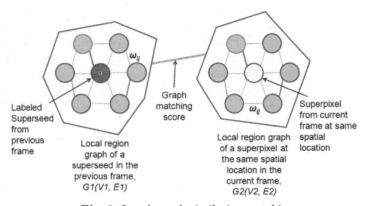

Fig. 3. Local graph similarity matching

Local Graph Matching. Local region graphs are constructed surrounding each superseed in the previous frame and surrounding corresponding pixels (having same spatial locations as that of the superseeds in the previous frame) in the current frame. This is illustrated in Fig. 3. These two graphs are compared to propagate the label from the previous frame to the current frame. Let $G1(V1, E1)$ and $G2(V2, E2)$ respectively represent the local region graph surrounding a superseed in the previous frame and the local region graph surrounding the pixel with same spatial location (as that of the superseed in the previous frame) in the current frame. We use graph Laplacian's eigenvalue-based score for matching [15]. Let $A1$ and $A2$ be the adjacency matrices, $D1$ and $D2$ be the diagonal matrices and $L1$ and $L2$ be the Laplacian matrices of the graphs $G1$ and $G2$ respectively. Then, we can write:

$$L1 = D1 - A1 \tag{10}$$

$$L2 = D2 - A2 \tag{11}$$

We use the similarity matching score $Sim_{G1,G2}$ between $G1$ and $G2$ by computing the top k eigenvalues of Laplacians $L1$ and $L2$, that contain 90% of energy, as given by:

$$Sim_{G1,G2} = \sum_{i=1}^{k}(\lambda_{1i} - \lambda_{2i})^2 \tag{12}$$

where k is chosen as shown below:

$$\min_{j} \left(\frac{\sum_{i=1}^{k} \lambda_{ji}}{\sum_{i=1}^{n} \lambda_{ji}} > 0.9 \right) \tag{13}$$

Low values of $Sim_{G1,G2}$ indicate that the graphs are very similar and vice-versa.

Temporal Consistency and Label Propagation. If the matching score (see equation (12)) is less than an experimentally chosen threshold (T_1), then the two co-located regions under consideration have temporal coherence. So, we simply copy the label of the superseed of the previous frame to the next frame. If this score is higher, then there is no such temporal consistency between the two corresponding regions. This may occur due to an exit or a new entry in the current frame. To further differentiate between these two situations, we check the spatial affinity (ω_{ij}) of the superpixel in the current frame with its neighbors in the local region graph. If the spatial affinity is more than an experimentally chosen threshold (T_2), it signifies an exit and no new label is required in that case. If the spatial affinity is less, it signifies an entry and we assign a new label to the superpixel in the current frame. In this manner, we ensure temporal coherence between each successive pair of frames under different situations (with or without entry and/or exit).

2.4   Watershed for Final Segmentation

We next employ the sequential unordered watershed algorithm with respect to topographical distance function [16], derived from the shortest path algorithm, to label the remaining pixels in the current frame to achieve the final segmentation. The basics of watershed transform following [16, 19] is included for the sake of completeness. Let f be a gray value of the morphologically processed input frame (image). The lower slope $LS(p)$ at pixel p is defined as the maximal slope linking p to any of its neighbors of lower altitude. Thus,

$$LS(p) = \max_{q \in N_G(p) \cup \{q\}} \left(\frac{f(p) - f(q)}{d(p, q)} \right) \tag{14}$$

where $N_{G(p)}$ is the set of neighbors of pixel p on the grid graph $G = (V, E)$ built on f and $d(p, q)$ is the distance associated with the edge (p, q). The cost of walking from a pixel p to its neighboring pixel q is defined as:

$$cost(p, q) = \begin{cases} LS(p) \cdot d(p, q) & if \ \ f(p) > f(q) \\ LS(q) \cdot d(p, q) & if \ \ f(p) < f(q) \\ \frac{1}{2}(LS(p) + LS(q)) \cdot d(p, q) & if \ \ f(p) = f(q) \end{cases} \tag{15}$$

The topographical distance along a path π between p and q is defined as:

$$T_f^\pi(p,q) = \sum_{i=0}^{l-1} d(p_i, p_{i+1}) \cdot cost(p_i, p_{i+1}) \qquad (16)$$

The topographical distance between p and q is the minimum of the topographical distances along all paths between p and q and is defined as:

$$T_f(p,q) = \min_{\pi \in [p \longrightarrow q]} T_f^\pi(p,q) \qquad (17)$$

Let $(m_i)_{i \in I}$ be the collection of minima (markers) of f. The catchment basins $CB(m_i)$ of f correspond to a minimum m_i is defined as the basin of the lower completion of f:

$$CB(m_i) = \{p \in D \mid \forall j \in I \backslash \{i\} : f^*(m_i) + T_{f^*}(p, m_i) < f^*(m_j) + T_{f^*}(p, m_j)\} \qquad (18)$$

where f^* is the lower completion of f. The watershed of f with 2D grid D are the points which do not belong to any catchment basin and is defined in the following manner:

$$Wshed(f) = D \cap (\cup_{i \in I} \cdot CB(m_i))^c \qquad (19)$$

The superseeds generated in the earlier stage of our solution pipeline act as the markers (regional minima). Thus construction of the catchment basins (segments) of the frame becomes a problem of finding a path of minimal cost between each pixel and a marker (regional minima). Note that for the second frame onwards, the watershed-based final segmentation provides the labels of the superpixels in the current frame. We then propagate the labels of the superseeds in the current frame to the next frame using the graph matching technique.

3   Experimental Results

Experiments are carried out over two different types of datasets, one acquired with a static camera (NYU depth dataset) [14] and the other acquired with a moving camera (NYU Scene Dataset) [3,4]. To evaluate the performance, we use the overall pixel accuracy (OP) [18] metric. We have implemented the proposed method in MATLAB $R2013b$ environment on a desktop PC with $3.4GHz$ Intel Core i7 CPU with $8GB$ RAM. SLIC for superpixels extraction is used from [11] and DBSCAN from [13]. The average execution time of the proposed method is 3.5 sec. out of which SLIC itself takes 3 sec. [20]. The values of the thresholds T_1 and T_2 are experimentally chosen as 0.45 and 0.50. To demonstrate the robustness of our method in terms of spatial consistency we compare our results with that of [3] and [5] in Fig. 4. For our experiment, we use 500 superpixels (an experimentally chosen value) for each frame. In case of the NYU scene dataset, the results are shown in Table 1. In this table, we compare our method with the results of frame by frame method, [4] and [3]. Table 1 clearly demonstrates the OP of our method (85.63) is superior as compared to that of the frame by frame (71.11), [4] (75.31), and [3] (76.27). We also show in Table 1 that the the modified DBSCAN (OP: 85.63) yield better results than the standard DBSCAN (OP: 78.26). In Fig. 5, we present the comparison of our semantic segmentation with the ground truth and with that of [3] for five intermediate frames 55 - 59 of the NYU Scene dataset. The labeled images are overlaid on the original frames for better representation. The results clearly show that our output frames resemble the ground truth much better as compared to that of [3]. The quantitative results in terms of overall pixel accuracy (OP) for the NYU Depth dataset are presented in Table 2. We experiment with four videos from the NYU Depth dataset, namely, Dining room, Living room, Classroom and Office. Our proposed method (using modified DBSCAN) with an average OP of 72.32 surpasses both the frame-by-frame approach with an OP of 60.5 and that of [3] with an average OP of 61.6.

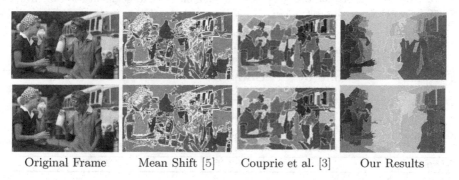

Original Frame   Mean Shift [5]   Couprie et al. [3]   Our Results

Fig. 4. Comparison of spatially consistent segments on different frames of Two women dataset [5] with independent segmentation

Table 1. OP values for the semantic segmentation task on the NYU Scene dataset

	frame by frame	Miksik et al. [4]	Couprie et al. [3]	Proposed method	
				DBSCAN [13] for initial frame	*modified DBSCAN for initial frame*
Accuracy	71.11	75.31	76.27	**78.26**	**85.63**

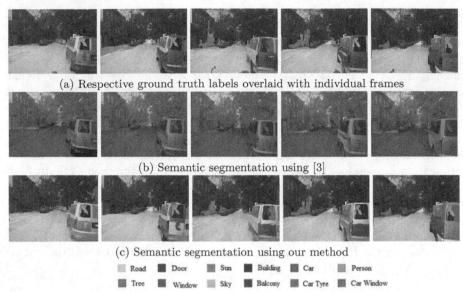

(a) Respective ground truth labels overlaid with individual frames

(b) Semantic segmentation using [3]

(c) Semantic segmentation using our method

| Road | Door | Sun | Building | Car | Person |
| Tree | Window | Sky | Balcony | Car Tyre | Car Window |

Fig. 5. Comparison of temporally consistent semantic video segmentation on frames 55 - 59 of NYU Scene dataset

Table 2. OP for the semantic segmentation task on the NYU Depth dataset

Dataset	Frame by frame	Couprie et al. [3]	Proposed Method With Modified DBSCAN
Dining room	63.8	58.5	**78.80**
Living room	65.4	72.1	**83.28**
Classroom	56.5	58.3	**65.55**
Office	56.3	57.4	**61.63**
Mean :	60.5	61.6	**72.32**

4   Conclusions

In this paper, we have presented a solution for the problem of causal video segmentation using superseeds and local graph matching. The superseeds are selected from the superpixels extracted using the SLIC algorithm. The labels of the superseeds are propagated using local graph matching. Finally, watershed algorithm is used to obtain the complete segmentation. In future, we will work on improving the execution time of our method. We will also explore how the segmentation accuracy can be further improved.

References

1. Comaniciu, D., Meer, P.: Mean Shift: A Robust Approach Toward Feature Space Analysis. IEEE TPAMI 24, 603–619 (2002)
2. Lee, Y.J., Kim, J., Grauman, K.: Key-Segments for Video Object Segmentation. In: ICCV, pp. 1995–2002 (2011)

3. Couprie, C., Farabet, C., LeCun, Y., Najman, L.: Causal Graph-Based Video Segmentation. In: ICIP, pp. 4249–4253 (2013)
4. Miksik, O., Munoz, D., Bagnell, J.A.D., Hebert, M.: Efficient Temporal Consistency for Streaming Video Scene Analysis. Tech. Report CMU-RI-TR-12-30, Robotics Institute, Pittsburgh, PA (2012)
5. Paris, S.: Edge-preserving smoothing and mean-shift segmentation of video streams. In: Forsyth, D., Torr, P., Zisserman, A. (eds.) ECCV 2008, Part II. LNCS, vol. 5303, pp. 460–473. Springer, Heidelberg (2008)
6. Galasso, F., Cipolla, R., Schiele, B.: Video Segmentation with Superpixels. In: Lee, K.M., Matsushita, Y., Rehg, J.M., Hu, Z. (eds.) ACCV 2012, Part I. LNCS, vol. 7724, pp. 760–774. Springer, Heidelberg (2013)
7. Kumar, M.P., Torr, P., Zisserman, A.: Learning Layered Motion Segmentations of Video. In: ICCV, pp. 301–319 (2012)
8. Galasso, F., Iwasaki, M., Nobori, K., Cipolla, R.: Spatio-temporal Clustering of Probabilistic Region Trajectories. In: ICCV, pp. 301–319 (2011)
9. Grundmann, M., Kwatra, V., Han, M., Essa, I.: Efficient Hierarchical Graph-Based Video Segmentation. In: ICPR, pp. 2141–2148 (2010)
10. Ferreira de Souza, K.J., Arajo, A.A., Patrocnio Jr., Z.K.G., Guimares, S.J.F.: Graph-based Hierarchical Video Segmentation Based on a Simple Dissimilarity Measure. Pattern Recognition Letters 47, 85–92 (2014)
11. Achanta, R., Shaji, A., Smith, K., Lucchi, A., Fua, P., Susstrunk, S.: SLIC Superpixels Compared to State-of-the-art Superpixels Methods. IEEE TPAMI 34, 2274–2281 (2012)
12. Ojala, T., Pietikainen, M., Maenpaa, T.: Multiresolution Gray-scale and Rotation Invariant Texture Classification with Local Binary Patterns. IEEE TPAMI 24, 971–987 (2002)
13. Ester, M., Kriegel, H.P., Sander, J., Xu, X.: A Density-Based Algorithm for Discovering Clusters, pp. 226–231. AAAI Press (1996)
14. Silberman, N., Hoiem, D., Kohli, P., Fergus, R.: Indoor segmentation and support inference from RGBD images. In: Fitzgibbon, A., Lazebnik, S., Perona, P., Sato, Y., Schmid, C. (eds.) ECCV 2012, Part V. LNCS, vol. 7576, pp. 746–760. Springer, Heidelberg (2012)
15. Koutra, D., Parikh, A., Ramdas, A., Xiang, J.: Algorithms for Graph Similarity and Subgraph Matching. Tech. Report CMU (2011)
16. Meyer, F.: Topographic Distance and Watershed Lines. Signal Processing 38, 113–125 (1994)
17. Cousty, J., Bertrand, G., Najman, L., Couprie, M.: Watershed Cuts: Minimum Spanning Forests and The Drop of Water Principle. IEEE TPAMI 31(8), 1362–1374 (2009)
18. Csurka, G., Larlus, D., Perronnin, F.: What Is a Good Evaluation Measure for Semantic Segmentation? BMVC, 2013/027 (2013)
19. Roerdink, J.B.T.M., Meijster, A.: The Watershed Transform: Definitions, Algorithms and Parallelization Strategies. Fundamenta Informaticae 41, 187–228 (2001)
20. http://www.csse.uwa.edu.au/~pk/research/matlabfns/Spatial/slic.m
21. Zhou Y., Bai X., Liu W., and Latecki L.J.: Fusion With Diffusion for Robust Visual Tracking. The Neural Information Processing Systems (NIPS), 2987–2995 (2012)

Fast Minimum Spanning Tree Based Clustering Algorithms on Local Neighborhood Graph

R. Jothi[✉], Sraban Kumar Mohanty, and Aparajita Ojha

Indian Institute of Information Technology, Design and Manufacturing Jabalpur,
Madhya Pradesh, India
{r.jothi,sraban,aojha}@iiitdmj.ac.in
http://www.iiitdmj.ac.in

Abstract. Minimum spanning tree (MST) based clustering algorithms have been employed successfully to detect clusters of heterogeneous nature. Given a dataset of n random points, most of the MST-based clustering algorithms first generate a complete graph G of the dataset and then construct MST from G. The first step of the algorithm is the major bottleneck which takes $O(n^2)$ time. This paper proposes two algorithms namely MST-based clustering on K-means Graph and MST-based clustering on Bi-means Graph for reducing the computational overhead. The proposed algorithms make use of a centroid based nearest neighbor rule to generate a partition-based Local Neighborhood Graph (LNG). We prove that both the size and the computational time to construct the graph (LNG) is $O(n^{3/2})$, which is a $O(\sqrt{n})$ factor improvement over the traditional algorithms. The approximate MST is constructed from LNG in $O(n^{3/2} \lg n)$ time, which is asymptotically faster than $O(n^2)$. The advantage of the proposed algorithms is that they do not require any parameter setting which is a major issue in many of the nearest neighbor finding algorithms. Experimental results demonstrate that the computational time has been reduced significantly by maintaining the quality of the clusters obtained from the MST.

Keywords: Clustering · MST · K-means · Bi-means · Local neighborhood graph

1 Introduction

Graph-based clustering algorithms have been used extensively in cluster analysis due to their efficient functionality in a wide range of problem domains [1][2]. Graph clustering identify similar subgraphs based on the topological properties of the graph. Several graph construction methods have been widely studied in the context of clustering, to name a few [10][11]. The motivation of many of the graph learning methods is to obtain a robust and sparse affinity graph and applying clustering algorithms such as spectral clustering on the affinity graph.

In recent years, Minimum Spanning Tree (MST) based graph clustering algorithms have drawn much attention, as they are capable of identifying clusters irrespective of their shapes and sizes [8].

© Springer International Publishing Switzerland 2015
C.-L. Liu et al. (Eds.): GbRPR 2015, LNCS 9069, pp. 292–301, 2015.
DOI: 10.1007/978-3-319-18224-7_29

The MST-based clustering method comprises of the following steps [3]:

1. Given a set of points, compute pairwise dissimilarity matrix which defines the adjacency matrix of the graph G.
2. Construct MST of G.
3. Remove the inconsistent edges until the required number of connected components are found. Each resulting component corresponds to a cluster.

Most of the previous work on MST-based clustering algorithms focuses on improving the quality of the clusters [4][5][6][7][8]. But computational efficiency is also an important issue to be considered. Very few algorithms were proposed in this direction [9][12][13]. Wang et al. proposed a MST based clustering algorithm using a divide-and-conquer scheme [9]. Using cut and cycle properties, the long edges are identified at an early stage, so as to save distance computations. However, the worst-case complexity of the algorithm remains as $O(n^2)$.

A fast approximate MST algorithm was proposed by C.Zhong et al. in [12]. By dividing the dataset into k subsets using K-means algorithm and applying exact MST algorithm on each of these subsets separately, they obtain approximate MST in $O(n^{3/2})$ complexity. However, the actual running time of the algorithm and the accuracy of the MST mainly depend on the distribution of the partitions from K-means [12].

X.Chen proposed two graph clustering algorithms namely clustering based on a near neighbor graph (CNNG) and clustering based on a grid cell graph (CGCG) [13]. While CNNG algorithm construct MST based on δ-nearest neighbors, the CGCG algorithm determines the nearest neighbors by dividing the attribute space into grid cells. The worst-case complexity of these algorithms is $O(n^2)$ and they speed up the clustering process using multidimensional grid partition and index searches. But the algorithms are sensitive to the parameters such as δ-near neighbors and interval length in each dimension of grid cells [13].

Contribution. Most of the previous algorithms assume complete graph of the dataset [7]. If we could represent the underlying structure of the dataset using a local neighborhood graph instead of a complete graph, the cost of MST-based clustering algorithms can be reduced. With this motivation, we present two efficient algorithms to improve the run time overhead caused in clustering based on MST. A centroid based nearest neighbor rule is proposed in this paper for identifying the local neighborhood of the points. We show that the number of edges in the local neighborhood graph generated by the proposed algorithms as well as the cost of graph generation is $O(n^{3/2})$. Also the proposed algorithms do not require any parameters.

The rest of the paper is organized as follows. The details of the proposed algorithms are explained in Section 2. The complexity of the proposed algorithms is discussed in Section 3. The experimental analysis is shown in Section 4. The conclusion and future scope are given in Section 5.

2   Proposed Method

Let $X = \{x_1, x_2, \cdots, x_n\}$ be a dataset of n d-dimensional points. A weighted undirected graph $G = (V, E)$ is constructed from X, where the vertex set $V = \{X\}$ and the edge set $E = \{(x_i, x_j) \mid x_i, x_j \epsilon V\}$. The edges are weighted using Euclidean distance $d(x_i, x_j)$. In most of the previous algorithms, as all-pair edges are assumed, $|E| = n(n-1)/2$, resulting in a complete graph.

As choosing the edges for MST comes from local neighborhood principle, the points which are far apart are not connected by an edge in the MST. Hence an initial edge pruning strategy is required in order to rule out the longest edges which play no role in the construction of MST. Fig. 1 illustrates this.

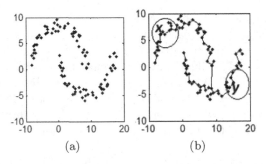

(a)   (b)

Fig. 1. MST connects points in the neighborhood:(a) A given dataset. (b) The points x and y are located far apart and the distance computation between x and y is not necessary.

In order to reduce the size of nearest neighbor search during MST construction, the proposed algorithms carry out a preliminary partition which gives us an intuition about the approximate nearest neighbors for each point. This saves much of the pair-wise distance computations and as a result the size of the graph is reduced from $O(n^2)$ to $O(n^{3/2})$. This is the key idea used in our proposed algorithms namely MST-based clustering on K-means Graph (KMGClust) and MST-based clustering on Bi-means Graph (BMGClust).

Let X be divided into a set of partitions $S = \{S_1, S_2, \cdots, S_k\}$, where k is the number of partitions. Let μ_i be the center of the partition S_i, where $1 \leq i \leq k$. Based on the partitions and their neighboring nature, we obtain local neighborhood of the points. A brute-force search on distance between centers of all pairs of partitions would easily reveal their adjacent nature. Let ω_{ij} be the distance between centers of a pair of partitions S_i and S_j. Let ω be the average distance between centers of all such pairs. ω is computed as follows.

$$\omega = \frac{1}{T_p} \sum_{S_i \epsilon S} \sum_{S_j \epsilon S} d(\mu_i, \mu_j)$$

where $T_p = k \times (k-1)/2$ and $d(a, b)$ denotes the Euclidean distance between a and b. The two partitions S_i and S_j are said to be neighbors iff $d(\mu_i, \mu_j) \leq \omega$.

With the above information, we define Local Neighborhood (LN) and Local Neighborhood Graph (LNG) as follows.

Definition 1. [Local Neighborhood] Consider a point $x_i \epsilon X$. Let S_i be the partition containing x_i. The local neighborhood of the point x_i is defined as:

$$LN(x_i) = \begin{cases} x_j, & \forall x_j \epsilon S_i \\ x_j, & \forall x_j \epsilon S_j, i \neq j, \ S_i \ \& \ S_j \text{ are neighboring partitions} \end{cases}$$

The above definition states that the points are likely to be in the neighborhood if either they belong to the the same partition or they fall in the boundary of the two partitions. Such boundary points are recognized using the Centroid Based Rule (CBR) which is stated as follows:

Definition 2. [Centroid Based Rule (CBR)] Let S_i and S_j be any two adjacent partitions with μ_{S_i} and μ_{S_j} as their centers respectively. Let x_i and x_j be any two points in the dataset X, such that $x_i \epsilon S_i$ and $x_j \epsilon S_j$. Let us define D_{ii}, D_{ij}, D_{jj} and D_{ji} as: $D_{ii} = d(x_i, \mu_{S_i})$; $D_{ij} = d(x_i, \mu_{S_j})$; $D_{jj} = d(x_j, \mu_{S_j})$; $D_{ji} = d(x_j, \mu_{S_i})$. Then, the points x_i and x_j are said to be boundary points iff $(D_{ij} \leq 2D_{ii})$ OR $(D_{ji} \leq 2D_{jj})$.

Definition 3. [Local neighborhood graph] $G_{LN} = (V, E)$ is a weighted undirected graph, where $V = \{X\}$ and E is defined as follows:

$$E = \{(x_i, x_j) \mid x_i, x_j \epsilon V \ \& \ (x_i \epsilon LN(x_j) \ OR \ x_j \epsilon LN(x_i))\} \tag{1}$$

Once the local neighborhood graph G_{LN} is constructed with respect to the above definition, MST can be generated from this graph. As only the points in the closer proximity are considered in the edge set of G_{LN}, much of the distance computations are saved.

2.1 MST-Based Clustering on K-means Graph

The KMGClust algorithm is briefed as follows. First, the given dataset X is divided into k partitions using K-means algorithm, where k is set to $\sqrt{n}$ [7]. Then, the local neighborhood graph G_{LN} is obtained by considering the points in the neighboring partitions. Finally, the MST is constructed from $G_{LN} = (V, E)$.

We divide the edge set E of G_{LN} as intra-partition edges (E_{intra}) and inter-partition edges (E_{inter}). Let $S = \{S_1, S_2, \cdots, S_k\}$ be the set of partitions produced by K-means with μ_i as the center of each partition S_i , where $1 \leq i \leq k$. Each partition S_i is a complete subgraph and hence each pair of points within a partition S_i will be connected by an intra-partition edge. The inter-partition edges are computed only between the adjacent partitions so as to rule out the comparisons between partitions which lie significantly far apart.

Once the adjacent partitions are determined, the adjoining edges are computed between these partitions. The points which lie on the boundary of a partition S_i will likely to have nearest neighbors from the boundary points of another

---

Algorithm 1. *KMGClust Algorithm.*

Input: *Dataset X.*
Output: *Approximate MST of X.*

---

1 **Partition Phase.**
   1.1 *Divide the dataset X using K-means.*
   1.2 *Let S be the set of partitions with μ_i as their respective centroids.*
2 **MST Generation Phase**
   2.1 *Let $G_{LN} = (V, E)$ be the local neighborhood graph to be constructed, where $V = \{X\}$ and $E = \{\phi\}$.*
   2.2 *Compute intra-partition all pair-edges as follows:*
         For each point x_j in subset S_i,
             For each point x_k in subset S_i, where $j \neq k$,
                 Compute $d(x_j, x_k)$ and add this edge (x_j, x_k) to E.
   2.3 *Compute inter-partition edges based on CBR as follows:*
         For each pair of neighboring partitions S_i and S_j, where $S_i \neq S_j$,
             For each pair of elements x_i and x_j, where $x_i \epsilon S_i$ and $x_j \epsilon S_j$,
                 Compute D_{ii}, D_{ij}, D_{jj} and D_{ji} as follows:
                 $D_{ii} = d(x_i, \mu_{S_i}); D_{ij} = d(x_i, \mu_{S_j}); D_{jj} = d(x_j, \mu_a); D_{ji} = d(x_j, \mu_b)$
                 If $(D_{ij} < 2D_{ii})$ || $(D_{ji} < 2D_{jj})$
                     Compute $d(x_i, x_j)$ and add this edge (x_i, x_j) to E.
   2.4 *Run Kruskal's algorithm on the graph G_{LN} to obtain MST.*

---

partition S_j, where S_i and S_j are adjacent partitions. The boundary points are recognized using CBR. The details of the KMGClust algorithm is given in Algorithm 1.

Issues with KMGCLust Algorithm. The purpose of creating an initial partition of the dataset is to minimize the number of edges in the nearest neighbor search during MST construction phase. The purpose can be achieved only if the partitions are of approximately equal size and the inter-partition distance, the distance between centers of the two partitions, is maximized. K-means algorithm may not produce balanced partitions, as a consequence, the neighborhood graph G_{LN} tends to have more number of intra-partition edges. This is explained in Fig. 2.

2.2 MST-Based Clustering on Bi-means Graph

As the initial centers for K-means partition are chosen randomly, the two or more centers may collide in a nearest region. This is shown Fig. 2c with solid stars representing the center of the partitions. In an effort to maximize the distance

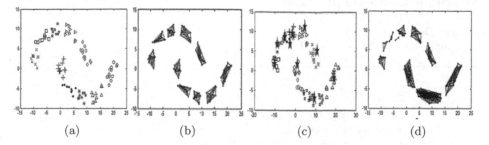

(a)   (b)   (c)   (d)

Fig. 2. Relation between equilibrium of partition and size of graph: (a) Approximately equal sized partitions. (b) Subgraphs of the partitions in (a). (c) An unbalanced partition with colliding centers. (d) Subgraphs of the partitions in (c).

between the centers of two partitions, BMGClust algorithm makes use of Bi-means algorithm which is a recursive bi-partitioning of the dataset with two centers chosen in a wise manner.

The Bi-partitioning method recursively split the dataset into a binary tree of partitions with different partitioning criteria [14]. The Bi-means algorithm used in this paper makes use of bi-partitioning method but in a different manner. Partitioning of the dataset continues as long as the size of the subset to be partitioned is greater than $\sqrt{n}$. The subsets, which cannot be partitioned further, are stored in the set S. The Bi-means algorithm is explained in Algorithm 2.

The local neighborhood graph G_{LN} can be generated from the set of partitions S returned by Bi-means Algorithm. Then, approximate MST is constructed by applying Kruskal's algorithm on the graph G_{LN}. The complete steps involved in BMGClust algorithm is described in Algorithm 3.

In the context of generating local neighborhood graph, the Bi-means algorithm has few advantages over K-means algorithm. First, unlike K-means, the Bi-means algorithm does not require to preset the number of initial partitions. Second, as the centers chosen from each of partition level are widely separated, the chance of getting colliding centers is very less.

3   Theoretical Analysis

The number of edges in the local neighborhood graph $G_{LN} = (V, E)$ resulting from KMGClust and BMGClust is bounded by the relation $|E| \leq O(n^{3/2})$. Thus, the complexity of constructing MST by KMGClust algorithm is $O(n^{3/2} \lg n)$, considering the average case. The worst case of BMGClust is $O(n^{3/2} \lg n)$.

4   Experimental Results

In order to demonstrate the efficiency of our algorithms on various clustering problems, we consider four types of artificial datasets as described in Table 1.

Algorithm 2. *Bi-means Algorithm.*

Input: *Dataset X.*
Output: *Binary partition tree of X.*

---

1 *$n' = |X|$.*
2 *If $n' > \sqrt{n}$*
   2.1 Find the center μ of the dataset X.
   2.2 Choose a point $o \epsilon X$ such that $d(o, \mu)$ is minimum.
   2.3 Compute the distance from o to all other points.
   2.4 Choose two centers $p \epsilon X$ and $q \epsilon X$ such that p is farthest from o and q is farthest from p.
   2.5 Split the dataset X into two subsets X_p and X_q according to centers p and q.
   2.6 Bi-partitioning(X_p).
   2.7 Bi-partitioning(X_q).
3 *else Append X to the table S.*
4 *return S.*

Algorithm 3. *BMGClust Algorithm.*

Input: *Dataset X.*
Output: *Approximate MST of X.*

---

1 *Let S be the set of partitions returned by Bi-means algorithm (ref. Algorithm 2).*
2 *Find Intra-partition edges as in KMGClust Algorithm (ref. Algorithm 1).*
3 *Find Inter-partition edges as in KMGClust Algorithm (ref. Algorithm 1).*
4 *Run Kruskal's algorithm on the graph G_{LN} to obtain MST.*

Experiments were conducted on a computer with an Intel Core2 Duo Processor 2GHz CPU and 2GB memory running Ubuntu Linux. We have implemented all the algorithms on C++.

The efficiency of the proposed algorithms are assessed in terms of size of the graph they generate, time required to construct MST and the validity of the clusters obtained from the MST. For the sake of clarity, let us denote the complete graph as CG and the clustering algorithm on the MST of CG as CGClust. Also We denote the local neighborhood graphs generated by our proposed algorithms KMGClust and BMGclust as KMLNG(K-means local neighborhood graph) and BMLNG(Bi-means local neighborhood graph) respectively.

Table 2 demonstrate the experimental results of the proposed algorithms on the datasets mentioned in Table 1. Fig.3 illustrates the comparison of time to construct MST (*Time*) from CGClust, KMGClust and BMGClust. In case of KMGClust and BMGClust algorithms, *Time* includes both the time spent in initial partitioning and the time spent in constructing MST. It is clear from the

Table 1. Details of the Dataset: No. of points (n), No. of clusters (k)

Dataset	n	k	Description
Data11	5000	15	Gaussian clusters
Data12	5000	15	
Data21	11522	2	Half-moon clusters
Data22	11522	2	
Data31	5000	2	Cluster inside another cluster
Data32	10106	2	
Data41	10000	2	Well-separated clusters
Data42	10000	2	

Table 2. Comparison of size of the graph ($|E|$), *Time* to construct MST and *Weight* of MST on different datasets

Dataset	Method	$|E|$	Time	Weight
Data11	CGClust	12497500	25.32	2.3430e+07
	KMGClust	824130	4.77	2.3859e+07
	BMGClust	782189	4.33	2.3430e+07
Data12	CGClust	12497500	26.77	2.7858e+07
	KMGClust	891126	15.15	2.7952e+07
	BMGClust	842218	5.13	2.7858e+07
Data21	CGClust	66372481	180.3	1026.42
	KMGClust	4783583	38.22	1027.2
	BMGClust	1933932	24.61	1026.44
Data22	CGClust	66372481	216.9	1118.45
	KMGClust	1204878	43.72	1193
	BMGClust	1893911	26.83	1113.21
Data31	CGClust	12497500	26.42	156.633
	KMGClust	607361	5.72	156.634
	BMGClust	447691	2.87	156.633
Data32	CGClust	51060565	133.76	158.017
	KMGClust	1692949	36.11	158.018
	BMGClust	1341682	11.42	158.017
Data41	CGClust	49995000	135.28	10501.8
	KMGClust	4792022	35.63	13782.2
	BMGClust	2845821	17.41	10412.9
Data42	CGClust	49995000	134.54	9711.24
	KMGClust	6118279	36.94	13501
	BMGClust	1581607	16.76	9459.83

figure that the running time of the proposed algorithms is significantly smaller as compared to CGClust.

In order to demonstrate that the local neighborhood graphs KMLNG and BMLNG will not miss any information that is significant for clustering, we test the performance of the clustering results on the approximate MST obtained from KMLNG and BMLNG using Zahn's clustering algorithm [3]. The results

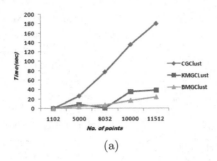

(a)

Fig. 3. Comparison of proposed algorithms

Table 3. Comparison of quality indices of clusters obtained from proposed methods with CGClust on synthetic datasets

Dataset	Method	Rand	Jaccard	FM	ARand
	CGClust	0.87565	0.34796	0.58910	0.46329
Data11	KMGClust	0.90200	0.40371	0.63461	0.53074
	BMGClust	0.87565	0.34796	0.58910	0.46329
	CGClust	1.0000	1.0000	1.0000	1.0000
Data21	KMGClust	1.0000	1.0000	1.0000	1.0000
	BMGClust	1.0000	1.0000	1.0000	1.0000
	CGClust	1.0000	1.0000	1.0000	1.0000
Data31	KMGClust	1.0000	1.0000	1.0000	1.0000
	BMGClust	1.0000	1.0000	1.0000	1.0000
	CGClust	0.83962	0.55453	0.74447	0.61432
Data41	KMGClust	0.81266	0.48295	0.67378	0.53446
	BMGClust	0.83970	0.55465	0.74455	0.61447

are validated using the external quality indices such as Rand, FM, Jaccard and Adjusted Rand [7]. Table 3 shows the quality indices of clustering on various datasets.

5   Conclusion

This paper proposed two efficient algorithms namely KMGClust and BMGClust for obtaining MST in the context of clustering in a less than quadratic time. The theocratical analysis proved that the size of the graph generated by the proposed algorithms is bounded by $O(n^{3/2})$ and thus the time for constructing MST has been reduced, while maintaining the quality of the clusters. Experimental results on various datasets demonstrated the efficiency of the proposed algorithms. As a future work, we will carry out an extensive analysis of local neighborhood graph generated by our proposed algorithms on various graph clustering algorithms.

References

1. Jain, A.K., Murty, M.N., Flynn, P.J.: Data clustering: a review. ACM Computing Surveys (CSUR) 31(3), 264–323 (1999)
2. Schaeffer, S.E.: Graph clustering. Computer Science Review 1(1), 27–64 (2007)
3. Zahn, C.T.: Graph-theoretical methods for detecting and describing gestalt clusters. IEEE Transactions on Computers 100(1), 68–86 (1971)
4. Xu, Y., Olman, V., Xu, D.: Minimum spanning trees for gene expression data clustering. GENOME INFORMATICS SERIES, pp. 24–33 (2001)
5. Laszlo, M., Mukherjee, S.: Minimum spanning tree partitioning algorithm for microaggregation. IEEE Transactions on Knowledge and Data Engineering 17(7), 902–911 (2005)
6. Luo, T., Zhong, C.: A neighborhood density estimation clustering algorithm based on minimum spanning tree. In: Yu, J., Greco, S., Lingras, P., Wang, G., Skowron, A. (eds.) RSKT 2010. LNCS, vol. 6401, pp. 557–565. Springer, Heidelberg (2010)
7. Zhong, C., Miao, D., Fränti, P.: Minimum spanning tree based split-and-merge: A hierarchical clustering method. Information Sciences 181(16), 3397–3410 (2011)
8. Wang, X., Wang, X.L., Chen, C., Wilkes, D.M.: Enhancing minimum spanning tree-based clustering by removing density-based outliers. Digital Signal Processing 23(5), 1523–1538 (2013)
9. Wang, X., Wang, X., Wilkes, D.M.: A divide-and-conquer approach for minimum spanning tree-based clustering. IEEE Transactions on Knowledge and Data Engineering 21(7), 945–958 (2009)
10. Cheng, B., Yang, J., Yan, S., Fu, Y., Huang, T.S.: Learning with L1-graph for image analysis. IEEE Transactions on Image Processing 19(4), 858–866 (2010)
11. Liu, H., Yan, S.: Robust graph mode seeking by graph shift. In: Proceedings of the 27th International Conference on Machine Learning (ICML 2010), pp. 671–678 (2010)
12. Zhong, C., Malinen, M., Miao, D., Fränti, P.: Fast approximate minimum spanning tree algorithm based on K-means. In: Wilson, R., Hancock, E., Bors, A., Smith, W. (eds.) CAIP 2013, Part I. LNCS, vol. 8047, pp. 262–269. Springer, Heidelberg (2013)
13. Chen, X.: Clustering based on a near neighbor graph and a grid cell graph. Journal of Intelligent Information Systems 40(3), 529–554 (2013)
14. Chavent, M., Lechevallier, Y., Briant, O.: DIVCLUS-T: A monothetic divisive hierarchical clustering method. Computational Statistics and Data Analysis 52(2), 687–701 (2007)

Graph-Based Applications

From Bags to Graphs of Stereo Subgraphs in Order to Predict Molecule'S Properties

Pierre-Anthony Grenier[1]([✉]), Luc Brun[1], and Didier Villemin[2]

[1] GREYC UMR CNRS 6072
[2] LCMT UMR CNRS 6507
Caen, France
{pierre-anthony.grenier,luc.brun,didier.villemin}@ensicaen.fr

Abstract. Quantitative Structure Activity and Property Relationships (QSAR and QSPR), aim to predict properties of molecules thanks to computational techniques. In these fields, graphs provide a natural encoding of molecules. However some molecules may have a same graph but differ by the three dimensional orientation of their atoms in space. These molecules, called stereoisomers, may have different properties which cannot be correctly predicted using usual graph encodings. In a previous paper we proposed to encode the stereoisomerism property of each atom by a local subgraph. A kernel between bags of such subgraphs then provides a similarity measure incorporating stereoisomerism properties. However, such an approach does not take into account potential interactions between these subgrahs. We thus propose in this paper, a method to take these interactions into account hence providing a global point of view on molecules's stereoisomerism properties.

Keywords: Graph kernel · Chemoinformatics · Stereoisomerism

1 Introduction

QSAR and QSPR methods are based on a basic principle which states that: "two similar molecules should have similar properties". An usual way to encode molecules is to use their molecular graphs. A molecular graph is a simple graph $G = (V, E, \mu, \nu)$, where each node $v \in V$ encodes an atom, each edge $e \in E$ encodes a bond between two atoms and the labeling functions μ and ν associate to each vertex and each edge a label encoding respectively the nature of the atom (carbon, oxygen,...) and the type of the bond (single, double, triple or aromatic).

However, molecular graphs have a limitation: they do not encode the spatial configuration of atoms. Indeed, some molecules, called stereoisomers, are associated to a same molecular graph but differ by the relative positioning of their atoms. We can imagine for example, a carbon atom, with four neighbors, each of them located on a summit of a tetrahedron. If we permute two of the atoms, we obtain a different spatial configuration (Figure 1a). An atom is called a stereocenter if a permutation of two atoms belonging to its neighborhood produces a different stereoisomer. Two connected atoms also define a stereocenter if a permutation of the positions of two atoms belonging to the union of

© Springer International Publishing Switzerland 2015
C.-L. Liu et al. (Eds.): GbRPR 2015, LNCS 9069, pp. 305–314, 2015.
DOI: 10.1007/978-3-319-18224-7_30

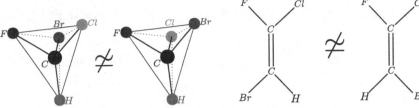

(a) Two different spatial configurations of the neighbors of a carbon

(b) Two different spatial configurations of two carbons linked by a double bond

Fig. 1. Two types of stereocenters

their neighborhoods produces a different stereoisomer (Figure 1b). According to chemical experts [8], within molecules currently used in chemistry, 98% of stereocenters correspond either to carbons with four neighbors, called asymmetric carbons (Figure 1a) or to couples of two carbons adjacent through a double bond (Figure 1b). We thus restrict the present paper to such cases.

Graph kernels [9–11, 3], allow us to combine a graph encoding of molecules with usual machine learning methods. Up to now, only few methods have attempted to incorporate stereoisomerism within the graph kernel framework. Brown et al. [1] have proposed to incorporate this information through an extension of the tree-pattern kernel [10]. In this last method, similarity between molecules is deduced from the number of common tree-patterns between two molecules. When several stereocenters are close to each other, one pattern may implicitly encode a walk which connect them. However the size of patterns being limited, in some cases the influence of a permutation around stereocenters is not detected.

Intuitively, stereoisomerism property is related to the fact that permuting two neighbors of a stereocenter produces a different spatial configuration. If those two neighbors have a same label, the influence of the permutation should be searched beyond the direct neighborhood of this stereocenter. Based on this ascertainment, we have proposed in [7] to characterized a stereocenter by a subgraph, big enough to highlight the influence of each permutation of the neighbors of this stereocenter but sufficiently small to provide a local characterization of it. We then proposed a kernel based on those subgraphs.

One drawback of our previous approach is that each subgraph, and thus each stereocenter, is considered independently. We thus present in this paper a method based on [7], which explicitly encode all minimal stereo subgraphs associated to stereocenters together with several types of interactions between these subgraphs.

In Section 2 we remind the two main points of [7], the encoding of molecules by ordered graphs, and the construction of minimal stereo subgraphs which characterize stereocenters. Then in Section 3 we present new graph models taking into account relationships between minimal stereo subgraphs. Results obtained with those graphs are provided in Section 4.

2 Ordered Graphs and Minimal Stereo Subgraphs

2.1 Encoding of Molecules by Ordered Graphs

The spatial configuration of the neighbors of each atom may be encoded through an ordering of its neighborhood [7]. In order to encode this information, we introduce the notion of ordered graph. An ordered graph $G = (V, E, \mu, \nu, ord)$ is a molecular graph $G_m = (V, E, \mu, \nu)$ together with a function $ord : V \to V^*$ which maps each vertex to an ordered list of its neighbors. Two ordered graphs G and G' are isomorphic $(G \underset{o}{\simeq} G')$ if there exists an isomorphism f between their respective molecular graphs G_m and G'_m such that $ord'(f(v)) = (f(v_1) \ldots f(v_n))$ with $ord(v) = (v_1 \ldots v_n)$ (where $N(v) = \{v_1, \ldots, v_n\}$ denotes the neighborhood of v). In this case f is called an ordered isomorphism between G and G'.

However, different ordered graphs may encode a same molecule. We thus have to define an equivalence relationship between ordered graphs, such that two ordered graphs are equivalent if they represent a same molecular configuration.

To do so, we introduce the notion of re-ordering function σ, which associates to each vertex $v \in V$ of degree n a permutation $\sigma(v)$ on $\{1, \ldots, n\}$, which allows to re-order its neighborhood. The graph with re-ordered neighborhoods $\sigma(G)$ is obtained by mapping for each vertex v its order $ord(v) = v_1 \ldots . v_n$ onto the sequence $v_{\sigma(v)(1)} \ldots . v_{\sigma(v)(n)}$ where $\sigma(v)$ is the permutation applied on v.

The set of re-ordering functions, transforming an ordered graph into another one representing the same configuration is called a valid family of re-ordering functions Σ [4]. We say that it exists an equivalent ordered isomorphism f between G and G' according to Σ if it exists $\sigma \in \Sigma$ such that f is an ordered isomorphism between $\sigma(G)$ and G' $(\sigma(G) \underset{o}{\simeq} G')$. The equivalent order relationship defines an equivalence relationship [4] and two different stereoisomers are encoded by non equivalent ordered graphs. We denote by $\text{IsomEqOrd}(G, G')$ the set of equivalent ordered isomorphism between G and G'.

Carbons with four neighbors, and double bonds between carbons, are not necessarily stereocenters. If they are not stereocenters, any permutation in their neighbourhood would lead to an equivalent ordered graph. We thus define for an ordred graph $G = (\widehat{G} = (V, E, \mu, \nu), ord)$ and one of its vertex $v \in V$ a set of ordrered isomorphism $\mathcal{F}_G^v$:

$$\mathcal{F}_G^v = \bigcup_{\substack{(i,j) \in \{1, \ldots, |N(v)|\}^2 \\ i \neq j}} \{f \mid f \in \text{IsomEqOrd}(G, \tau_{i,j}^v(G)) \text{ with } f(v) = v\}$$

where $\tau_{i,j}^v$ is a re-ordering function equals to the identity on all vertices except v for which it permutes the vertices of index i and j in $ord(v)$.

We then define a stereo vertex as a vertex for which any permutation of two of its neighbors produces a non-equivalent ordered graph:

Definition 1 (Stereo vertex). Let $G = (V, E, \mu, \nu, ord)$ be an ordered graph. A vertex $v \in V$ is called a stereo vertex iff $\mathcal{F}_G^v = \varnothing$.

Two carbons linked by a double bond form a stereocenter and we have proved in [6] that if a carbon of a double bond is a stereo vertex then the other one is also a stereo vertex. Therefore we denote by $kernel(s)$ the set of stereo vertices corresponding to a stereocenter ($kernel(s) = \{s\}$ if s is an asymmetric carbon and $kernel(s) = \{s, u\}$ if s is a carbon of a double bond, where u is the other carbon of the double bond). We further denote by $StereoStar(s)$ the set composed of a stereocenter and its neighbourhood: $StereoStar(s) = kernel(s) \cup N(kernel(s))$.

2.2   Minimal Stereo Subgraphs

Definition 1 is based on the whole graph G to test if a vertex v is a stereo vertex. However, given a stereo vertex s, one can observe that on some configurations, the removal of some vertices far from s should not change its stereo property. In order to obtain a more local characterization of a stereo vertex, we should thus determine a vertex induced subgraph H of G, including s, large enough to characterize the stereo property of s, but sufficiently small to encode only the relevant information characterizing the stereo property of s. Such a subgraph is called a minimal stereo subgraph of s.

We now present a constructive definition of a minimal stereo subgraph of a stereo vertex. Let s denote a stereo vertex and let H_s be a subgraph of G containing $kernel(s)$. We say that the stereo property of s is not captured by H_s if (Definition 1):

$$\mathcal{F}_{H_s}^s \neq \varnothing \tag{1}$$

To define a minimal stereo subgraph of s, we consider a finite sequence $(H_s^k)_{k=1}^n$ of vertex induced subgraphs of G. The first element of this sequence H_s^1 is the smallest vertex induced subgraph for which we can test (1): $V(H_s^1) = StereoStar(s)$.

If the current vertex induced subgraph H_s^k does not capture the stereo property of s, we know by (1), that it exists some isomorphisms $f \in \mathcal{F}_{H_s^k}^s$. We denote by $\mathcal{E}_f^k$ the set of vertices of H_s^k inducing the isomorphism f in H_s^k:

$$\mathcal{E}_f^k = \{v \in V(H_s^k) \mid \exists p = (v_0, \ldots, v_q) \in H_s^k \text{ with } v_0 \in kernel(s) \text{ and } v_q = v$$
$$\text{s.t. } f(v_1) \neq v_1\} \tag{2}$$

In [6], we show that for any f in $\mathcal{F}_{H_s^k}^s$, $\mathcal{E}_f^k$ is not empty. A vertex v belongs to $\mathcal{E}_f^k$ if neither its label nor its neighborhood in H_s^k allow to differentiate it from $f(v)$. The basic idea of our algorithm consists in enforcing constraints on each $v \in \mathcal{E}_f^k$ at iteration $k+1$ by adding to H_s^k the neighborhood of v in G. The set of vertices of the vertex induced subgraph H_s^{k+1} is thus defined by:

$$V(H_s^{k+1}) = V(H_s^k) \cup \bigcup_{f \in \mathcal{F}_{H_s^k}^s} N(\mathcal{E}_f^k) \tag{3}$$

where $N(\mathcal{E}_f^k)$ denote the neighborhood of $\mathcal{E}_f^k$.

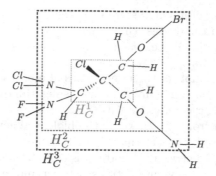

Fig. 2. An asymmetric carbon and its associated sequence $(H_C^k)_{k=1}^3$

The algorithm stops when the set $f \in \mathcal{F}_{H_s^k}^s$ becomes empty. We proved in [6] that the subgraph obtained by this algorithm captures the stereo property of s. Figure 2 illustrates our algorithm. Remarks that the computation of the minimal stereo subgraph requires the computation of graph isomorphisms and is thus nearly NP-complete. However, minimal stereo subgraphs correspond to a local characteristic of vertex and have consequently a limited size [7].

Thus for each stereo vertex we can construct its minimal stereo subgraph to characterize it. We consider two stereo vertices as similar if they have a same minimal stereo subgraph, and to test it efficiently, we transform our minimal stereo subgraphs S into codes c_S thanks to the method described in [13].

3   Graph of Interactions

In the previous section we have defined a way to encode molecules and construct an oriented subgraph which characterizes a stereocenter. We associate to an

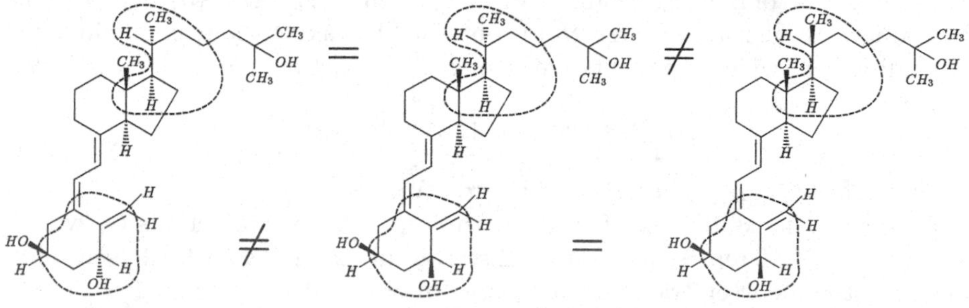

(1) : 194   (2) : 160   (3) : 554

Fig. 3. 3 molecules with the value of their biological activities. Minimal stereo subgraphs which differs between them are surrounded by dotted lines

ordered graph G its bag of minimal stereo subgraph $\mathcal{H}(G)$. In [7], we have proposed a kernel between these bags which encodes a similarity between ordered graphs. However, using this kernel, each minimal stereo subgraph is considered independently.

Figure 3 shows an example of three molecules of a dataset used in Section 4. In this figure, (2) have only one minimal stereo subgraph different from (1) and (3). Thus by considering the notion of distance associated to the kernel of [7], (1) and (3) are equidistant from (2). However the biological activity of (2) is closer from the biological activity of (1) than from the one of (3). As the different minimal stereo subgraph between (2) and (3) is close from other minimal stereo subgraphs, taking into account interactions between minimal stereo subgraphs allows to obtain a smaller distance between (1) and (2) than between (3) and (2), which may help to obtain a better prediction of the property.

Unfortunately, the amount of interactions between two minimal stereo subgraphs which may influence a molecular property is yet unknown both in chemoinformatics and chemistry fields. Hence we propose to define different functions of interactions, encoding different degrees of information about the interactions between stereo vertices.

Functions of interactions are defined according to a sequence of conditions $(c_1, \ldots, c_n)$. These conditions are increasingly constraining:

$$\forall i \in \{1, \ldots, n-1\}\, c_{i+1} \Rightarrow c_i$$

Let H_1 and H_2 be two minimal stereo subgraphs, such that s_1 is the stereo vertex of H_1 and s_2 is the stereo vertex of H_2. We propose the following set of conditions:

$$
\begin{aligned}
&c_1 : H_1 \bigcap H_2 \neq \emptyset && c_2 : kernel(s_1) \subset H_2 \\
&c_3 : StereoStar(s_1) \subset H_2 && c_4 : H_1 \subset H_2
\end{aligned}
\tag{4}
$$

We consider in this paper three functions of interactions F_i. Each function F_i is designed in order to be more restrictive than F_j (with $j < i$). To do so each F_i is defined by only using conditions c_j with j in $\{i, \ldots, 4\} \cup \{0\}$, where c_0 is defined as $\neg c_i$. The value $F_i(H_1, H_2)$ is obtained by taking the maximal index j of conditions c_j which represents the strongest interaction between H_1 and H_2:

$$F_i(H_1, H_2) = \max\{j \in \{i, \ldots, 4\} \cup \{0\} \mid c_j\}$$

Note that $(F_i)_{i \in \{1,2,3\}}$ are non symmetric functions.

We define thanks to those functions, three graphs of interactions G_i where each vertex $v \in V_i$ represents a minimal stereo subgraph and each edge encodes an interaction between two minimal stereo subgraphs deduced from F_i :

Definition 2 (Graph of interactions). A graph of interactions $G_i = (V_i, E_i, \mu_i, \nu_i)$ is a graph built from an ordered graph $G = (G_m = (V, E, \mu, \nu), ord)$ and a function of interaction F_i such that :

- $\forall u \in V_i,\ \exists! H(u) \in \mathcal{H}(G)$.
- $\forall u \in V_i,\ \mu_i(u) = c_{H(u)}$, where c_H is the code defined in [13] (Section 2).
- $\exists (u_1, u_2) \in E_i \iff F_i(H(u_1), H(u_2)) \neq 0$ or $F_i(H(u_2), H(u_1)) \neq 0$.
- $\forall e = (u_1, u_2) \in E_i,\ \nu_i(e) = \min(F_i(H(u_1), H(u_2)), F_i(H(u_2), H(u_1))) \odot \max(F_i(H(u_1), H(u_2)), F_i(H(u_2), H(u_1)))$.

where $\odot$ denotes the concatenation.

We can check if a vertex is in a minimal stereo subgraph in constant time. Thus, the complexity to check each condition (equation 4) considering two minimal stereo subgraphs H_1 and H_2 is $\mathcal{O}(max(|H_1|, |H_2|))$ for c_1, $\mathcal{O}(|kernel(s_1)|)$ for c_2, $\mathcal{O}(|StereoStar(s_1)|)$ for c_3 and $\mathcal{O}(|H_1|)$ for c_4. The worst case complexity for computing graphs of interactions is thus equal to $\mathcal{O}(|\mathcal{H}(G)|^2 \max_{H \in \mathcal{H}(G)} |H|)$. In practice this value is small (for the vitamin dataset presented in Section 4, we have at most $|\mathcal{H}(G)| = 9$ and $\max_{H \in \mathcal{H}(G)} |H| = 24$).

Figure 4 shows the graphs of interactions obtained from an ordered graph using the different functions of interactions. The graph G_1 is built by taking all conditions. However we may suppose that the weakest interaction c_1 may not be relevant. Indeed, an intersection between two minimal stereo subgraphs may not be a sufficiently relevant information. Thus the graph G_2 is designed by considering that two stereo vertices are related if we have at least $kernel(s_1) \subset H_2$ or $kernel(s_2) \subset H_1$. Moreover, a vertex s_1 is a stereo vertex because of the relative positioning of its neighbour. So we may suppose that, if a stereo vertex is present in a minimal stereo subgraph ($kernel(s_1) \subset H_2$), but not its neighbourhood

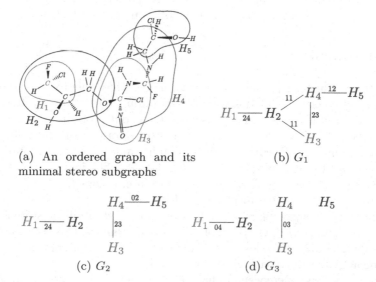

(a) An ordered graph and its minimal stereo subgraphs

(b) G_1

(c) G_2

(d) G_3

Fig. 4. One ordered graph and its different graphs of interactions G_i, obtained using F_i with $i \in \{1, 2, 3\}$

Table 1. Average values of numbers of vertices ($\overline{|V|}$), number of edges ($\overline{|E|}$), number of labels($\overline{|\mathcal{L}_V|}$,$\overline{|\mathcal{L}_E|}$), and mean degree ($\overline{d}$) of graph of interactions

(a) ACE dataset

	$\overline{|V|}$	$\overline{|E|}$	$\overline{|\mathcal{L}_V|}$	$\overline{|\mathcal{L}_E|}$	$\overline{d}$
Graph 1	5	7	4.5	3	2.8
Graph 2	5	2	4.5	2	0.8
Graph 3	5	1	4.5	1	0.4

(b) Vitamin dataset

	$\overline{|V|}$	$\overline{|E|}$	$\overline{|\mathcal{L}_V|}$	$\overline{|\mathcal{L}_E|}$	$\overline{d}$
Graph 1	8.55	17.4	8.38	5.71	4.07
Graph 2	8.55	11.3	8.38	4.71	2.62
Graph 3	8.55	6.14	8.38	2.71	1.43

Table 2. Classification of the ACE inhibitory activity of perindopirilates stereoisomers

Method	Accuracy				
Brown [1]	96.875				
Stereo Kernel [7]	87.5				
Stereo + Extended subgraphs [5]	96.875				
Graph of interactions with	[3]	[9]	[10]	[11]	[3] with MKL
Graph of interactions 1	93.75	84.375	93.75	93.75	**100**
Graph of interactions 2	93.75	62.5	62.5	62.5	87.5
Graph of interactions 3	84.375	62.5	62.5	62.5	90.625

($StereoStar(s_1) \not\subset H_2$), the stereo vertex may have a similar influence in H_2 than a non-stereo vertex. Thus G_3 is built by considering that two stereo vertices are related if we have at least $StereoStar(s_1) \subset H_2$ or $StereoStar(s_2) \subset H_1$.

As graphs of interactions are graphs without order, we may apply any graph kernel (e.g. [9–11, 3]) to measure their similarities. Note that for the treelet kernel [3], treelets of size 1 are vertices of graphs of interactions, and thus encode the same notion of similarity than the bags of minimal stereo subgraphs [7].

4   Experiments

We have tested our method on two datasets. For both of them we use the same protocol: a nested cross-validation which select parameters and estimate the performance. The outer cross-validation is a leave-one-out procedure, used to compute an error of prediction for each molecule of the dataset. For each fold, we use another leave-one-out procedure on the remaining molecules, to compute a validation error. We use standard SVM methods for classification and regression of molecules. Basic statisistics about the graphs of interactions $G_i = (V_i, E_i, \mu_i, \nu_i)$ deduced from each dataset is displayed in Table 1.

Our first experiment is based on a dataset composed of all the stereoisomers of the perindoprilate [2]. As this molecule has 5 stereocenters, the dataset is composed of $2^5 = 32$ molecules. In this dataset, we try to predict if a molecule inhibit the angiotensin-converting enzyme (ACE).

In this dataset two stereocenters have a same minimal stereo subgraph, but different surrounding. The stereo kernel [7] and one of the graph of interactions (G_3), can not differentiate those two stereocenters, which have different influence

Table 3. Prediction of the biological activity of synthetic vitamin D derivatives

Method	RMSE			
1 - Tree patterns Kernel [10]	0.251			
2 - Treelet Kernel [3]	0.271			
3 - Brown [1]	0.184			
4 - Stereo Kernel [7]	0.194			
5 - Stereo + Extended subgraphs [5]	0.180			
Graph of interactions with	[3]	[9]	[10]	[11]
6 - Graph of interactions 1	0.177	0.184	0.185	0.201
7 - Graph of interactions 2	0.169	0.189	*0.162*	0.166
8 - Graph of interactions 3	0.172	0.172	**0.161**	*0.162*

on the property, this explains why other method ([1, 5] and the two other graphs of interactions) obtain a better accuracy. However, for our graphs of interactions, treelet of size one have a negative effect on the classification, this explains why we do not obtain better results than [1, 5]. By using a multiple kernel learning algorithm [12], we can learn a weight for each treelet, that allow us to discard treelet of size 1 and to obtain the best results with the first graph of interactions. The second graph of interactions have very few edges and a low degree (Table 1a). This last point explains why some kernel [9–11] and the treelet kernel with multiple kernel learning obtains poor results on this graph.

The second dataset is a dataset of synthetic vitamin D derivatives, used in [1]. This dataset is composed of 69 molecules, with an average of 8.55 stereocenters per molecule. This dataset is associated to a regression problem, which consists in predicting the biological activity of each molecule.

Methods which do not encode stereoisomerism information [10, 3] obtain poor results as we can see in Table 3 (lines 1-2). The adaptation of the tree pattern kernel to stereoisomerism [1] and our previous kernels [7, 5] (lines 3-5) improves the results over the two previous methods hence showing the insight of adding stereoisomerism information. Taking into account relationships between minimal stereo subgraphs (lines 6-8) allows us to obtain better results than our previous method [5]. Unlike the previous dataset, the graphs of interactions G_2 and G_3 have higher degree (≈ 2) and thus obtains good results. Graph G_1 have a high degree (4) and a high number of different labels on vertices, which induces a lot of unique patterns in each graph. This last point decreases the number of patterns common to two graphs and explains why on this dataset G_1 does not obtain results as good as G_2 and G_3.

In conclusion, the treelet kernel applied on graphs of interactions seems to obtain equivalent or better results than alternative kernel methods. Moreover, given a data set of molecules related to stereoisomerism, our experiments show that the choice of a particular graph of interactions should be based on the mean degree of vertice of these graphs. A mean degree of 2 seems to correspond to a good compromise between a very low degree corresponding to a node set without graph structure and a too high degree which also hidden the graph structure.

5   Conclusion

In this paper we have proposed an extension of our previous method [7] based on a new graph model, where each node represents a stereo subgraph and each edge encodes an interaction between two stereo subgraphs. This graph allows us to take into account interactions between stereo subgraphs. The relevance of this approach is demonstrated through experiments on two datasets.

Acknowledgements. This work has been performed using computing resources partially funded by the CPER Normandie.

References

1. Brown, J., Urata, T., Tamura, T., Arai, M.A., Kawabata, T., Akutsu, T.: Compound analysis via graph kernels incorporating chirality. Journal of Bioinformatics and Computational Biology 8(1), 63–81 (2010)
2. Castillo-Garit, J.A., Marrero-Ponce, Y., Torrens, F., Rotondo, R.: Atom-based stochastic and non-stochastic 3d-chiral bilinear indices and their applications to central chirality codification. Journal of Molecular Graphics and Modelling 26(1), 32–47 (2007)
3. Gaüzère, B., Brun, L., Villemin, D.: Two New Graphs Kernels in Chemoinformatics. Pattern Recognition Letters 33(15), 2038–2047 (2012)
4. Grenier, P.-A., Brun, L., Villemin, D.: Incorporating stereo information within the graph kernel framework. Technical report, CNRS UMR 6072 GREYC (2013), http://hal.archives-ouvertes.fr/hal-00809066
5. Grenier, P.-A., Brun, L., Villemin, D.: Incorporating molecule's stereisomerism within the machine learning framework. In: Fränti, P., Brown, G., Loog, M., Escolano, F., Pelillo, M. (eds.) S+SSPR 2014. LNCS, vol. 8621, pp. 12–21. Springer, Heidelberg (2014)
6. Grenier, P.-A., Brun, L., Villemin, D.: Taking into account interaction between stereocenters in a graph kernel framework. Technical report, CNRS UMR 6072 GREYC (2014), https://hal.archives-ouvertes.fr/hal-01103318
7. Grenier, P.-A., Brun, L., Villemin, D.: A graph kernel incorporating molecule's stereisomerism information. In: Proceedings of ICPR (2014)
8. Jacques, J., Collet, A., Wilen, S.: Enantiomers, racemates, and resolutions. Krieger Pub. Co. (1991)
9. Kashima, H., Tsuda, K., Inokuchi, A.: Marginalized kernels between labeled graphs. In: ICML, vol. 3, pp. 321–328 (2003)
10. Mahé, P., Vert, J.-P.: Graph kernels based on tree patterns for molecules. Machine Learning 75(1), 3–35 (2008)
11. Shervashidze, N., Schweitzer, P., Van Leeuwen, E.J., Mehlhorn, K., Borgwardt, K.M.: Weisfeiler-lehman graph kernels. The Journal of Machine Learning Research 12, 2539–2561 (2011)
12. Varma, M., Babu, B.R.: More generality in efficient multiple kernel learning. In: Proceedings of the 26th Annual International Conference on Machine Learning, pp. 1065–1072. ACM (2009)
13. Wipke, W.T., Dyott, T.M.: Stereochemically unique naming algorithm. Journal of the American Chemical Society 96(15), 4834–4842 (1974)

Thermodynamics of Time Evolving Networks

Cheng Ye[1(✉)], Andrea Torsello[2], Richard C. Wilson[1], and Edwin R. Hancock[1]

[1] Department of Computer Science, University of York, York, YO10 5GH, UK,
{cy666,richard.wilson,edwin.hancock}@york.ac.uk
[2] Dept. Environmental Sciences, Informatics and Statistics,
Ca' Foscari University of Venice, Dorsoduro 3246 - 30123 Venezia, Italy

Abstract. In this paper, we present a novel and effective method for better understanding the evolution of time-varying complex networks by adopting a thermodynamic representation of network structure. We commence from the spectrum of the normalized Laplacian of a network. We show that by defining the normalized Laplacian eigenvalues as the microstate occupation probabilities of a complex system, the recently developed von Neumann entropy can be interpreted as the thermodynamic entropy of the network. Then, we give an expression for the internal energy of a network and derive a formula for the network temperature as the ratio of change of entropy and change in energy. We show how these thermodynamic variables can be computed in terms of node degree statistics for nodes connected by edges. We apply the thermodynamic characterization to real-world time-varying networks representing complex systems in the financial and biological domains. The study demonstrates that the method provides an efficient tool for detecting abrupt changes and characterizing different stages in evolving network evolution.

Keywords: Thermodynamics · Time-varying networks · Von Neumann entropy · Internal energy · Temperature

1 Introduction

The study of complex networks has been shown to play a crucial role in natural sciences. Examples of such networks include the World Wide Web, metabolic reaction networks, financial market stock correlations, scientific collaboration, coauthorship and citation relations, and social interactions [10]. One of the most important characteristics of complex networks is their internal structure, i.e., how the components and connections between components are arranged or organized. The reason for this is that the structure of complex networks has a significant influence on their function and performance [6].

The normalized Laplacian spectrum has provided a convenient route to a complexity level characterization of networks via the von Neumann entropy (or quantum entropy) associated with a density matrix [11]. By mapping between discrete Laplacians and quantum states, provided that the discrete Laplacian is scaled by the inverse of the volume of the graph, a density matrix is obtained whose entropy can be computed using the spectrum of the discrete Laplacian [3].

© Springer International Publishing Switzerland 2015
C.-L. Liu et al. (Eds.): GbRPR 2015, LNCS 9069, pp. 315–324, 2015.
DOI: 10.1007/978-3-319-18224-7_31

Moreover, the von Neumann entropy can be efficiently approximated in terms of simple degree statistics hence avoiding the need to compute the normalized Laplacian eigenvalues, for both undirected and directed graphs [8] [14].

Statistical thermodynamics can be combined with both network theory and kinetics to provide a practical framework for handling highly structured and highly interactive complex systems [9]. By using a random walk that maximizes the Ruelle-Bowens free-energy rate on weighted graphs, a novel centrality measure can be computed, and this has been successfully applied to both connected and disconnected large-scale networks [5]. Recently, it has been demonstrated that the subgraph centrality can be interpreted as a partition function of a network [7], and as a result the entropy, internal energy and the Helmholtz free energy can be defined using spectral graph theory. Moreover, various relations between these thermodynamic variables and both graph structure and graph dynamics can be obtained.:

Although the bulk of existing network theory is concerned with static networks, most realistic networks are in reality dynamic in nature. Specifically, networks grow and evolve with the addition of new components and connections, or the rewiring of connections from one component to another [1]. Motivated by the need to fill this gap in the literature and to augment the methods available for understanding the evolution of time-varying networks, in this paper we establish a thermodynamic framework for analyzing the time-varying networks. We first explore whether the approximate von Neumann entropy can be used as a thermodynamic entropy of a network. Then, we derive expressions for additional thermodynamic variables of networks, including the internal energy and temperature.

The remainder of the paper is organized as follows. In Sec. II, we provide a detailed account of the development of the thermodynamic variables of networks, i.e., the entropy, internal energy and temperature. In Sec. III, we apply the resulting characterization to a number of real-world time-varying networks, including the New York Stock Exchange (NYSE) data and fruit fly life cycle gene expression data. Finally, we conclude the paper and make suggestions for future work.

2   Thermodynamic Variables of Complex Networks

In this section, we provide a detailed development of expressions for thermodynamic network variables, including the entropy, internal energy and temperature. To do this, we first show that the recently developed approximate von Neumann entropy can also be interpreted as the thermodynamic entropy of a network, when we associate the microscopic configurations of a network with the eigenstates of the normalized Laplacian spectrum. We then develop an expression for network internal energy and determine the network temperature by measuring fluctuations in entropy and internal energy. Each of these thermodynamic variables can be computed using simple graph statistics, including the number of nodes and edges together with node degree statistics.

2.1   Initial Considerations

Let $G(V, E)$ be an undirected graph with node set V and edge set $E \subseteq V \times V$. The adjacency matrix A of graph $G(V, E)$ is defined as

$$A_{uv} = \begin{cases} 1 & \text{if } (u, v) \in E \\ 0 & \text{otherwise.} \end{cases} \tag{1}$$

The degree of node u is $d_u = \sum_{v \in V} A_{vu}$. The normalized Laplacian matrix $\tilde{L}$ is $\tilde{L} = D^{-1/2}(D - A)D^{-1/2}$ where D is the degree diagonal matrix whose elements are given by $D(u, u) = d_u$ and zeros elsewhere. The elementwise expression of $\tilde{L}$ is

$$\tilde{L}_{uv} = \begin{cases} 1 & \text{if } u = v \text{ and } d_v \neq 0 \\ -\frac{1}{\sqrt{d_u d_v}} & \text{if } u \neq v \text{ and } (u, v) \in E \\ 0 & \text{otherwise.} \end{cases} \tag{2}$$

The normalized Laplacian matrix $\tilde{L}$ and its spectrum yield a number of very useful graph invariants for a finite graph. For example, the eigenvalues for the graph normalized Laplacian are real numbers, bounded between 0 and 2 [4].

According to [11], the normalized Laplacian matrix $\tilde{L}$ can be interpreted as the density matrix of an undirected graph. With this choice of density matrix, the von Neumann entropy of the undirected graph is defined as the Shannon entropy associated with the normalized Laplacian eigenvalues, i.e.,

$$H_{VN} = -\sum_{i=1}^{|V|} \frac{\tilde{\lambda}_i}{|V|} \ln \frac{\tilde{\lambda}_i}{|V|} \tag{3}$$

where $\tilde{\lambda}_i$, $i = 1, \ldots, |V|$, are the eigenvalues of $\tilde{L}$.

In this paper, we aim at developing a thermodynamic characterization of network structure. We commence by assuming that at any instant in time a network $G(V, E)$, is statistically distributed across an ensemble of $|V|$ microstates. The probability that the system occupies a microstate indexed s is given by $p_s = \tilde{\lambda}_s / \sum_{s=1}^{|V|} \tilde{\lambda}_s$, where $\tilde{\lambda}_s$, $s = 1, 2, \ldots, |V|$ are the eigenvalues of the normalized Laplacian matrix of graph G. Noting that the trace of a matrix is the sum of its eigenvalues, we have $\sum_{s=1}^{|V|} \tilde{\lambda}_s = Tr[\tilde{L}] = |V|$, so the microstate occupation probability is simply $p_s = \tilde{\lambda}_s / |V|$.

We define the thermodynamic entropy of a network using the Shannon formula that is exclusively dependent on the probabilities of the microstates:

$$S = -k \sum_{s=1}^{|V|} p_s \ln p_s = -k \sum_{s=1}^{|V|} \frac{\tilde{\lambda}_s}{|V|} \ln \frac{\tilde{\lambda}_s}{|V|}, \tag{4}$$

where k is the Boltzmann constant and is set to be 1 to simplify matters.

It is clear that the thermodynamic entropy Eq.(4) and the von Neumann entropy Eq.(3) take the same form. Both depend on the graph size and the eigenvalues of the normalized Laplacian matrix. It is reasonable to suggest that the von Neumann entropy can be interpreted as the thermodynamic entropy of a complex network.

2.2   Approximate Von Neumann Entropy for Complex Networks

In prior work, we have shown how the von Neumann entropy of a network Eq.(3) can be simplified by making use of the quadratic approximation (i.e., $-x \ln x \approx x(1-x)$),

$$H_Q = \sum_{i=1}^{|V|} \frac{\tilde{\lambda}_i}{|V|} (1 - \frac{\tilde{\lambda}_i}{|V|}). \tag{5}$$

For undirected graphs this quadratic approximation allows the von Neumann entropy to be expressed in terms of the trace of the normalized Laplacian and the trace of the squared normalized Laplacian, with the result that

$$H_{VN} = \frac{Tr[\tilde{L}]}{|V|} - \frac{Tr[\tilde{L}^2]}{|V|^2}. \tag{6}$$

The two traces appearing in the above expression are given in terms of node degree statistics [8], leading to

$$H_{VN} = 1 - \frac{1}{|V|} - \frac{1}{|V|^2} \sum_{(u,v) \in E} \frac{1}{d_u d_v}. \tag{7}$$

This formula contains two measures of graph structure, the first is the number of nodes of graph, while the second is based on degree statistics for pairs of nodes connected by edges. Moreover, the expression for the approximate entropy has computational complexity that is quadratic in graph size, which is simpler than the original von Neumann entropy that is cubic since it requires enumeration of the normalized Laplacian spectrum.

In order to obtain a better understanding of the entropic measure of graphs, it is interesting to explore how the von Neumann entropy is bounded for graphs of a particular size, and in particular which topologies give the maximum and minimum entropies. From Eq.(7) it is clear that when the term under the summation is minimal, the von Neumann entropy reaches its maximal value. This occurs when each pair of graph nodes is connected by an edge, and this means that the graph is complete. On the other hand, when the summation takes on its maximal value, the von Neumann entropy is minimum. This occurs when the structure is a string.

The maximum and minimum entropies corresponding to these cases are as follows. For a complete graph K_n, in which each node has degree $n-1$, it is straightforward to show that

$$H_{VN}(K_n) = 1 - \frac{1}{n} - \frac{1}{n^2} \cdot \frac{n(n-1)}{2(n-1)^2} = 1 - \frac{2n-1}{2n(n-1)}.$$

In the case of a string P_n ($n \geq 3$), in which two terminal nodes have degree 1 while the remainder have degree 2, we have

$$H_{VN}(P_n) = 1 - \frac{1}{n} - \frac{1}{n^2} \cdot \frac{n+1}{4} = 1 - \frac{5n+1}{4n^2}.$$

As a result, the graph von Neumann entropy is bounded as follows:

$$1 - \frac{5|V| + 1}{4|V|^2} \leq H_{VN}(G) \leq 1 - \frac{2|V| - 1}{2|V|(|V| - 1)}$$

where the lower boundary is obtained for strings, which are the simplest regular graph, and the upper bound is reached for complete graphs.

2.3   Internal Energy and Temperature of Complex Networks

The internal energy of a network is defined as the mean value of the total energy, i.e., the sum of all microstate energies, each weighted by its occupation probability:

$$U = \sum_{s=1}^{|V|} p_s U_s, \tag{8}$$

where U_s is the energy of microstate s. Here we take the internal energy to be the total number of edges in the graph i.e., $U = |E|$. From the properties of the Laplacian and normalized Laplacian matrices, we have $|E| = Tr[L] = Tr[D^{1/2}\tilde{L}D^{1/2}] = Tr[D\tilde{L}]$. This can be achieved if we set the microstate energies to be $U_s = |V|d_s$, i.e., proportional to the node degrees.

Suppose that the graphs $G = (V, E)$ and $G' = (V', E')$ represent the structure of a time-varying complex network at two consecutive epochs t and t' respectively. The reciprocal of the thermodynamic temperature T is the rate of change of entropy with internal energy, subject to the condition that the volume and number of particles are held constant, i.e., $1/T = dH_{VN}/dU$. This definition can be applied to evolving complex networks which do not change size during their evolution.

We write the change of the von Neumann entropy H_{VN} between graphs G and G' as

$$dH_{VN} = H_{VN}(G') - H_{VN}(G) = \sum_{(u,v) \in E, E'} \frac{d_u \Delta_v + d_v \Delta_u + \Delta_u \Delta_v}{d_u(d_u + \Delta_u)d_v(d_v + \Delta_v)},$$

where Δ_u is the change of degree of node u: $\Delta_u = d'_u - d_u$; Δ_v is similarly defined. The change in internal energy, is equal to the change in the total number of edges: $dU = U(G') - U(G) = |E'| - |E| = \Delta|E|$. Hence the reciprocal temperature T is:

$$\frac{1}{T(G,G')} = \sum_{(u,v) \in E, E'} \frac{d_u \Delta_v + d_v \Delta_u + \Delta_u \Delta_v}{\Delta|E|d_u(d_u + \Delta_u)d_v(d_v + \Delta_v)}. \tag{9}$$

The temperature measures fluctuations in the internal structure of the time evolving network, and depends on two properties of the network. The first is the overall or global change of the number of edges, while the second property is a local one which measures the change in degree for pairs of nodes connected by edges. Both quantities measure fluctuations in network structure, but at different levels of detail. The temperature is greatest when there are significant differences in the global number of edges, and smallest when there are large local variations in edge structure which do not result in an overall change in the number of edges.

To summarize, in this section we have proposed a novel method for characterizing the evolution of complex networks by employing thermodynamic variables. Specifically, we show that the network von Neumann entropy can be used as a thermodynamic characterization, provided that the eigenstates of the normalized Laplacian matrix define the network's microstates together with their occupation probabilities. Moreover, the internal energy depends on the number of edges in the network. The thermodynamic temperature measures fluctuations via the change in the number of edges and node degree changes.

3   Experiments and Evaluations

We have derived expressions for the thermodynamic entropy, internal energy and temperature of time-evolving complex networks. In this section, we explore whether the resulting measures can be employed to provide a useful tool for better understanding the evolution of dynamic networks. In particular, we aim at applying the novel thermodynamic method to a number of real-world time-evolving networks in order to analyze whether abrupt changes in structure or different stages in network evolution can be efficiently characterized.

We commence by giving a brief overview of the two datasets used for experiments, which are both extracted from real-world complex systems.

Dataset 1: Is extracted from a database consisting of the daily prices of 3799 stocks traded on the New York Stock Exchange (NYSE). To construct the dynamic network, the 347 stock that have historical data from January 1986 to February 2011 are selected [12]. Then, we use a time window of 28 days and move this window along time to obtain a sequence (from day 29 to day 6004) in which each temporal window contains a time-series of the daily return stock values over a 28 day period. We represent trades between the different stocks as a network. For each time window, we compute the cross correlation coefficients between the time-series for each pair of stocks, and create connections between them if the absolute value of the correlation coefficient exceeds a threshold. This yields a time-varying stock market network with a fixed number of 347 nodes and varying edge structure for each of 5976 trading days.

Dataset 2: Is from the domain of biology and records the dynamic interactions between genes of the fruit fly (Drosophila melanogaster) during its complete life cycle. The data is sampled at 66 developmental time points. The fruit fly life cycle is divided into four stages, namely the embryonic (samples 1-30), larval (samples 31-40) and pupal (samples 41-58) periods together with the first 30 days of adulthood (samples 59-66). To represent this data using a time evolving network, the following steps are followed [13]. At each developmental point the 588 genes that are known to play an important role in the development of the Drosophila are selected. These genes are the nodes of the network, and edges are established based on the microarray gene expression measurements reported in [2]. This dataset yields a time-evolving Drosophila gene-regulatory network with a fixed number of 588 nodes, sampled at 66 developmental time points.

3.1   Thermodynamic Measures for Analyzing Network Evolution

We apply our thermodynamic characterization method to the dynamic networks in Dataset 1 and Dataset 2. At each time step we compute the entropy, internal energy and temperature according to Eq.(7), Eq.(8) and Eq.(9) respectively. This allows us to investigate how these network thermodynamic variables evolve with time and whether some critical events can be detected in the network evolution. These include financial crises or crashes in the stock market, and the essential morphological transformations that occur in the development of the Drosophila.

The leftmost plot in Fig. 1 is a 3-dimensional scatter plot showing the thermodynamic variables for the time-evolving stock correlation network. It represents a thermodynamic space spanned by entropy, internal energy and temperature. Also shown in Fig. 1 are the individual times series for the different thermodynamic variables. The most important feature here is that the thermodynamic distribution of networks in Fig. 1(a) shows a strong manifold structure with different phases of network evolution occupying different volumes of the thermodynamic space. There are though outliers, and these appear as peaks and troughs in the individual time series (in Fig. 1(b), (c) and (d)). The outliers indicate significant global events. Examples include Black Monday, the outbreak of the Persian Gulf War (17 January 1991) and the 24 October 2008 stock market crash. Another interesting feature in Fig. 1(a) is that the Dot-com bubble period (approximately from 1997 to 2000) which is represented by red dots, is separated from the background data points and occupies a distinct region in the thermodynamic space. The reason for this is that during the Dot-com bubble period, a significant number of Internet-based companies were founded, leading to a rapid increase of both stock prices and market confidence. This considerably changed both the inter-relationships between stocks and the resulting structure of the entire market.

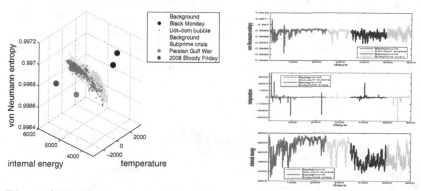

Fig. 1. a)The 3D scatter plot of the dynamic stock correlation network in the thermodynamic space. Yellow dots: 1986 - 1996 background data; red dots: Dot-com bubble; green dots: 2001 - 2006 background data; cyan dots: Subprime crisis; black: Black Monday; magenta: Outbreak of Persian Gulf War; blue: 2008 Bloody Friday stock market crash. b) The von Neumann entropy versus time for the dynamic stock correlation network. c) The internal energy versus time for the dynamic stock correlation network. d) The temperature versus time for the dynamic stock correlation network.

We now study three financial crises in detail, and explore how the stock market network structure changes with time according to the thermodynamic variables. In Fig. 2 we show the trace of the stock network on the entropy-energy plane during Black Monday, the Asian Financial Crisis and the Lehman Brothers bankruptcy respectively. The number beside each data point represents the day number in the time series. From the left-hand panel we observe that before Black Monday, the network structure remains stable, neither the network entropy nor the internal energy change significantly. However, during Black Monday (day 115 and 116), the network experiences a considerable change in structure since the entropy increases dramatically. After the crisis, the network entropy slowly decreases, and the stock correlation network gradually returns to its normal state. A similar observation can also be made concerning the 1997 Asian Financial Crisis which is shown in the middle panel. The stock network again undergoes a significant crash in which the network structure undergoes a significant change, as signalled by a large drop in network entropy. The crash is followed by a slow recovery. For the Lehman Brothers bankruptcy in the right-hand panel, it is interesting to note that as the time series evolves, both the network entropy and the internal energy continue to grow gradually. This illustrates a very different pattern. Hence, our thermodynamic representation can be used to both characterize and distinguish between different financial crises.

We now turn our attention to the fruit fly network, i.e., the Drosophila gene regulatory network contained in Dataset 2. In Fig. 3, we again show the 3-dimensional scatter plot of the time-varying thermodynamic variables space, together with the entropy, energy and temperature times series. The four developmental stages are shown in different colours. The different stages of evolution are easily distinguished by the thermodynamic variables. For instance from Fig. 3(a), since the early development of an embryo, the red dots (embryonic period) show significant fluctuations. This is attributable to strong and rapidly changing gene interactions, because of the need for rapid development. Moreover, in the pupal stage, the data are relatively sparsely distributed in the thermodynamic space. This is attributable to the fact that during this period, the pupa undergoes a number of significant pupal-adult transformations. As the organism evolves into an adult, the gene interactions

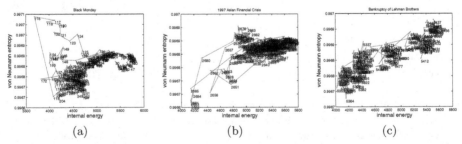

(a)   (b)   (c)

Fig. 2. Trace of the time-evolving stock correlation network in the entropy-energy plane during financial crises (the number beside data point is the day number). Left: Black Monday (from day 30 to 300 in the time series); middle: Asian Financial Crisis (from day 2500 to 2800); right: Bankruptcy of Lehman Brothers (from day 5300 to 5500).

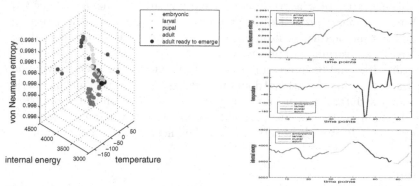

Fig. 3. a) The 3D scatter plot of the Drosophila melanogaster gene regulatory network in the thermodynamic space. Red dots: embryonic period; cyan dots: larval period; blue dots: pupal period: green dots: adulthood; black dot: adult ready to emerge. b) The von Neumann entropy versus time for the Drosophila melanogaster gene regulatory network. c) The internal energy versus time for the Drosophila melanogaster gene regulatory network. d) The temperature versus time for the Drosophila melanogaster gene regulatory network.

which control its growth begin to slow down. Hence the green points (adulthood) remain stable. Finally, the black data points are well separated from the remainder of the developmental samples, and correspond to the time when the adult emerges.

In this section we have undertaken experiments on a number of realistic time-varying complex systems in order to analyze whether the thermodynamic characterizations we have developed are efficient in studying the evolution of dynamic networks. The experimental results of both the stock correlation network of NYSE and the Drosophila gene regulatory network demonstrate that the thermodynamic entropy, internal energy together with temperature provide a powerful tool for detecting abrupt events and characterizing different stages in the network evolution.

4   Conclusions

In this paper, we adopt a thermodynamic representation of network structure in order to visualise and understand the evolution of time-varying networks. We provide expressions for thermodynamic network variables, including the entropy, internal energy and temperature. This analysis is based on statistical thermodynamics and commences from a recently derived expression for the von Neumann entropy of a network. The analysis then defines the microscopic configurations of a network to be the normalized Laplacian eigenstates. The internal energy depends on the number of edges in the network. The thermodynamic temperature measures fluctuations via changes in the number of edges and individual node degree changes.

We have evaluated the method experimentally using data representing a variety of real world complex systems, taken from the financial and biological domains. The experimental results demonstrate that the thermodynamic variables are efficient in analyzing the evolutionary properties of dynamic networks, including the detection of abrupt changes and phase transitions in structure or other distinctive periods in the evolution of time-varying complex networks.

In the future, it would be interesting to see what features the thermodynamic network variables reveal in additional domains, such as human functional magnetic resonance imaging data. Another interesting line of investigation would be to explore if the thermodynamic framework can be extended to the domains of dynamic directed networks, edge-weighted networks, labelled networks and hypergraphs.:

References

1. Albert, R., Barabási, A.L.: Topology of evolving networks: Local events and universality. Physical Review Letters 85(24), 5234–5237 (2000)
2. Arbeitman, M., Furlong, E.E., Imam, F., Johnson, E., Null, B.H., Baker, B.S., Krasnow, M.A., Scott, M.P., Davis, R.W., White, K.P.: Gene expression during the life cycle of drosophila melanogaster. Science 297(5590), 2270–2275 (2002)
3. Braunstein, S., Ghosh, S., Severini, S.: The laplacian of a graph as a density matrix: A basic combinatorial approach to separability of mixed states. Annals of Combinatorics 10(3), 291–317 (2006)
4. Chung, F.R.K.: Spectral Graph Theory. AMS (1997)
5. Delvenne, J.C., Libert, A.S.: Centrality measures and thermodynamic formalism for complex networks. Phys. Rev. E. 83(046117) (2011)
6. Estrada, E.: The Structure of Complex Networks: Theory and Applications. Oxford University Press (2011)
7. Estrada, E., Hatano, N.: Statistical-mechanical approach to subgraph centrality in complex networks. Chem. Phys. Lett. 439, 247–251 (2007)
8. Han, L., Escolano, F., Hancock, E.R., Wilson, R.C.: Graph characterizations from von neumann entropy. Pattern Recognition Letters 33, 1958–1967 (2012)
9. Mikulecky, D.C.: Network thermodynamics and complexity: a transition to relational systems theory. Computers & Chemistry 25, 369–391 (2001)
10. Newman, M.: The structure and function of complex networks. SIAM Review 45(2), 167–256 (2003)
11. Passerini, F., Severini, S.: Quantifying complexity in networks: The von neumann entropy. Inthernational Journal of Agent Technologies and Systems 1, 58–67 (2008)
12. Peron, T.K.D., Rodrigues, F.A.: Collective behavior in financial markets. EPL 96(48004) (2011)
13. Song, L., Kolar, M., Xing, E.P.: Keller: estimating time-varying interactions between genes. Bioinformatics 25(12), 128–136 (2009)
14. Ye, C., Wilson, R.C., Comin, C.H., Costa, L.D.F., Hancock, E.R.: Approximate von neumann entropy for directed graphs. Phys. Rev. E. 89(052804) (2014)

Isometric Mapping Hashing

Yanzhen Liu[1]([✉]), Xiao Bai[1], Haichuan Yang[1], Zhou Jun[2], and Zhihong Zhang[3]

[1] School of Computer Science and Engineering,
Beihang University XueYuan Road No.37, Beijing, HaiDian District, China
{lyzeva,baixiao,yanghaichuan}@buaa.edu.cn
[2] School of Information and Communication Technology,
Griffith University Nathan, Nathan, QLD 4111, Australia
jun.zhou@griffith.edu.au
[3] Xiamen University, Software School
General Office, Software School of Xiamen University, Xiamen, 361005, Fujian,
People's Republic of China
zhihong@xmu.edu.cn

Abstract. Hashing is a popular solution to Approximate Nearest Neighbor (ANN) problems. Many hashing schemes aim at preserving the Euclidean distance of the original data. However, it is the geodesic distance rather than the Euclidean distance that more accurately characterizes the semantic similarity of data, especially in a high dimensional space. Consequently, manifold based hashing methods have achieved higher accuracy than conventional hashing schemes. To compute the geodesic distance, one should construct a nearest neighbor graph and invoke the shortest path algorithm, which is too expensive for a retrieval task. In this paper, we present a hashing scheme that preserves the geodesic distance and use a feasible out-of-sample method to generate the binary codes efficiently. The experiments show that our method outperforms several alternative hashing methods.

Keywords: Hashing · Manifold · Isomap · Out-of-sample extension

1 Introduction

Nearest Neighbor (NN) search is a basic and important step in image retrieval. For a large scale dataset of size n, the complexity of NN search is $O(n)$, which is still too high for big data processing. To solve this problem, sublinear approximate nearest neighbor methods have been proposed. Hashing methods are among such well-performed methods.

The basic idea of hashing is to use binary code to represent high-dimensional data, with similar data pair having smaller Hamming distance of binary codes. In other words, we should construct a mapping from high-dimension Euclidean space to Hamming space and preserve the original Euclidean distance of data in the Hamming space. We can see that hashing accelerates NN search on two aspects: one is that binary XOR operation to calculate the Hamming distance is much faster than calculating the Euclidean distance; the other is that binary code can be used as address to index the original data.

C.-L. Liu et al. (Eds.): GbRPR 2015, LNCS 9069, pp. 325–334, 2015.
DOI: 10.1007/978-3-319-18224-7_32

A simple and basic hash method is Locality-Sensitive Hashing(LSH) [1]. LSH is based on generating random hyperplane in Euclidean space. A hyperplane generates a bit of code. Data points on one side of the hyperplane are coded 0 at this bit, while at the other side are coded 1. Critically, points on the hyperplane can be coded either 0 or 1, but the probability is very low. LSH successfully reduces the hashing process to sublinear query time with acceptable accuracy. However, because of its randomness, reaching higher accuracy needs longer code, and there is redundancy to some degree.

To get more compact binary code, some researchers started to analyze the inner structure of dataset. These efforts led to data-dependent hashing methods while LSH is data-independent method. Among the data-dependent methods, some are linear such as Iterative Quantization (ITQ) [6] and some are non-linear such as Spectral Hashing (SH) [14], and Anchor Graph Hashing (AGH) [8]. By learning the inner structure of dataset, the hash codes gain higher accuracy with shorter code length.

Although all the above hashing methods are concentrating on preserving Euclidean distance, however, the most similar data pair may not be the nearest neighbor in Euclidean space, but the nearest neighbor on the same manifold. Thus if we want to improve the precision of hashing method, we must take manifold structure into consideration as well. For linear manifold, linear methods such as PCAH, PCA-ITQ is applicable. But for non-linear manifold such as Swiss Roll, further exploration is needed to catch the manifold structure in hashing.

Graph methods play an important role in image analysis [15–17]. Many non-linear manifold learning methods are motivated by graph approaches that create low-dimension embedding of data with manifold distribution in Euclidean space, such as Laplace Eigenmap (LE) [2], Locally Linear Embedding (LLE) [11], and Isometric Mapping (ISOMAP) [13]. Some manifold based hashing methods utilize these ideas and extend them to embedding in Hamming space, such as Inducitve Manifold Hashing (IMH) [12] and Locally Linear Hashing (LLH) [5]. IMH-tSNE extracts manifold structure with tSNE [9] and gives an out-of-sample extension scheme by locally linear property of manifold. LLH learns the locally-linear structure of manifold and preserves it in hash code.

In this paper, we utilize a widely used non-linear dimensionality reduction method Isometric Mapping (ISOMAP) to catch the geodesic distance structure of manifold and preserve it in hash code. We name the method Isometric Mapping Hashing (IsoMH). The experiments show that our IsoMH have comparable precision with the state-of-the-art methods.

2   Isometric Mapping Hashing

Given a set of data points $X = [\mathbf{x}_1, \mathbf{x}_2, ..., \mathbf{x}_n]^T \left(\mathbf{x}_k \in \mathbb{R}^d, X \in \mathbb{R}^{n \times d} \right)$, we want to get a mapping from X to a binary hash code set $B \in \{-1, 1\}^{n \times c}$ that the hamming distance in B best approximates the data similarity in X.

Here, we explore the geodesic distance on manifold rather than the Euclidean distance as the quantitative measurement of data similarity. So in the process of mapping, we choose to preserve geodesic distance between data points in X and then use a non-linear dimension reduction method ISOMAP (Section 2.1). After dimension reduction with ISOMAP, we got the low-dimensional embedding $Y \in \mathbb{R}^{n \times c}$, and we quantize Y into a binary hash code B. Direct quantization with sign function $B = sign(Y)$ may lead to great quantization loss. Therefore, we employ the iteration quantization method ITQ [6] to calculate B, which significantly reduces the quantization loss (Section 2.2). Finally we use an efficient out-of-sample extension method rather than doing ISOMAP again for a single data query (Section 3). The experiments to evaluate our method is introduced in Section 4.

Here is a list of mathematical symbols appearing in the following sections:

b: number of bits of the final hash code.

ξ: eigenvector of the kernel matrix.

V: eigenvector matrix, $V = (\xi_1, \xi_2, ..., \xi_b)^T$.

λ: eigenvalue of the kernel matrix

Λ: diagonal matrix of eigenvalues.

$\mathbf{y}$: embedding vector.

Y: data matrix, $Y = (\mathbf{y}_1, \mathbf{y}_2, ..., \mathbf{y}_b)^T$.

B: binary code matrix.

2.1 Dimensionality Reduction with ISOMAP

Isomap is a non-linear dimensionality reduction method which aims at preserving the geodesic distance of the manifold structure. Real geodesic distance is difficult to compute, so Isomap takes shortest path distance to approximate the geodesic distance in large dataset.

With the dataset X, we can construct a neighborhood graph G, whose edges connect only neighboring nodes with the Euclidean distance of neighboring nodes as the weights. To properly define the neighborhood, we set a threshold ϵ. Only node pairs whose Euclidean distances are below the threshold are considered as neighboring nodes. Alternately, we can choose a number k, only node pairs being k-nearest-neighbor to each other are regarded as neighboring. With this neighborhood graph G, we compute the shortest path distance between all node pairs in G to approximate the geodesic distance of those node pairs. We denote the approximate geodesic distance matrix as D.

When using the Euclidean distance to reduce dimensionality, we come up with Multidimensional Scaling (MDS) [4]. MDS is a linear dimension reduction method aiming at preserving the pairwise distance. According to [4], we can get the low-dimensional embedding by decomposing the kernel matrix

$$K = -\frac{1}{2}HDH \qquad (1)$$

where $H = E - \frac{1}{t}ee^T$ (e is a column vector of all 1). We pick eigenvectors ξ_k ($1 \leqslant k \leqslant b$) of the top b eigenvalues λ_k ($1 \leqslant k \leqslant b$) of K. Then the low-dimensional embedding can be calculated as

$$Y = \Lambda^{\frac{1}{2}}V \tag{2}$$

2.2 Quantize Low-Dimension Embedding with ITQ

After getting the low-dimension embedding Y, we are able to quantize it into binary code. In order to get the binary code, we need to quantize each dimension in the low-dimensional embedding into 0 and 1 according to the formula

$$B = sign(Y) \tag{3}$$

Direct quantization may cause great loss as explained in [6]. So we try to calculate an optimized rotation matrix to minimize the quantization loss Q.

$$Q = \|B - YR\|_F \tag{4}$$

Tackling this optimization problem, an iterative method was proposed to solve B and R one after the other [6]. When solving B, we fix R and calculate B with sign function $B = sign(YR)$. When solving R, we fix B and the problem becomes the classic Orthogonal Procrustes Problem. It can be solved by computing the Singular Value Decomposition (SVD) of $B^T V$ as $S\Omega\hat{S}$, and we get $R = \hat{S}S^T$. Finally, we get a local minimum of the quantization loss Q and its corresponding optimal rotation matrix R. Then binary hash code of training set can be represented as $B = sign(YR)$.

The whole process of training is described in pseudo-code in Algorithm 1.

Algorithm 1. Training phase of Isometric Mapping Hashing

Data: training data X, code length b
Result: shortest path matrix D, eigenvector matrix V, eigenvalue matrix Λ,
           Euclidean embedding E, rotation matrix R, hash code B
Compute Euclidean distances of all data points;
Sort Euclidean distances to identify neighborhoods;
Construct adjacency matrix A of neighborhood graph G;
Calculate shortest path distance D of graph G;
Execute MDS on D and get V, Λ and Y;
Execute ITQ on E to get the best quantization code B, with rotation matrix R;

---

3 Out-of-Sample Extension

By applying the above described method on the training set, we can get the binary codes of the whole training set. If we get another single testing data, we must calculate all other training data together with this single testing data. With the shortest path distance and matrix decomposition, the Isomap takes $O(n^3)$ to execute, which is difficult to meet the need of online query. To solve this problem, we adopt an out-of-sample extension method [3] to calculate the approximate Isomap embedding.

Given a query $\mathbf{q}$, we first need to calculate its approximate geodesic distance to all other training set points, and that is, to calculate its shortest path distance to all others. In fact, there is no need to run Dijkstra one more time. If the neighborhood of $\mathbf{q}$ is $\mathbf{x}_{l_1}, \mathbf{x}_{l_2}, ..., \mathbf{x}_{l_k}$ $(1 \leqslant l_i \leqslant n)$, then the shortest path distance of $\mathbf{q}$ to all other point in X can be calculated by the shortest path distance of its neighbors. Denoting by $D(\mathbf{x}_i, \mathbf{x}_j)$, the shortest path distance from $\mathbf{x}_i$ to $\mathbf{x}_j$, it is obvious that

$$D(\mathbf{x}, \mathbf{q}) = \min_i \{ \|\mathbf{q} - \mathbf{x}_{l_i}\| + D(\mathbf{x}_{l_i}, \mathbf{x}) \} \tag{5}$$

To calculate a pair of shortest path distance, we just need to compare k-1 times and add k times.

With the newly calculated $D(\mathbf{x}, \mathbf{q})$, we can calculate the embedding $\mathbf{y}$ of an out-of-sample $\mathbf{x}$ according to the expression

$$\mathbf{y}_k(\mathbf{q}) = \frac{1}{2} \frac{1}{\sqrt{\lambda_k}} \sum_{i=1}^{n} v_{ki} \left(E_{\mathbf{x}'} \left[D^2 \left(\mathbf{x}', \mathbf{x}_i \right) \right] - D^2 (\mathbf{x}_i, \mathbf{q}) \right) \tag{6}$$

given in [3]. And it can be represented in matrix as

$$\mathbf{y} = \frac{1}{2} \Lambda^{\frac{1}{2}} V \eta \tag{7}$$

where $\eta = E_{\mathbf{x}'} \left[D^2 \left(\mathbf{x}', \cdot \right) \right] - D^2 (\mathbf{q}, \cdot)$ is a column vector.

The pseudo-code of out-of-sample extension is written as follow.

Algorithm 2. Out-of-sample Extension

Data: training data X, code length b, shortest path matrix D, eigenvector matrix V, eigenvalue matrix Λ, Euclidean embedding Y, rotation matrix R, hash code B, out-of-sample $\mathbf{q}$
Result: hash code of out-of-sample data $\mathbf{b}_q$
Compute Euclidean distances of $\mathbf{x}_q$ to all training data;
Decide the k neighborhood point set of $\mathbf{q}$: $\mathbf{x}_{l_1}, \mathbf{x}_{l_1}, ..., \mathbf{x}_{l_k}$;
Calculate shortest path distance $\mathbf{d}_q$ from $\mathbf{q}$ to all other training data according to equation(5);
Calculate Euclidean embedding $\mathbf{y}_q$ of out-of-sample point according to equation(6);
$\mathbf{b}_q = R^T sign(\mathbf{y}_q)$;

---

4   Experiments

We ran experiments to test our IsoMH method on two frequently-used image datasets: **MNIST** and **CIFAR-10** [7].

MNIST. The **MNIST** dataset has 70000 28*28 small grayscale images of handwritten digits from '0' to '9'. Each small image is represented by a 784-dimensional vector and a single digit label. We used 2000 random images as the training set, and 500 random images as the query set.

Fig. 1. Images in the left column are query images selected from 10 classes, and the rest images are the retrieval results using IsoMH. Images in green boxes represent the correct results.

CIFAR-10. There are 60000 32*32 small color images in **CIFAR10**. These small images are from 10 classes: airplane, automobile, bird, cat, deer, dog, frog, horse, ship, and truck. Each class has 6000 images. Every image in **CIFAR10** is represented by a 320-dimensional GIST feature [10] vector and a label from '1' to '10'. Fig. 1 shows sample results on **CIFAR-10**. We selected 10 images from different classes as the query and search the top 9 nearest neighbors in hamming distance of hash codes.

Besides IsoMH, we also tested other 5 unsupervised methods: LSH [1], PCA-ITQ [6], AGH [8], IMH-LE, and IMH-tSNE [9]. LSH and PCA-ITQ are linear while AGH, IMH-LE and IMH-tSNE are nonlinear. Specifically, two IMH methods are also manifold based. We used codes released by the original authors during the experiments.

Manifold based methods aim at getting semantically similar results. Though unsupervised, we used class labels as the ground truth, and evaluated the performance of those methods by Precision-Recall (P-R) curves and time consumption of training and query. We calculated the precision and recall by:

$$precision = \frac{\text{Number of retrieved true neighbor pairs}}{\text{Number of retrieved neighbor pairs}} \tag{8}$$

$$recall = \frac{\text{Number of retrieved true neighbor pairs}}{\text{Total number of true neighbor pairs}} \tag{9}$$

All our experiments were run on a 64-bit PC with 3.50 GHz Intel i7-4770K CPU and 16.0GB RAM.

4.1 Parameters Analysis

We set the hash code length to 12bits, 32bis, 64bits, 96bits and 128bits and performed the experiments, respectively. The results are shown in Fig. 2. The curves show that when the number of bits is small, results will be better with the increasing of code length, and if the code length is large enough, adding hash bits cannot bring much advance in retrieval precision.

For the construction of nearest neighbor graph, different methods and parameters have almost the same results with shortest path distance algorithm, because the training set is large. In our experiments, we set $k = 7$.

4.2 Comparison to Other Methods

Fig. 3 shows the P-R curves of the above-mentioned methods on MNIST. Our IsoMH performs better than all other methods. Linear method PCA-ITQ and manifold method IMH-tSNE are two most competitive methods. As the figure showing, IMH-tSNE performs better with longer code. For the training time

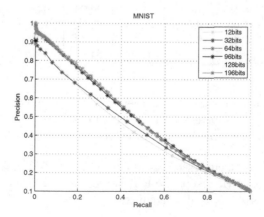

Fig. 2. The Precision-Recall curve of IsoMH with different hash code length on the MNIST dataset

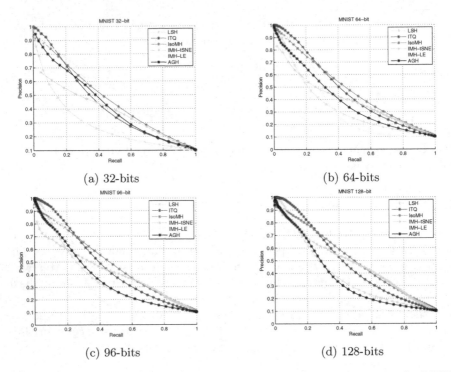

(a) 32-bits

(b) 64-bits

(c) 96-bits

(d) 128-bits

Fig. 3. Precision-Recall curves of different methods with experiments on the MNIST dataset

Table 1. Time consumption (milliseconds) of different methods on the MNIST dataset

Methods	32 bits		64 bits		96 bits		128 bits	
	Training	Indexing	Training	Indexing	Training	Indexing	Training	Indexing
IsoMH	1853.44	189.03	1901.18	193.63	1954.06	211.14	2048.67	232.61
PCA-ITQ	10.61	0.18	18.27	0.35	29.78	0.50	42.52	0.69
IMH-tSNE	35.01	1.09	69.02	1.13	85.92	1.22	95.07	1.18
IMH-LE	96.38	2.20	90.61	2.20	87.13	2.52	87.99	2.28
AGH	84.02	10.92	87.42	13.35	84.97	13.20	85.47	13.09

consumption, as shown in Table 1, our method is not very competitive because the time complexity of shortest path algorithm is $O(n^3)$. In P-R curve, our IsoMH performs better in middle and lower precision region while be weaker in higher precision region.

Fig. 4 shows the P-R curves of results on CIFAR-10. CIFAR-10 has more complex images than MNIST so these P-R curves are lower than MNIST's. And our IsoMH is still better than all the curves.

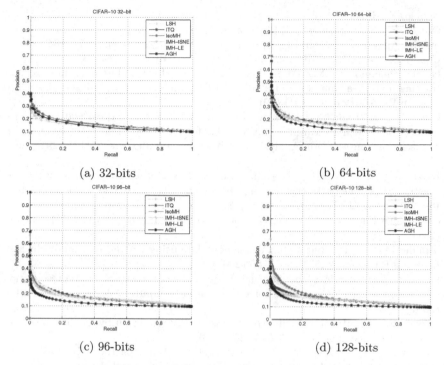

(a) 32-bits

(b) 64-bits

(c) 96-bits

(d) 128-bits

Fig. 4. Precision-Recall curves of different methods with experiments on the CIFAR-10 dataset

5 Conclusion

Manifold based hashing methods have advantages on retrieving semantically similar neighbors. In this paper, we preserve geodesic distance of manifold with binary code and utilize ISOMAP to implement our mapping. IsoMH outperforms many state-of-the-arts methods in precision but consume more time in computing the shortest distance of neighborhood graph. In the future work, we will concentrate on improving the time efficiency of IsoMH.

Acknowledgement. This research is supported by National Natural Science Foundation of China (NSFC) projects No. 61370123 and No.61402389.

References

1. Andoni, A., Indyk, P.: Near-optimal hashing algorithms for approximate nearest neighbor in high dimensions. In: 47th Annual IEEE Symposium on Foundations of Computer Science, FOCS 2006, pp. 459–468. IEEE (2006)
2. Belkin, M., Niyogi, P.: Laplacian eigenmaps and spectral techniques for embedding and clustering. In: NIPS, vol. 14, pp. 585–591 (2001)

3. Bengio, Y., Paiement, J.F., Vincent, P., Delalleau, O., Roux, N.L., Ouimet, M.: Out-of-sample extensions for LLE, Isomap, MDS, eigenmaps, and spectral clustering. In: Advances in Neural Information Processing Systems, pp. 177–184 (2004)
4. Cox, T.F., Cox, M.A.: Multidimensional Scaling. CRC Press (2010)
5. Go, I., Zhenguo, L., Xiao-Ming, W., Shih-Fu, C.: Locally linear hashing for extracting non-linear manifolds. In: 2014 IEEE Conference on Computer Vision and Pattern Recognition (CVPR) (2014)
6. Gong, Y., Lazebnik, S., Gordo, A., Perronnin, F.: Iterative quantization: A procrustean approach to learning binary codes for large-scale image retrieval. IEEE Transactions on Pattern Analysis and Machine Intelligence 35(12), 2916–2929 (2013)
7. Krizhevsky, A., Hinton, G.: Learning multiple layers of features from tiny images. Master's thesis, Department of Computer Science, University of Toronto (2009)
8. Liu, W., Wang, J., Kumar, S., Chang, S.F.: Hashing with graphs. In: Proceedings of the 28th International Conference on Machine Learning (ICML 2011), pp. 1–8 (2011)
9. Van der Maaten, L., Hinton, G.: Visualizing data using t-SNE. Journal of Machine Learning Research 9(2579-2605), 85 (2008)
10. Oliva, A., Torralba, A.: Modeling the shape of the scene: A holistic representation of the spatial envelope. International Journal of Computer Vision 42(3), 145–175 (2001)
11. Roweis, S.T., Saul, L.K.: Nonlinear dimensionality reduction by locally linear embedding. Science 290(5500), 2323–2326 (2000)
12. Shen, F., Shen, C., Shi, Q., Van Den Hengel, A., Tang, Z.: Inductive hashing on manifolds. In: 2013 IEEE Conference on Computer Vision and Pattern Recognition (CVPR), pp. 1562–1569. IEEE (2013)
13. Tenenbaum, J.B., De Silva, V., Langford, J.C.: A global geometric framework for nonlinear dimensionality reduction. Science 290(5500), 2319–2323 (2000)
14. Weiss, Y., Torralba, A., Fergus, R.: Spectral hashing. In: Advances in Neural Information Processing Systems, pp. 1753–1760 (2009)
15. Xiao, B., Hancock, E.R., Wilson, R.C.: Graph characteristics from the heat kernel trace. Pattern Recognition 42(11), 2589–2606 (2009)
16. Xiao, B., Hancock, E.R., Wilson, R.C.: Geometric characterization and clustering of graphs using heat kernel embeddings. Image and Vision Computing 28(6), 1003–1021 (2010)
17. Zhang, H., Bai, X., Zhou, J., Cheng, J., Zhao, H.: Object detection via structural feature selection and shape model. IEEE Transactions on Image Processing 22(11-12), 4984–4995 (2013)

Skeletal Graphs from Schrödinger Magnitude and Phase

Francisco Escolano[1(✉)], Edwin R. Hancock[2], and Miguel A. Lozano[3]

[1] Department of Computer Science and AI, University of Alicante, Alicante, Spain
sco@dccia.ua.es
[2] Department of Computer Science, University of York, York, UK
erh@cs.york.ac.uk
[3] Department of Computer Science and AI, University of Alicante, Alicante, Spain
malozano@ua.es

Abstract. Given an adjacency matrix, the problem of finding a simplified version of its associated graph is a challenging one. In this regard, it is desirable to retain the essential connectivity of the original graph in the new representation. To this end, we exploit the magnitude and phase information contained in the Schrödinger Operator of the Laplacian. Recent findings based on continuous-time quantum walks suggest that the long-time averages of both magnitude and phase are sensitive to long-range interactions. In this paper, we depart from this hypothesis to propose a novel representation: skeletal graphs. Using the degree of interaction (or "long-rangedness") as a criterion we propose a structural level set (i.e. a sequence of graphs) from which it emerges the simplified graph. In addition, since the same theory can be applied to weighted graphs, we can analyze the implications of the new representation in the problems of spectral clustering, hashing and ranking from computer vision. In our experiments we will show how coherent transport phenomenon implemented by quantum walks discovers the long-range interactions without breaking the structure of the manifolds on which the graph or its connected components are embedded.

Keywords: Graph simplification · Schrödinger Operator · Quantum walks

1 Introduction

Given a set of points $\mathcal{X} = \{x_1, \ldots, x_N\} \subset \mathbb{R}^d$ to be clustered, ranked or hashed, and a metric $d : \mathcal{X} \times \mathcal{X} \to \mathbb{R}$, the spectral approach consists of mapping the x_i to the vertices V, with $N = |V|$, of an undirected weighted graph $G(V, E)$ so that the edge weight $W(i,j) = e^{-d^2(x_i, x_j)/t}$ is a similarity measure between x_i and x_j.

The analysis of the similarity/affinity matrix W is key to the success of the pattern recognition task. This is usually done through the study of the Laplacian matrix $L = D - W$ or of its normalized counterpart. Then, despite W contains

© Springer International Publishing Switzerland 2015
C.-L. Liu et al. (Eds.): GbRPR 2015, LNCS 9069, pp. 335–344, 2015.
DOI: 10.1007/978-3-319-18224-7_33

pairwise relations, the spectrum and eigenvectors of L contain global information. However, it has been recently pointed out [1] that the limit analysis of a graph Laplacian, for instance setting $t \to 0$ as a consequence of $N \to \infty$, reveals some flaws or degenerate behaviors. For instance, ranking functions, which are usually implemented by Green's functions, diverge when $N \to \infty$. Zhout et al. propose to solve this problem by computing Green's functions of "higher-order" Laplacians, i.e. L^m and $m \geq 0$.

In clustering, the well known approach of normalized cuts method [2] has been improved by introducing topological distances which are consistent with a metric (commute-times) [3]. State-of-the-art methods for image segmentation do not only combine different eigenvectors of the Laplacian (normalized in this case) [4] but also incorporate better dissimilarity measures to make the weight matrix W more discriminative.

Finally, spectral hashing [5][6] consists of fusing the information contained in several eigenvectors of L in order to define uncorrelated and efficient codes. Anchor graphs [7] provide a scalable approach for spectral hashing through the simplification of the original similarity matrix W.

Most of the exisiting methods for improving ranking, semi-supervised learning, clustering or hashing rely on the concept of a random walk. For instance, in [8] pixels are labeled in terms of the probability that a random walk will reach them from a given seed. In this regard, the underpinning principle is to minimize the combinatorial Dirichlet integral associated with the weighted Laplacian of a graph. Seeds are assumed to be the boundary conditions of a Dirichlet problem and minimization seeks the values of the unknown labels so that the Dirichlet integral is the smoothest one. This means that a harmonic solution is preferred (the probability of an unknown label for a pixel must be average of the probabilities of neighboring pixels). The method relies on solving a linear system. The resulting method is more robust to noise than normalized cuts.

Recent findings from the study of quantum walks suggest alternative ways of incorporating "high-order" information. For instance, in [9] a quantum version of the Jensen-Shannon divergence is used to compute a graph kernel, whereas in [10] it is exploited to detect both symmetric and anti-symmetric structures in graphs. In [11] and the references therein, it is suggested that quantum walks provide information about long-range interactions. For instance, in dendrimers (trees structured in strata or "generations") a quantum random walk starting at the root reaches (in the limit) nodes lying in the same generation with similar probability.

Our long-term aim is to address the question whether the above ideas (simplification of W, incorporation of long-range interactions in L, evaluation of the similarity measure) which have contributed to the improvement of spectral methods for clustering, ranking and hashing can be driven from quantum walks. In this paper, we will focus on the improvement of spectral clustering and ranking.

2   Quantum Walks for Analyzing Transport

2.1   Unitary Evolution and the Schrödinger Operator

Let $\mathcal{H} = span\{|j\rangle \mid j = 1, \ldots, N\} = \mathbb{C}^N$ be a N–dimensional Hilbert space where $\langle j| = (0 \ldots 1 \ldots 0)$ with a 1 at the $j - th$ position. We use the Dirac bra-ket notation where: $|a\rangle = \boldsymbol{a}$, $\langle a| = \boldsymbol{a}^*$, $\langle a|b\rangle = \boldsymbol{a}^*\boldsymbol{b}$ is the inner product and therefore $\langle j|k\rangle = \boldsymbol{j}^*\boldsymbol{k} = \delta_{jk}$ and $\sum_{j=1}^N |j\rangle\langle j| = \mathbf{1}$. Then, a point in the Hilbert space is given by $|\psi\rangle = \sum_{j=1}^N c_j|j\rangle$ with $c_j \in \mathbb{C}$ so that $|c_1|^2 + |c_2|^2 + \ldots + |c_N|^2 = 1$ and $|c_j|^2 = \bar{c}_j c_j$.

The Schrödinger equation describes how the complex state vector $|\psi(t)\rangle \in \mathbb{C}^n$ of a continuous-time quantum walk varies with time:

$$\frac{\partial|\psi(i)\rangle}{\partial t} = -i\boldsymbol{L}|\psi_t\rangle. \tag{1}$$

Given an initial state $|\psi(0)\rangle = \sum_{j=1}^N c_j^0|j\rangle$ the latter equation can be solved to give $|\psi(t)\rangle = \boldsymbol{\Psi}(t)|\psi(0)\rangle$, where $\boldsymbol{\Psi}(t) = e^{-iLt}$ is a complex $n \times n$ *unitary matrix*. In this paper we refer to $\boldsymbol{\Psi}(t)$ as the *Schrödinger operator*. In this regard, Stone's theorem [12] establishes a one-to-one correspondence between a time parameterized unitary matrix $\boldsymbol{U}(t)$ and a self-adjoint (Hermitian) operator $\boldsymbol{H} = \boldsymbol{H}^*$ such that there is a unique Hermitian operator satisfying $\boldsymbol{U}(t) = e^{it\boldsymbol{H}}$. Such an operator $\boldsymbol{H}$ is the *Hamiltonian*. In the case of graphs $\boldsymbol{H} = -\boldsymbol{L}$ and then we have that $\boldsymbol{\Psi}(t) = e^{-it\boldsymbol{L}}$ is a unitary matrix for $t \in \mathbb{R}$. Therefore, given an initial state $|\psi(0)\rangle$, the Schrödinger Operator characterizes the evolution of a Continuous-Time Quantum Walk (CTQW). The probability that the quantum walk is at node j is given by $|\langle j|\psi\rangle|^2 = |c_j|^2$. The $|c_j|^2$ are known as the amplitudes of the wave traveling through the graph.

2.2   Long-Time Averages from Magnitude

Different choices of the initial state $|\psi(0)\rangle = \sum_{j=1}^N c_j^0|j\rangle$ lead to different ways of probing the graph by exploiting properties of quantum superposition and quantum interference. For instance, in [9], initial amplitudes are set to $c_j^0 = \sqrt{\frac{d_j}{\sum_{k=1}^N d_k}}$, in order to compute de quantum version of the Jensen-Shannon divergence. However, in [10], where the focus is on identifying whether the vertices i and j are symmetrically placed in the graph we have that either $c_j^0 = 1/\sqrt{2}$ and $c_k^0 = 1/\sqrt{2}$ (in phase) or $c_j^0 = 1/\sqrt{2}$ and $c_k^0 = -1/\sqrt{2}$ (in antiphase). Actually, the Quantum Jensen-Shannon divergence has a low value when pairs or vertices are located anti-symmetrically and a high value when they are symmetrically placed.

In this paper, we use the classical choice proposed by Farhi and Gutman for studying transport properties of quantum walks in trees [13]. In such approach, states $|j\rangle$ are associated with excitations at the nodes j. Therefore, the evolution of a CTQW commencing at node $|j\rangle$ is given by $|j(t)\rangle = e^{-it\boldsymbol{L}}|j\rangle$. In this regard, the amplitude of a transition between nodes j and k at time t is given by $c_{jk}(t) = \langle k|j(t)\rangle = \langle k|e^{-itL}|j\rangle$, and the quantum-mechanical probability of such a transition is $\pi_{jk}(t) = |c_{jk}(t)|^2 = |\langle k|e^{-itL}|j\rangle|^2$.

Since the spectral decomposition of the Laplacian is $L = \Phi\Lambda\Phi^T$, where $\Phi = [\phi_1|\phi_2|\ldots|\phi_n]$ is the $N \times N$ matrix of ordered eigenvectors according to the corresponding eigenvalues $0 = \lambda_1 \leq \lambda_2 \leq \ldots \leq \lambda_n$, and $\Lambda = diag(\lambda_1\,\lambda_2\,\ldots\,\lambda_n)$, we have that $e^{-itL} = \Phi e^{-it\Lambda}\Phi^T$ where where $e^{-it\Lambda} = diag(e^{-it\lambda_1}\,e^{-it\lambda_2}\,\ldots\,e^{-it\lambda_n})$. Then, the probability of a transition between j and k at time t is given by

$$\pi_{jk}(t) = \left|\langle k|e^{-itL}|j\rangle\right|^2 = \left|\langle k|\Phi e^{-it\Lambda}\Phi^T|j\rangle\right|^2$$

$$= \left|\langle k|\sum_{u=1}^{N} e^{-it\lambda_u}|\phi_u\rangle\langle\phi_u|j\rangle\right|^2 = \left|\sum_{u=1}^{N} e^{-it\lambda_u}\langle k|\phi_u\rangle\langle\phi_u|j\rangle\right|^2, \qquad (2)$$

where $\langle k|\phi_u\rangle = \phi_u(k)$ and $\langle\phi_u|j\rangle = \phi_u(j)$ respectively. Therefore we have

$$\pi_{jk}(t) = \sum_{u=1}^{N}\sum_{v=1}^{N} e^{-it(\lambda_u-\lambda_v)} z_u(k,j) z_v(k,j), \qquad (3)$$

where $z_u(k,j) = \phi_u(k)\phi_u(j)$ and $z_v(k,j) = \phi_v(k)\phi_v(j)$ account for the correlations between the k–th and j–th components of the eigenvectors ϕ_u and ϕ_v. Then, since $e^{-it(\lambda_u-\lambda_v)} = \cos(t(\lambda_u - \lambda_v)) - i\sin(t(\lambda_u - \lambda_v))$ we have that the long-time limit of $\pi_{jk}(t)$ does not exist, whereas the corresponding long-time limit of a classical continuous-time random walk is $1/N$. However, in the quantum mechanical case it is possible to compute the long-time average:

$$\chi_{jk} = \lim_{T\to\infty} \frac{1}{T}\int_0^T \pi_{jk}(t)dt$$

$$= \lim_{T\to\infty} \frac{1}{T}\int_0^T \sum_{u=1}^{N}\sum_{v=1}^{N} e^{-it(\lambda_u-\lambda_v)} z_u(k,j)z_v(k,j)$$

$$= \lim_{T\to\infty} \sum_{u=1}^{N}\sum_{v=1}^{N} z_u(k,j)z_v(k,j)\frac{1}{T}\int_0^T e^{-it(\lambda_u-\lambda_v)}dt$$

$$= \sum_{u=1}^{N}\sum_{v=1}^{N} \delta_{\lambda_u,\lambda_v} z_u(k,j)z_v(k,j), \qquad (4)$$

where $\delta_{\lambda_u,\lambda_v} = 1$ if $\lambda_u = \lambda_v$ and 0 otherwise. Therefore, the long-time averaged probabilities do not depend directly on the eigenvalues of the Laplacian but on their multiplicity. Mülken and Blumen have recently related the multiplicity to the transport efficiency of CTQWs [11]. More precisely, the averaged return probability $\bar{\pi}(t) = (1/N)\sum_{k=1}^{N} \pi_{k,k}(t)$ decays faster with time than that of a classical continuous-time random walk under conditions of low multiplicity. This means that in the long-time limit more probabilistic mass is allocated to nodes lying far away from the origin of the quantum walk, provided that there is low degeneracy (i.e. multiplicity).

2.3   Long-Time Averages for Phase

The consequences of the above analysis can be summarized as follows. The long-time averages of the amplitude encode transport information and this information is contained in the interactions (correlations) between the eigenvectors of the Laplacian. An asymptotic analysis of the Schrödinger Operator leads to understanding the coherent transport efficiency of the graph. However, the analysis is incomplete if we do not also consider phase in addition to amplitude (magnitude). In order to compute long-time averages of phase we must pay attention to the complex nature of the operator. More precisely, the phase of the transition is $\kappa_{jk}(t) = \arg(c_{jk}(t)) = \arg(\langle k|e^{-itL}|j\rangle)$. From $e^{-itL} = \Phi e^{-it\Lambda}\Phi^T$ we have that $c_{jk}(t) = \sum_{u=1}^{N} e^{-it\lambda_u}\langle k|\phi_u\rangle\langle\phi_u|j\rangle = \sum_{u=1}^{N} e^{-it\lambda_u} z_u(j,k)$. Therefore a transition $c_{jk}(t)$ is a sum of time-dependent phasors. The magnitude of each phasor is the correlation between the j−th and k−th components of the corresponding eigenvector ϕ_u, and the phase relies on the corresponding eigenvalue λ_u. Actually the values $e^{-it\lambda_u}$ are the eigenvalues of the Schrödinger Operator $\Psi(t) = e^{-itL}$ as well as points in the Argand circle unit radius. We define

$$
\begin{aligned}
\xi_{jk} &= \lim_{T\to\infty} \arg\left(\frac{1}{T}\int_0^T c_{jk}(t)dt\right) \\
&= \lim_{T\to\infty} \arg\left(\sum_{u=1}^{N} z_u(j,k)\frac{1}{T}\int_0^T e^{-it\lambda_u}dt\right) \\
&= \lim_{T\to\infty} \arg\left(\sum_{u=2}^{N} z_u(j,k)\frac{(1-e^{-iT\lambda_u})}{iT\lambda_u}\right) \\
&= \lim_{T\to\infty} \arg\left(\frac{1}{T}\left\{-iG(j,k) + \sum_{u=2}^{N} g_u(j,k)(\cos(T\lambda_u) - i\sin(T\lambda_u))\right\}\right) ,(5)
\end{aligned}
$$

which accounts for the limiting phase of the transition $c_{jk}(t)$. The limiting phase is obtained from $\arg(\xi_{jk})$, where $\arg(z) = \frac{1}{2i}\log\frac{z}{z^*}$. We have also that $G(j,k) = \sum_{u=2}^{N}\frac{1}{\lambda_u}\phi_u(j)\phi_u(k)$ is the Green's function of the Laplacian and $g_u(j,k) = \frac{1}{\lambda_u}\phi_u(j)\phi_u(k)$. The Green's function is the pseudo-inverse of the Laplacian, that is, $GL^+ = 1$. Such function, or more precisely the inverse $(\beta + L)^{-1}$ with $\beta > 0$, it is typically used for ranking the nodes in the graph [14], that is to estimate topological distances in the graph. For instance, if we define the vector $y = (y_1 y_2 \ldots y_N)^T$ so that $y_j = 1$ if node j is assumed to be "known", and $y_k = 0$ for $j \neq k$, then the function $f = (f_1 f_2 \ldots f_N)^T = (\beta+L)^{-1}y$ yields higher values f_l at nodes l closer to j. Then, the facts that: i) the Green's function appears in the imaginary part of the limit, and ii) the terms of the Green's function appear both in the real and imaginary part, partially explains the structure of the matrix ξ. It is patched with several clusters of components ξ_{jk} in phase, provided that those components link nodes with similar topological distances.

3   Skeletal Similarity Matrices

The simplification of W is a key ingredient of our approach. Such simplification results from computing the long-time averages from magnitude. The diagonal of the symmetric matrix χ contains the long-time probabilities that a CTQW returns to each node. The off-diagonal elements $\chi(j,k)$ are the probabilities that a CTQW commencing at the $j-$th node reaches the $k-$th one in the limit. Then, since $\sum_{k=1}^{T}\chi(j,k)=1$, $\forall j$ we can associate a probability density function (pdf) to each node j. The fraction of off-diagonal probability mass $e(W)=\sum_{j=1}^{N}\sum_{k\neq j}^{N}\chi(j,k)/N$ measures the transport efficiency of the weighted

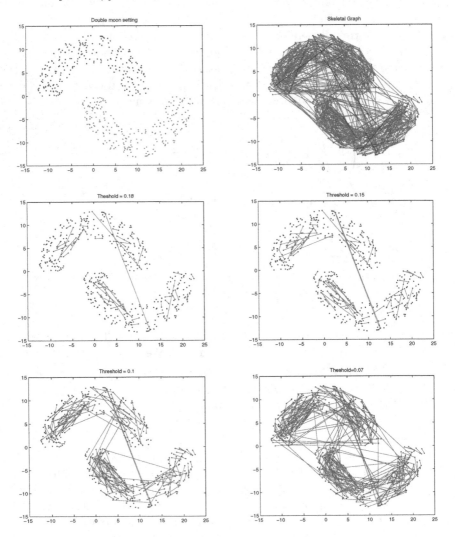

Fig. 1. Graph simplification: Double moon setting (top-left), skeletal graph (top-right), structural level sets for decreasing thresholds

graph. Since the CTQWs have a coherent behavior, $e(W)$ increases when there is enough similarity support, provided that the eigenvectors of the Laplacian matrix L are not degenerated. For instance, if W encodes a complete graph we have $\chi = 1$, that is, $e(W) = 0$ since the eigenvectors of the Laplacian are: 0 with multiplicity one, and N, with multiplicity $N - 1$.

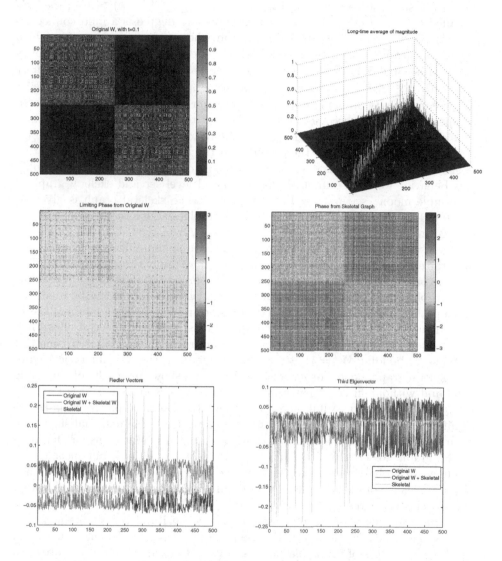

Fig. 2. Graph clustering: Similarity matrix for $t = 0.1$ (top-left), long-time averages of magnitude (top-left), long-time average of phase from the original similarity matrix (second row, left) and from the long-time average of magnitude (second row, right), comparison between the Fiedler vectors (third row, left) and between the third eigenvector (third row, right).

Since such support is not homogeneous we have that χ is sparser than $\boldsymbol{W}$. We also have that off-diagonal probabilities typically correspond to long-range interactions, i.e. to transitive links. If we sort the off-diagonal entries $\chi(j, k)$ in descending order, the sequence of graphs generated by incrementally decreasing the threshold γ used for retaining an edge if $\chi(j, k) \geq \gamma$ leads to a structural level set $\boldsymbol{S}_1, \boldsymbol{S}_2, \ldots, \boldsymbol{S}_r, \ldots$ dominated by long-range interactions. In addition, if we do not stop the sequence until we have $d_j = \sum_{k=1}^{N} S_r(j, k) > 0 \,\forall j$, then $\boldsymbol{S}_r$ is called a *skeletal graph*. In this context, non-zero degree is a mild connectivity criteria but it is enough to form a graph whose links connect points which are topologically close in the manifold, i.e. there is a minimal number of links between isolated clusters (specially when the bandwidth parameter t defining $W(i, j) = e^{-d^2(\boldsymbol{x}_i, \boldsymbol{x}_j)/t}$ is close to zero, as it happens when $N \to \infty$). Actually if our connectivity stopping criterion is stronger (e.g. find a unique connected component) then $\boldsymbol{S}_r$ becomes close to a complete graph. Therefore, the coherence of the CTQWs is exploited to form a sequence of graphs with minimal edge density, whereas if we build the graph from the original similarity information of short-range interactions dominate at the cost of a high density. To exemplify this behavior we have computed the structural level set and skeletal graph of a "double moon" setting (see Fig. 1-top left). The skeletal graph is given at $\gamma = 0.041$ (see Fig. 1-top right, where the last edges added to the structure are colored in blue).

The skeletal graph representation provides graph simplification by retaining only the edges in it. We can also discard nodes if we apply thresholds higher than the one providing the skeletal graph. But the representation is useful insofar the new graph retains good properties of the original $\boldsymbol{W}$, i.e., if it can be used to a given pattern recognition task (e.g. clustering or ranking) with a minimal loss of performance. For instance, when we apply this strategy to spectral clustering, it is better to combine the original affinity matrix with the "filtered" one (for instance by building $\boldsymbol{W}' = \boldsymbol{W} + \alpha\chi$) than to use χ or the weights associated to the skeletal graph alone. For instance, in Fig. 2 we show the results of clustering with and without filtered/sparse information. Fig. 2-top right shows that the CTQWs do not tend to link nodes of different clusters despite such links are significant in $\boldsymbol{W}$ due to the choice of the bandwidth t. The diagonal shows the return probabilities and also shows to what extent coherent transport happens in the graph: we have a high efficiency rate $e(\boldsymbol{W}) = 0.7414$, but most of the off-diagonal probabilities are retained by the links connecting nodes of the same cluster. This is not enough to let the skeletal graphs (or the weights of ξ) obtain a good clustering result. However, if we use $\boldsymbol{W}' = \boldsymbol{W} + \alpha\chi$ with $\alpha = 1$ we obtain a Rand index of 0.8609 which improves the Rand index obtained with the normalized cut method (0.824). The reason is explained by the structure of the Fiedler vectors of the Laplacians associated to each case (Fig. 2-third row, left). The Fiedler vector of $\boldsymbol{W}'$ (red) is very similar (in terms of partitioning information) to that of $\boldsymbol{W}$, but the amplitudes of their oscillations are slightly differ. This effect is better observed in the third eigenvectors (Fig. 2-third row,

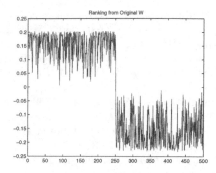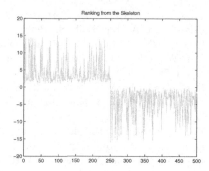

Fig. 3. Graph ranking: Ranking from the pseudo-inverse of the Laplacian. From the original similarity (first column), and from the skeletal similarity (second column)

right). Finally, the eigenvectors associated to the Skeletal graphs are unable to perform the partition.

Finally, in Fig. 3 we show that the skeletal version improves the original one despite having a more sparse pseudo-inverse (it is built from fragmentary evidence).

4   Conclusions

In this paper we have investigated the impact of continuous-time quantum walks in graph simplification, graph-based clustering and graph-based ranking. Future work includes intensive experimentation and the development of an information-theoretic interpretation of coherent transport.

Acknowledgements. Funding. F. Escolano: Project TIN2012-32839 (Spanish Gov.). E. R. Hancock: Royal Society Wolfson Research Merit Award.

References

1. Zhou, X., Belkin, M., Srebro, N.: An iterated graph laplacian approach for ranking on manifolds. In: Proceedings of the 17th ACM SIGKDD International Conference on Knowledge Discovery and Data Mining, San Diego, CA, USA, August 21-24, pp. 877–885 (2011)
2. Shi, J., Malik, J.: Normalized cuts and image segmentation. IEEE Trans. Pattern Anal. Mach. Intell. 22(8), 888–905 (2000)
3. Qiu, H., Hancock, E.R.: Clustering and embedding using commute times. IEEE Trans. Pattern Anal. Mach. Intell. 29(11), 1873–1890 (2007)
4. Arbelaez, P., Maire, M., Fowlkes, C., Malik, J.: Contour detection and hierarchical image segmentation. IEEE Trans. Pattern Anal. Mach. Intell. 33(5), 898–916 (2011)
5. Weiss, Y., Torralba, A., Fergus, R.: Spectral hashing. In: Advances in Neural Information Processing Systems 21, Proceedings of the Twenty-Second Annual Conference on Neural Information Processing Systems, Vancouver, British Columbia, Canada, December 8-11, pp. 1753–1760 (2008)

6. Weiss, Y., Fergus, R., Torralba, A.: Multidimensional spectral hashing. In: Fitzgibbon, A., Lazebnik, S., Perona, P., Sato, Y., Schmid, C. (eds.) ECCV 2012, Part V. LNCS, vol. 7576, pp. 340–353. Springer, Heidelberg (2012)
7. Liu, W., Wang, J., Kumar, S., Chang, S.: Hashing with graphs. In: Proceedings of the 28th International Conference on Machine Learning, ICML 2011, Bellevue, Washington, USA, June 28 - July 2, pp. 1–8 (2011)
8. Grady, L.: Random walks for image segmentation. TPAMI 28(11), 1768–1783 (2006)
9. Bai, L., Rossi, L., Torsello, A., Hancock, E.R.: A quantum jensen-shannon graph kernel for unattributed graphs. Pattern Recognition 48(2), 344–355 (2015)
10. Rossi, L., Torsello, A., Hancock, E.R., Wilson, R.C.: Characterizing graph symmetries through quantum jensen-shannon divergence. Phys. Rev. E 88, 032806 (2013)
11. Mülken, O., Blumen, A.: Continuous-time quantum walks: Models for coherent transport on complex networks. Physics Reports 502(2-3), 37–87 (2011)
12. Stone, M.: On one-parameter unitary groups in hilbert space. Annals of Mathematics 33(3), 643–648 (1932)
13. Farhi, E., Gutmann, S.: Quantum computation and decision trees. Phys. Rev. A 58, 915–928 (1998)
14. Zhou, D., Weston, J., Gretton, A., Bousquet, O., Schlkopf, B.: Ranking on data manifolds. In: Advances in Neural Information Processing Systems 16. MIT Press (2004)

Graph Based Lymphatic Vessel Wall Localisation and Tracking

Ehab Essa, Xianghua Xie, and Jonathan-Lee Jones

Swansea University, Swansea SA2 8PP, UK
{e.m.m.essa,x.xie,csjj}@swansea.ac.uk
http://csvision.swan.ac.uk

Abstract. We present a novel hidden Markov model (HMM) based approach to segment and track the lymph vessel in confocal microscopy images. The vessel borders are parameterised by radial basis functions (RBFs) so that the number of tracking points are reduced to a very few. The proposed method tracks the hidden states that determine the border location along a set of normal lines obtained from the previous frame. The border observation is derived from edge-based features using steerable filters. Two Gaussian probability distributions for the vessel border and background are used to infer the emission probability. The transition probability is learnt by using the Baum-Welch algorithm. A new optimisation method for determining the best sequence of the hidden states is introduced. We transform the segmentation problem into a minimisation of s-excess graph cost. Each node in the graph corresponds to one state, and the weight for each node is defined using its emission probability. The inter-relation between neighbouring nodes is defined using the transition probability. Its optimal solution can be found in polynomial time using the s-t cut algorithm. Qualitative and quantitative analysis of the method on lymphatic vessel segmentation show superior performance of the proposed method compared to the traditional Viterbi algorithm.

1 Introduction

Object tracking has been widely used in many applications such as motion-based recognition [1], surveillance [2, 3], and medical imaging [4, 5]. The object can be represented in various forms, e.g., points, geometric shape, and contour. Geometric shapes, e.g., rectangle and ellipse, are suitable for representing rigid objects. Tracking object contour is usually carried out for complex and nonrigid shapes where a contour provides more accurate representation than a bounding box. In this paper, we proposed an approach to track lymph anatomical borders over the cross-sectional images in order to achieve more coherent segmentation.

Various techniques of contour detection methods have been used in tracking and could be integrated with the state space model; for example, active contour and hidden Markov model. Active contour is an energy minimisation method that utilises two forces: internal force describes the regularity of the contour shape, and external force describes the boundary features to evolve the contour

© Springer International Publishing Switzerland 2015
C.-L. Liu et al. (Eds.): GbRPR 2015, LNCS 9069, pp. 345–354, 2015.
DOI: 10.1007/978-3-319-18224-7_34

to reach the object boundary. For example, Terzopoulos *et al.* [6] represent the system state by B-spline control points and integrate the Kalman filter with the snake model. In [7], authors also used B-spline representation and proposed the Condensation algorithm to track affine motion parameters using a particle filter. The measurement is computed by extracting edge features along normal lines. In [5] authors use watershed segmentation to initialise the contour position and then evolve an edge-based snake to track the left ventricle in echocardiographic images. However, active contour heavily depends on the initialisation of the contour and may converge to the local minima.

The hidden Markov model is a stochastic model in which the Markov property is assumed to be satisfied in a finite set of states in which these states are hidden. Many applications have demonstrated the advantage of HMM to deal with the time-varying signals such as speech recognition [8], classification [1], and tracking [9–11]. In [1], authors use HMM to classify the local wall motion of stress echocardiography to normal or abnormal. They build two HMM models, one for each class and use a forward algorithm to compute the probability of the observation of each model. In [12], HMM is used in conjunction with a particle filter to track hand motion. A particle filter is used to estimate the hand region that is most likely to appear. HMM estimates the hand shape using the Viterbi algorithm where the state is a set of quantised pre-learnt exemplars [13]. However, the number of exemplars can grow exponentially depending on the complexity of the object. In [14], the authors used the Kalman filter with P2DHMM (pseudo 2-dimensional HMM) to perform a person tracking. P2DHMM is a nested 1D HMM in which a number of super hidden states modelling image columns, each of which contains another number of HMM hidden states. The Viterbi algorithm is used to find the best sequence of states that classify the image as object and background. These measurements are used by the Kalman filter to predicate the rectangle box containing the object in the next frame. However, the system becomes more complex and time-consuming with the increase of object size. In [9, 10], the authors incorporated region and edge features with HMM. The contour is sampled into a set of discrete points, and the features are extracted along the normal direction at each contour point. The ellipse shape is fitted based on the contour and the unscented Kalman filter is used for tracking. In [11], authors extended the previous idea to deal with variable length open contours and investigated using arc emission instead of traditional state emission for defining the observation probabilities of the HMM. The optimal contour is identified using the Viterbi algorithm.

In this paper, we propose a segmentation and tracking method based on a novel HMM, in order to efficiently delineate the vessel borders in lymphatic images with the presence of noise and occlusions. The transformation of vessel images from Cartesian coordinates to polar coordinates allows efficient parametrisation of the border using RBFs and thus reduces the tracking size from a large number of border points to a very few RBF centres. A new optimisation method is proposed to find the best sequence of the hidden states based on the minimisation of an s-excess graph. Instead of using the traditional Viterbi algorithm, we transform the problem into finding the optimal closure graph that corresponds to the desired border and can be efficiently solved in polynomial time. The emission probability is defined based on two probability distributions for the vessel border and background respectively that are derived directly from edge-based features. The training of the transition probability is achieved by using the Baum-Welch algorithm.

2   Proposed Method

As illustrated in Figure 1, the border of interest is approximated by using the RBF functions where the hidden states of the HMM model are referred to the potential RBF centres. The border is evenly sampled into M points and at each point a line segment is drawn perpendicular to the tangent line of the border and each line segment has K points. The index of contour RBF centres is denoted $\phi = 1, \ldots, M$ and the index of each normal line is $\psi = 1, \ldots, K$ where K is an odd number. The initial RBF centres are defined from the previous frame and located as the centre of normal line $\psi = (K + 1)/2$ as shown in Figure 1. The normal line restricts the search space for the predicated contour to be within $(K - 1)/2$ point distance from the initial contour.

We denote all possible sequences of hidden states as $Q = \{q\}$, where $q = \{q_1, \ldots, q_\phi, \ldots, q_M\}$ is one possible state sequence and q_ϕ is the state on the normal at ϕ. These sequences correspond to possible RBF centre locations. The HMM observations $O = \{O_1, \ldots, O_\phi, \ldots, O_M\}$ are extracted from the normal lines. HMM is specified by three probability measures $\lambda = (A, B, \pi)$, where A, B and π are the probabilities for the transition, emission and initial states. The transition between states q at two normals ϕ and $\phi + 1$ are governed by a set of probabilities, i.e. transition probabilities $P(q_\phi | q_{\phi+1})$, and any state can only be observed by an output event according to associated probability distribution, i.e. emission probabilities $P(O_\phi | q_\phi)$. The output event is the image features extracted from each state at the normal line.

Steerable filter is used to highlight the border and to define the image observation. Two Gaussian distributions are inferred from the these observation to compute the emission probability. The transition probability is learnt from a set of training examples. To find the optimal sequence of hidden states, we propose a novel optimisation using an s-excess graph. The minimum s-excess graph is a variation of minimum closure graph. Each hidden state on the normal line corresponds to a node in the graph and the cost of each node is inversely proportional to its emission probability. The transition probability defines the inter-columns arcs cost. The minimisation of s-excess graph is effectively solved by the graph cut in polynomial time.

2.1   Border Parametrisation

The images are unraveled from Cartesian coordinates to polar coordinates so that the border of interest becomes a height field in the polar coordinates. Note the

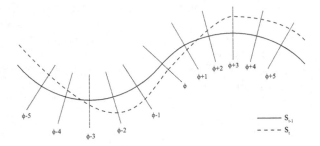

Fig. 1. An illustration of segmenting border S_t based on the border S_{t-1} obtained in the previous frame using the proposed HMM Model. The border is divided into M points and at each point a normal line is drawn with length K. The final border is interpolated using RBFs.

starting point and the end point is considered the same after unraveling to ensure continuity. Tracking all the border points individually in their polar coordinates is computationally expensive and unnecessary. We interpolate the border using radial basis functions. Thus, tracking the border of interest becomes localising those RBF centres at a set of angular positions in the polar coordinates. The thin plate local compact support RBF is used for interpolation. The movement of the border across frames is in effect determined by the locations of those RBF centres. The radial basis function is defined as a linear combination of the basis functions Φ and a low degree polynomial term $p(x)$, each basis function associated with a weighted coefficient ω_j, and centred at the constraint locations c_i. It can be formulated as $f(\mathbf{x}) = \sum_{i=1}^{M} \omega_i \Phi(\mathbf{x} - c_i) + p(\mathbf{x})$, where $\mathbf{x}$ is the border points [15]. The thin-plate function is given by $\Phi(x) = |x|^2 log(|x|)$.

2.2 Transition Probability

In [16], the authors used smoothness constraints to define the transition probability that favours small displacements from the initial contour, i.e., it limits the deformation ability for the contour. The object is assumed to be opaque and the segmented foreground and background regions have different appearances. Recently, the authors in [11] computed transition probabilities by modelling the amount of bending between the normal lines and the displacement of contour position with respect to the previous frame using two Gaussian distributions and defining weighted coefficients for each transition. A fixed point iterative method is used to find the optimal value of these coefficients.

Baum-Welch and Viterbi training are popular estimation methods for HMM parameters (A, B, π). The Viterbi training is an approximation of the Baum-Welch method and is computationally much faster, however, it may perform less well compared to the Baum-Welch method. In this work, we use the Baum-Welch (Forward-Backward) algorithm [8] to define both the transition and prior probabilities. The Baum-Welch algorithm is a special case of the Expectation-Maximisation method. We define the probability of traversing from state i at normal ϕ to state j at normal $\phi + 1$, where $1 \leq \phi \leq M$ and, $1 \leq i, j \leq K$ as:

$$\xi_\phi(i, j) = P(q_\phi = i, q_{\phi+1} = j | O, \lambda), \tag{1}$$

This definition can be written using forward and backward variables as:

$$\xi_\phi(i, j) = \frac{\alpha_\phi(i) a_{ij} b_j(O_{\phi+1}) \beta_\phi(j)}{\sum_{i=1}^{K} \sum_{j=1}^{K} \alpha_\phi(i) a_{ij} b_j(O_{\phi+1}) \beta_\phi(j)} \tag{2}$$

where $\alpha_\phi(i)$ and $\beta_\phi(j)$ are the forward and backward variables for state i and j respectively, $b_j(O_{\phi+1})$ is the emission probability and it is measured by $P(O_{\phi+1}|q_j)$. The probability of being in state i at normal ϕ is given by $\gamma_\phi(i) = \sum_{j=1}^{K} \xi_\phi(i, j)$. The summation of $\gamma_\phi(i)$ over all normals is interpreted as the expected number of transitions from state i, while the summation of $\xi_\phi(i, j)$ over all normals is the expected number of transitions from state i to state j. The expected initial state distribution at the first normal $\phi = 1$ is defined as $\hat{\pi}_i = \gamma_{\phi_1}(i)$, and the expected transition probability is given as $\hat{a}_{ij} = \sum_{\phi=1}^{M-1} \xi_\phi(i, j) / \sum_{\phi=1}^{M-1} \gamma_\phi(i)$. The process is iterated and each of iterations increases the likelihood of the data $P(O|\hat{\lambda}) \geq P(O|\lambda)$ until convergence.

2.3 Emission Probability

We use a similar method to describe the emission probability as presented in [11, 10]. Image observations are modelled by two probability density functions: one for the contour and the other for the background. Let $O_\phi = \{o_{\phi,1}, \ldots, o_{\phi,\psi}, \ldots, o_{\phi,K}\}$ be a set of features along the normal ϕ and $o_{\phi,\psi}$ a feature extracted from point ψ on the line. $P(o_{\phi,\psi}|FG)$ and $P(o_{\phi,\psi}|BG)$ represent the probabilities of that feature belonging to the border of interest and the background respectively. The state-emission probability is defined as:

$$P(O_\phi|q_\phi) \propto P(o_{\phi,\psi}|FG) \prod_{\psi \neq q_\phi} P(o_{\phi,\psi}|BG). \tag{3}$$

The likelihood of the observed variables O_ϕ from a state q_ϕ is computed by measuring the likelihood of each feature $o_{\phi,\psi}$ at index ψ of the line ϕ to belong to the contour and all the rest of features on that line belong to the background.

From a set of training data with a manually labelled borders of interest, we use first order derivatives of Gaussian (GD) filters [17] in six different orientations to highlight the edge features along the border, and extract features that correspond to the border and the background, in order to learn the parameters, i.e. mean and variance of two Gaussian distributions FG and BG.

2.4 Minimum s-Excess Graph Optimisation

Viterbi algorithm is commonly used to find the best sequence of hidden states [9, 11]. However, the search process can be hampered by occlusions in the image where the emission probability is not reliable and relying on the transition probably does not guarantee smoothness transition of the border. Moreover, the Viterbi algorithm cannot handle the close contour case when starting and ending point is the same. Here, we treat the problem as minimising an s-excess graph to return the minimum-cost closet set graph containing the best sequence of states on its envelope.

The minimum s-excess problem [18] is a relaxation of minimum closure set problem. A closed set is a subset of graph where all successors of any node in the set are also taken in the set. In s-excess problem, the successors of nodes may not be contained in the set but with cost equal to the edge cost heading to such successors.

Given a directed graph $G(V, E)$, each node $v \in V$ is assigned to a certain weight $w(v)$ and each edge $e \in E$ have a positive cost. The aim of the minimum s-excess problem [19] is to obtain a subset of graph nodes $S \subset V$ where the total nodes cost add to the cost of separating the set from the rest of the graph $S' = V - S$ is minimised:

$$\mathcal{E} = \sum_{v \in S} w(v) + \sum_{\substack{(u,v) \in E \\ u \in S, v \in S'}} c(u, v). \tag{4}$$

Minimisation of closure graph and s-excess problems have been used in image segmentation, e.g. [20–23]. In [20], the authors formulated the problem as a minimisation of a closed graph to find the optimal surfaces on d-Dimensional multi-column graphs ($d \geq 3$). However, only weighted graph nodes (first term in eq. (4)) is used to represent the segmentation problem by assuming the surface is changing within a global constant constraint which is difficult to define and may lead to abrupt changes of the surface due to the high influence of nodes weight. In [21], prior information is incorporated (in the second term in eq. (4)) by encoding the prior shapes as a set of convex functions defined between every pair of adjacent columns. Our work is similar to Song *et al.* [21] in that aspect. However, we formulate the problem differently as a tracking problem in consecutive frames, defining the pair-wise cost based on the transition probability learnt from training data and treat it as a stochastic process.

The dimension of the graph G is $M \times K$, where M is the number of RBF centres ($m = 1, \ldots, M$) and K is the length of each normal line ($k = 1, \ldots, K$). Each hidden state $\{q_{1,1}, \ldots, q_{m,k}, \ldots, q_{M,K}\}$ is equivalent to a graph node $V = \{v_{1,1}, \ldots, v_{m,k}, \ldots, v_{M,K}\}$. The nodes that are in one normal line K is treated as a column chain. We have two type of graph arcs: intra-chain and inter-chain arcs. For intra-chain, along each chain, every node $v(m, k)$ ($k > 1$) has a directed arc to the node $v(m, k - 1)$ with $+\infty$ weight assigned to the edge cost to ensure that the desired border intersects with each column once and once only. In the case of inter-chain, for every ordered pair of adjacent chains $Col(m_1, k_1)$ and $Col(m_2, k_2)$, a set of directed arcs are established to link every node $V(m_1, k_1)$ with nodes $V(m_2, k_2')$ in the adjacent chain $Col(m_2, k_2)$, where $1 \leq k_2' \leq k_1$. Similarly, from each node $V(m_2, k_2)$ to nodes $V(m_1, k_1')$ in the previous adjacent chain $Col(m_1, k_1)$, where $1 \leq k_1' \leq k_2$, are connected. Since each node refers to one hidden state, the cost of arcs between a pair of adjacent chains, $c(v_{m_1,k_1}, v_{m_2,k_2})$, is defined by the transition matrix a_{ij} discussed in Section 2.2. The last row of the graph is connected to each other with $+\infty$ to form the basis of the closed graph. Since the images are unraveled into polar coordinates, the first and last column chain is also connected in the same manner as the inter-chain arcs.

The emission probability for every node is converted to cost according to $C(m,k) = -log(P(o_{m,k}|q_{m,k}))$ and it is inversely correlated to the likelihood that the border of interest passes through the node (m,k). The weight for each node $w(m,k)$ on the directed graph is defined as $w(m,k) = C(m,k) - C(m,k-1)$, except for the bottom column, i.e. $w(m,1) = C(m,1)$.

Hochbaum [24] showed that the minimum s-excess problem is equivalent to solving the minimum $s - t$ cut defined on a proper graph. Hence, the s-t cut algorithm [25] can then be used to find the minimum closed set, based on the fact that the weight can be used as the basis for dividing the nodes into nonnegative and negative sets. The source s is connected to each negative node and every nonnegative node is connected to the sink t, both through a directed arc that carries the absolute value of the cost node itself. The running time $T(n,m)$ is thus bound to finding the minimum s-t cut, where n number of graph nodes and m number of edges. The cost of cut is the total number of edge cost separating the graph into source (i.e the minimum s-excess graph) and sink sets.

3   Application and Results

The proposed method is applied to segment lymphatic vessel images that are obtained using confocal microscopy. Confocal microscopy is a method that allows for the visualisation and imaging of 3D volume objects that are thicker than the focal plane of a conventional microscope. Four in vitro volumes were acquired using confocal microscopy after charged with fluorescent dye and pressure was applied to ensure valve opening. We randomly select 2 volumes to train the HMM and the the rest for testing. Anisotropic vessel enhancement diffusion [26] was applied to enhance image features and to improve vessel connectivity, as well as reduce the amount of image noise that is typical in this type of images. The size of each volume is $512 \times 512 \times Z$ where Z is the number of slices are varies. Each volume is converted to the trans-axial view and resized to $512 \times 512 \times 512$ then was unwrapped into polar coordinates of size 200×1257. the evaluation was carried out on every 10th frame where manual labelling was performed. Five different evaluation metrics are used, including absolute mean difference (AMD), Hausdorff distance (HD) in pixel, sensitivity (Sens. %), specificity (Spec. %), Jaccard measure (JM %). The inner region is considered as positive for Sens. and Spec. measurements. The normal line segments were measured at a length of 101 pixels, and 42 RBF centres in polar coordinates were used, i.e. one RBF centre every 30 pixels.

Tables 1,2 shows the quantitative results of segmenting both the inner and outer vessel walls. The result of the outer vessel border has lower distance error than the inner vessel wall with the absolute mean difference 2.42 and Hausdorff distance 10.59 pixels compared to 4.55 and 14.22 respectively as the lymph

Table 1. Inner border quantitative result. Mean value (standard deviation)

	AMD	HD	JM	Sens.	Spec.
Proposed method using Viterbi algorithm	13.56 (6.56)	51.31 (18.86)	81.17 (8.55)	82.14 (9.26)	**99.51** **(0.99)**
Proposed method using s-excess optimisation	**4.55** **(3.84)**	**14.22** **(8.63)**	**93.14** **(0.06)**	**95.45** **(5.74)**	98.84 (2.26)

Table 2. Outer border quantitative result. Mean value (standard deviation)

	AMD	HD	JM	Sens.	Spec.
Proposed method using Viterbi algorithm	4.56 (1.59)	30.46 (15.62)	94.93 (1.38)	97.48 (0.98)	97.88 (1.11)
Proposed method using s-excess optimisation	**2.42** **(0.67)**	**10.59** **(3.71)**	**97.18** **(0.83)**	**98.57** **(0.99)**	**98.90** **(0.57)**

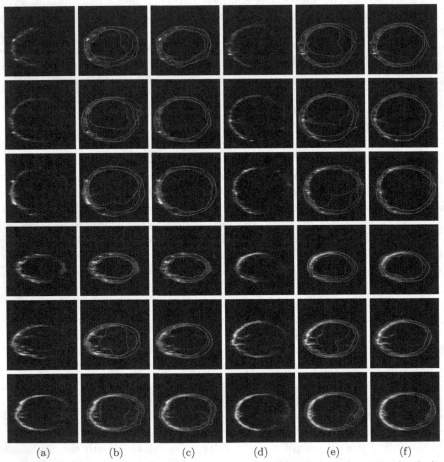

(a)   (b)   (c)   (d)   (e)   (f)

Fig. 2. Comparison between ground-truth (green) and segmentation results (red). (a,d) original image (b,e) Viterbi method. (c,f) proposed method.

valve intervenes with the inner border. The tables also present a comparison between using the Viterbi algorithm and the proposed s-excess optimisation in finding the best sequence of states. The proposed s-excess optimisation clearly outperformed Viterbi algorithm in tracking both borders. Viterbi based method is easily distracted by valve and produced high distance error and low sensitivity. It also requires a much larger amount of data to train the transition and prior probabilities. The proposed s-excess optimisation achieved much smoother border transition, particularly when there was an occlusion or missing feature. Figure 2 shows typical segmentation results when compared to the independent manual labelling, i.e. ground-truth.

4 Conclusions

We presented an HMM based segmentation method for delineating the inner and outer lymph vessel borders in confocal images. The method searches for the border along a set of normal lines based on the segmentation of the previous frame. Boundary based features were extracted to infer the emission probability. We show that the optimal sequence of the hidden states corresponds to RBFs of the border can be effectively obtained by minimising the s-excess problem. Qualitative and quantitative results on lymphatic images showed promising performance of the method.

Acknowledgement. We would like to thank James Cotton from Wolverhampton Hospital for providing the data. The work is funded by NISCHR.

References

1. Mansor, S., Noble, J.: Local wall motion classification of stress echocardiography using a hidden markov model approach. In: ISBI (2008)
2. Beymer, D., Konolige, K.: Real-time tracking of multiple people using continuous detection. In: ICCV Frame-Rate Workshop (1999)
3. Xie, L., Zhu, G., Wang, Y., Xu, H., Zhang, Z.: Real-time vehicles tracking based on kalman filter in a video-based ITS. In: ICCCS (2005)
4. Abolmaesumi, P., Sirouspour, M., Salcudean, S.: Real-time extraction of carotid artery contours from ultrasound images. In: CBMS (2000)
5. Cheng, J., Foo, S., Krishnan, S.M.: Watershed-presegmented snake for boundary detection and tracking of left ventricle in echocardiographic images. IEEE T-ITB 10(2), 414–416 (2006)
6. Terzopoulos, D., Szeliski, R.: Active vision. MIT Press, 3–20 (1993)
7. Isard, M., Blake, A.: Condensation–conditional density propagation for visual tracking. IJCV 29(1), 5–28 (1998)
8. Rabiner, L.: A tutorial on hidden markov models and selected applications in speech recognition. Proceedings of the IEEE 77(2), 257–286 (1989)
9. Chen, Y., Rui, Y., Huang, T.: JPDAF based HMM for real-time contour tracking. In: CVPR (2001)

354   E. Essa et al.

10. Chen, Y., Rui, Y., Huang, T.: Multicue HMM-UKF for real-time contour tracking. IEEE PAMI 28(9), 1525–1529 (2006)
11. Sargin, M., Altinok, A., Manjunath, B., Rose, K.: Variable length open contour tracking using a deformable trellis. IEEE TIP 20(4), 1023–1035 (2011)
12. Fei, H., Reid, I.D.: Joint bayes filter: A hybrid tracker for non-rigid hand motion recognition. In: Pajdla, T., Matas, J(G.) (eds.) ECCV 2004. LNCS, vol. 3023, pp. 497–508. Springer, Heidelberg (2004)
13. Toyama, K., Blake, A.: Probabilistic tracking with exemplars in a metric space. IJCV 48(1), 9–19 (2002)
14. Breit, H., Rigoll, G.: Improved person tracking using a combined pseudo-2D-HMM and kalman filter approach with automatic background state adaptation. In: ICIP (2001)
15. Turk, G., O'brien, J.F.: Modelling with implicit surfaces that interpolate. ACM ToG 21(4), 855–873 (2002)
16. Chen, Y., Huang, T., Rui, Y.: Parametric contour tracking using unscented kalman filter. In: ICIP (2002)
17. Freeman, W.T., Adelson, E.H.: The design and use of steerable filters. IEEE PAMI 13(9), 891–906 (1991)
18. Hochbaum, D.S.: Anniversary article: Selection, provisioning, shared fixed costs, maximum closure, and implications on algorithmic methods today. Manage. Sci. 50(6), 709–723 (2004)
19. Wu, X., Chen, D.Z.: Optimal net surface problems with applications. In: Widmayer, P., Triguero, F., Morales, R., Hennessy, M., Eidenbenz, S., Conejo, R. (eds.) ICALP 2002. LNCS, vol. 2380, pp. 1029–1042. Springer, Heidelberg (2002)
20. Li, K., Wu, X., Chen, D.Z., Sonka, M.: Optimal surface segmentation in volumetric images-a graph-theoretic approach. IEEE PAMI 28(1), 119–134 (2006)
21. Song, Q., Bai, J., Garvin, M., Sonka, M., Buatti, J., Wu, X.: Optimal multiple surface segmentation with shape and context priors. IEEE TMI 32(2), 376–386 (2013)
22. Essa, E., Xie, X., Sazonov, I., Nithiarasu, P.: Automatic IVUS media-adventitia border extraction using double interface graph cut segmentation. In: ICIP, pp. 69–72 (2011)
23. Essa, E., Xie, X., Sazonov, I., Nithiarasu, P., Smith, D.: Shape prior model for media-adventitia border segmentation in IVUS using graph cut. In: Menze, B.H., Langs, G., Lu, L., Montillo, A., Tu, Z., Criminisi, A. (eds.) MCV 2012. LNCS, vol. 7766, pp. 114–123. Springer, Heidelberg (2013)
24. Hochbaum, D.S.: An efficient algorithm for image segmentation, markov random fields and related problems. J. ACM 48(4), 686–701 (2001)
25. Boykov, Y., Kolmogorov, V.: An experimental comparison of min-cut/max-flow algorithms for energy minimization in vision. IEEE PAMI 26(9), 1124–1137 (2004)
26. Manniesing, R., Viergever, M.A., Niessen, W.J.: Vessel enhancing diffusion: A scale space representation of vessel structures. MIA 10(6), 815–825 (2006)

A Comic Retrieval System Based on Multilayer Graph Representation and Graph Mining

Thanh-Nam Le$^{(\boxtimes)}$, Muhammad Muzzamil Luqman,
Jean-Christophe Burie, and Jean-Marc Ogier

L3i Laboratory, University of La Rochelle, 17042 La Rochelle Cedex 1, France
{thanh_nam.le,muhammad_muzzamil.luqman,jean-christophe.burie,
jean-marc.ogier}@univ-lr.fr

Abstract. Comics analysis offers a lot of interesting Content-based Image Retrieval (CBIR) applications focusing in this special type of document images. In this paper, we propose a scheme to represent and to retrieve comic-page images using attributed Region Adjacency Graphs (RAGs) and their frequent subgraphs. We first extract the graphical structures and local features of each panel in the whole comic volume, then separate different categories of local features to different layers of attributed RAGs. After that, a list of frequent subgraphs is obtained using Frequent Subgraph Mining (FSM) techniques. For CBIR purpose, the recognition and ranking are done by checking for isomorphism between the graphs representing the query image versus the list of discovered frequent subgraphs. Our experimental results show that the proposed approach can achieve reliable retrieval results for comic images browsing using query-by-example (QBE) model.

Keywords: Comics · Attributed region adjacency graph · Spatial relation · Structural pattern recognition · CBIR · Query-by-example

1 Introduction and Related Work

Comics, a kind of visual medium to express ideas via drawing, has its large audience and market throughout the world. Besides a tremendous amount of paper-based comic books produced since their early age, *e-comics* has become more and more common recently thanks to its advantage of being electronically distributed. The explosion of smart reading devices is also pulling a noticeable amount of research interest to develop an interactive environment for reading-on-small-devices of comics.

Normally it is rather inconvenient for a reader to perform a search for a specific scene or comic character in large volumes, as he has to perform brute force search with only clues based on his own impression and/or memory of the content. In addition, it may take quite a long time between the releases of volumes, so the readers may find it hard to find out the correlation between chapters if they forget some detail. Searching through larger archives (*e.g.* online bookstores) is even more difficult and limited, as the search support is generally only keyword-based query by title, author or publisher. For the above reasons, developing a

CBIR (Content-based Image Retrieval) system focusing in comics, where the indexing and retrieval involve the drawing content, would potentially make the search experience more efficient, intuitive, and enjoyable.

The lack of such an integrated comics analysis and retrieval system until now is however understandable, due to a lot of challenges associated with content-based comic retrieval. As comic-page images are one kind of complex document image (unstructured combination of lines, text, stylistic text, curves and other sketch strokes), previous researches have been carried out to localize the "basic" elements to facilitate further analysis, such as: analyzing the layout of a page, extracting panels [1,2,3], localizing speech balloons [4,5], or detecting text [5,6]. Yet the biggest challenge remains in recognizing comic characters and understanding the scenery, which can be made up from different objects against complex background. The great variation in size, shape and pose of the characters, and severe occlusion problem account for the main difficulties of the recognition task. In attempt to detect and retrieve comic characters, several approaches have been proposed. In [7], the authors proposed a retrieval method by matching descriptors composed of the most representative colors in the query and local regions in comic panels. In [8], the authors proposed automatic indexing of comic-page images by modeling the problem as a subgraph spotting task. For human-like characters in *manga*, the authors in [9] tried to detect *manga* faces by features extraction and deformable part model.

A lot of CBIR systems employ feature vectors as statistical representations of images [10,11]. The choice of vectors is mainly motivated by their ease of handling in vector space (computational feasibility of vector distances). However, other than statistical method, document images can also be represented by a structural representation, such as trees or graphs. The fact that graphs are naturally well-suited to model objects in terms of parts and their relations makes them a strong candidate for various recognition applications, *e.g.* general object recognition, handwriting recognition, sketch recognition [12], etc. In such domains, the recognition problem turns into the task of graph matching, or evaluating the similarity between graphs. The same philosophy applies for comic analysis using graph approach.

In this paper, we present a system to represent comic-page images by multi-layer, attributed regional adjacency graphs (RAG), and to retrieve them with query-by-example as input. In our method, each panel is represented by several layers of attributed RAGs, where in each layer, the nodes correspond to the segmented regions of a panel; the nodes' attributes store the local characteristics of the corresponding regions; and the edges are constructed based on the proximity of the segmented regions.

The rest of paper is organized as follows: In Sec. 2, we focus on graph construction from the original comic-page images; Sec. 3 details our method of mining frequent subgraphs, and retrieval using matching between query and mined frequent patterns, followed by evaluation and discussion in Sec. 4. Finally the conclusion comes in Sec. 5.

2 Multilayer Graph Representation of Comic-Page Image

In this section we present the construction of multilayer attributed RAGs from comic-page images. An attributed graph is defined as follows: *Let A_V and A_E denote the domains of possible values for attributed nodes and edges. An attributed graph AG over (A_V, A_E) is a four-tuple $AG = (V, E, \mu^V, \mu^E)$, where: V is a set of nodes; $E \subseteq V \times V$ is a set of edges; $\mu^V : V \to A_V$ is a function assigning attributes to nodes; and $\mu^E : E \to A_E$ is a function assigning attributes to edges.* In an attributed graph, $|V|$ is the order of graph, $|E|$ is graph size, and the degree of a node is number of edges connected to it.

2.1 Preprocessing of Panels and Nodes Construction

A comic-page image usually consists of several smaller panels (or frames), each one captures a moment of the story. We first extract the panels in the comic-page image using the method proposed by Rigaud *et al.* [13]. After extracting the panels, each panel is segmented into smaller regions. The segmentation of panels is an important step as it directly affects the size and semantic structure of the resulting graphs. Filtering, which refers to eliminating areas that are not essential such as black lines delimiting the colored regions or tiny artifacts, is taken into consideration as well. Those regions are ignored for the above reason and to limit the size of the final graph. Among a lot of segmentation techniques that can be employed, we choose MSER (Maximally Stable Extremal Region) [14], as it has been shown to be superior in terms of stability and performance compared to other techniques which are also based on intensity extrema [15]. Due to the great variety of styles, or level of detail in different comic titles, no global or optimal settings for segmentation is sought. The parameters of MSER are set adaptively based on "style signature" of the author. In fact, it is observed that an author tends to generate rather consistent content in terms of drawing elaboration. The range of 25 to 50 regions per panel has been experimentally determined to limit the number of nodes (and so the size) of the resulting graphs.

2.2 Node Labeling

Node labeling is done in this step by assigning the local characteristics of each region to its node. Color, texture and shape are fundamental elements of object appearance used by human to differentiate objects. However, drawings in comics in general are poorly textured, so we rely mainly on color and shape features. Each node is characterized by the following attributes:

Color: The color information of a region is the average value of color values of all pixels within it, encoded in CIE Lab color space. L*a*b is a color-opponent space, which has been specially designed to be perceptually uniform based on human visual system. This means a change of the same amount in a color value should produce a change of about the same visual importance, thus Euclidean distances in this space correspond more closely and uniformly to the differences of the perceived colors.

Moment Invariants: The shape of a region is also an important discriminatory characteristic. The set of seven Hu moments [16], which was shown to be invariant under translation, changes in scale, mirror symmetry and rotation, are employed due to their invariance characteristics. As Hu moments values are generally very small in numerical value, the actual values used as node attribute are in their logarithmic scale, and the distance between descriptors A and B in terms of moments h_i ($i = 1..7$) is calculated as: $d(A, B) = \sum_{i=1}^{7} \left| \frac{1}{m_i^A} - \frac{1}{m_i^B} \right|$ where $m_i = sign(h_i) \log(|h_i|)$.

Compactness: Another shape descriptor of a region is the compactness measure, or circularity. It is the ratio of the area of the shape to the area of a circle having the same perimeter: $circularity = 4\pi \frac{area}{perimeter^2}$. This feature can also be used to describe a region: a circularity value of 1 indicates a perfect circle (the most compact shape), while smaller values indicate increasingly elongated shapes.

2.3   Edges Construction and Labeling

After the nodes together with their attributes are obtained, the edges of the graphs are constructed based on the criteria of adjacency and proximity. For any pair of regions in the set of regions, we set an edge linking the nodes representing them if the regions are determined as a proximate pair. We take into account both the distances between the centroid of regions and the distances between points on their contours.

The edge attribute assigned to each edge is to quantify the relationship between two connected regions. This measure should be invariant to survive rotation and scaling change. Based on the assumption that the proportion between parts of a certain character will be preserved, regardless it is drawn from near or far perspective, we propose to use the ratio between surface areas (in terms of number of pixels) of adjacent regions. Let R_1 and R_2 be two connected regions that $S_{R_1} < S_{R_2}$, where S_{R_1} and S_{R_2} denote the surface areas of R_1 and R_2. The attribute of the edge between R_1 and R_2 is the ratio $\frac{S_{R_1}}{S_{R_2}}$.

2.4   Multilayer Representation and Quantization of Node Labels

Evaluating the similarity between nodes by calculating the distances with all the obtained attributes as a single feature vector is not efficient, as not all types of the regions' attributes vary the same amount, or contribute the same weight in determining how a node is "different" from another. For example, in the presence of color feature, as in our case, retrieval based on color has been shown to outperform retrieval based on shape alone, both in terms of efficiency and robustness [17]. So we propose the following steps to separate different types of node information into different graph layers, as illustrated in Fig. 1:

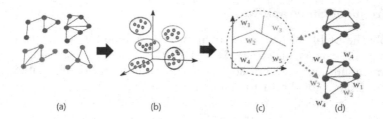

(a)   (b)   (c)   (d)

Fig. 1. (a) A single graph layer, (b) Value space of all the nodes' attribute in that layer, (c) A dictionary is built based on the observed values (d) Reassigning the attribute in each node by its corresponding label in the dictionary in (c)

Step 1: Each graph is triplicated but each copy only stores one of the three attributes: color information, moment invariants, and compactness value. From the original dataset D of graphs representing the panels in the volume, we obtain three sets D_{color}, $D_{moments}$, and $D_{compact}$.

Step 2: In each set, all the observed values of node attributed are gathered and clustered using k−means algorithm [18]. The number of cluster is determined based on the inherent cluster characteristic of the observed node values, in the beginning of dictionary building process.

Step 3: The original attribute in each node is replaced by the label value of its corresponding cluster from the built dictionary. After this step, we have three layers of graphs representing the comic volume, with the graphs in each layer having the same edge structure but different type of node attributes. The matching and sorting used in FSM procedure in the following section are based on this single label value in each node.

3   Querying and Retrieval

Besides complexity, an inherent problem with the use of graph, an exact matching algorithm also suffers from the *low repeatability* problem. It will reject the match between two "similar" graphs when there is discrepancy in even only one of the nodes. This problem arises as the encoding of an object is not perfect due to noise and errors introduced in low-level stages. Since it is essential that the exactness constraint has to be relaxed in a CBIR system to avoid low repeatability problem, we propose a search scheme based on partial matching and voting between subgraph matching only. The search is performed on a list of frequent subgraphs obtained from Frequent Subgraph Mining (FSM) process.

This approach is based on the assumption that the same character, or some parts of it, would be represented by similar subgraphs in the panels where it appears. Characters are the main focus of the storyline, so characters or parts of them should have a degree of redundancy and occurrence frequency higher than that of arbitrary objects.

3.1 Frequent Subgraph Mining

We apply the well-known algorithm gSpan [19] to mine the frequent subgraphs. Let $\mathbb{D}$ be the dataset of graphs, $\mathbb{D} = \{G_1, G_2, \ldots, G_n\}$ comprising n graphs, and a minimum support threshold $minSup$ (in absolute or percentage value), the goal of the algorithm is to enumerate all subgraphs G_f that are frequent, i.e. $sup(G_f) \geqslant$ minSup. For example, applying FSM with $minSup = 45\%$ on a graph dataset $\mathbb{D}$ containing 1000 graphs returns a set of k frequent subgraphs (or frequent patterns) $\mathbf{FP} = \{fp_1, fp_2, \ldots fp_k\}$ where each fp_i occurs *at least* 450 times. The algorithm starts with an empty DFS code and recursively grows it with valid, canonical extensions, so if fp_i is frequent, every substructures of it are frequent too. Fig. 2 shows a simplified example of the frequent subgraphs returned on the color layer dataset. We apply FSM to all 3 layers of graphs representing the comic volume, and obtain a list of frequent patterns for each layer, together with their frequency and original occurrence.

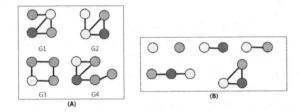

Fig. 2. Toy example of mining frequent subgraphs in graph data set (color layer): (A) dataset containing 4 graphs, (B) frequent subgraphs which occur at least 3 times ($minSup_{abs} = 3$ or $minSup_\% = 75\%$)

3.2 Retrieval with Query-by-Example

Querying is a difficult issue in comic retrieval. If the text in each comic-page image can be reliably localized and transcribed, we can gain some advantage of the contextual information such as dialog content or narration. However, as the OCR quality of textual content in comics is far from perfect in general, we suggest a visual-aid query type, such as sketch-based or example-based, for a natural and intuitive interface. While sketching requires a certain level of drawing skill, query-by-example (QBE) offers a simple yet interesting interface, as readers will choose from a set of main characters or characters of interest which are cropped from the original comic-page images.

As an adaptation of Bag-of-Words model to graph domain, instead of pairwise comparison between the query and all the graphs in the dataset, we only check if the frequent patterns is within the query, i.e. to see if the query has some degree of similarity with all the panels, and which panels collect the most signals of matching. The querying and retrieval steps are as follows:

Querying: The query image I_q is undergone the same processing step as the original comic-page image. We obtain the query graph G_q after this step.

Similarity Measure: We then check for occurrence of subgraph isomorphism between G_q and each discovered frequent pattern P_{freq} using VF2 algorithm [20]. A mapping M between G_q and P_{freq} is subgraph isomorphism iff M is an isomorphism between P_{freq} and a subgraph of G_q.

Edge Verification: In each edge match of mapping M, we further verify the match by comparing the edge's attribute. As an edge's attribute denotes the surface area ratio between regions, we only verify an edge match if their labels do not vary over a threshold: $\frac{|e_1 - e_2|}{\max(e_1, e_2)} \leqslant \omega$. Here ω is set to 0.3.

Retrieval: For each P_{freq} in the list of frequent patterns which appear in G_q (the graph representing the query), all the corresponding graphs in each layer containing these frequent patterns are collected. We then formulate a candidate set consist of graphs in the original dataset by merging the appearance information. For simplicity, the merging is done by sum pooling, i.e. each graph G_c in the candidate set is associated with the total number of occurrence of the frequent patterns which are contained in both them and in the query graph, forming a tuple (G_c, f_{G_c}), where $f_{G_c} = \sum_{P_{\text{freq}}} \delta(P_{\text{freq}}, G_c)$. Here the delta function $\delta(P_{\text{freq}}, G_c) = 1$ if P_{freq} is subgraph isomorphism with G_c, and $\delta(P_{\text{freq}}, G_c) = 0$ otherwise. However, as we consider color information gives more correct and reliable matching, each subgraph isomorphism with P_{freq} of layer "color" gives G_c twice the weight compare with a subgraph isomorphism with P_{freq} of layer "compact" or layer "moments". Finally, the candidate set is sorted accordingly to f_{G_c} and the panels represented by graphs that have highest f_{G_c} are returned.

4   Evaluation and Discussion

Dataset: We evaluate the proposed scheme on the first dataset and ground truth of comics for comic analysis research in the literature [21]. The dataset contains different comic titles from a variety of genres and styles. In each title, several characters were randomly cropped and chosen as the target object to test the retrieval. E.g. the title "Cosmozone" dataset consists of 94 pages, each page has 3-5 panels on average, resulting in the total of 371 panels. For each target offered to readers as QBE, we manually generate the ground truth by identifying its occurrence in all the panels. For example, to evaluate the retrieval of character *J. Cool* we identified 159 panels containing him as positive samples, and the remainder as negative ones. In this title, the characters may have relatively consistent size and orientation in a few consecutive panels. Table 1 shows the occurrence frequency of some selected characters in this title. Fig. 3 shows some samples directly cropped from random panels to be used as query. Table 2 gives the result in terms of the precision and recall rate of the retrieval of 4

Table 1. Occurrence Frequency of Charaters in COSMOZONE

Character	J. Cool	Admiral	Zuü	Apatchai	Goadec	Dokk	Sylvie
Appear Times	159	22	79	38	49	54	3
Frequency(%)	42.8	5.9	21.3	10.2	13.2	14.5	0.8

Fig. 3. Cropped areas of characters to be used as QBE. Characters' name, from left to right: J. Cool, Apatchai, Admiral, Dokk, Goadec, Zuü (©2009 Studio Cyborga)

characters in the same title for comparative study. The system yields the results by matching each of the layers of the graph representing the cropped area used as queries, versus the list of frequent subgraphs mined with minSup value of 40%. The node attributes are clustered into 10 groups of color values, and 5 groups of shape and compactness. Precision @10 and @20 measure how correct the top 10 and 20 returned results are, e.g. if *precision @10* of a target is 0.6, a reader would receive 6 panels containing him in the top 10 results.

In Fig 4, we show an instance of the retrieval result for character *J. Cool*. Note that the first two queries have rather impressive retrieval result: 8 out first 10 results are correct. As this system is one kind of comics browsing – which means a user is unlikely to browse through tens of pages results, or want to get hundreds of results even most of them are correct, (in that case he or she would want a more advanced or refined search),– the top results are the most interesting and important ones. The other two queries having the retrieval rate dropped down significantly can be explained by their occurrence frequencies in the title (quite lower than that of the first two); thus the matched frequent subgraphs are the ones scattering over the whole dataset, and even the top candidate graphs cannot gather the matching score to boost them to be prominent, resulting in more irrelevant results in the top page.

Table 2. Recall and Precision Results of Several QBE

Query	J. Cool	Lt. Zuü	Apachai	Dokk
Average Recall	54.86%	73.15%	50.31%	38.7%
Average Precision	74.1%	50.08%	34.24%	44.1%
Precision @10	0.8	0.8	0.4	0.3
Precision @20	0.85	0.7	0.25	0.35

Fig. 4. 12 out of top 20 retrieval results for character "J. Cool" by using the query on the left column. We obtained quite a lot of relevant results in the top returned images.

5   Conclusion

We have presented a method to represent comic-page images by region adjacency graphs to capture both local features and global structure in each panel. We have also presented how to retrieve interested characters using a sample area, known as query-by-example. The region's attributes are separated into different graph layers to flexibly assign score to each matching signal; and node values are quantized for faster matching. This graph representation scheme is robust to geometrical deformations; as it was shown the returned results consist of different poses, scales, orientations, and facial expressions of the same character. By mining frequent subgraphs and dealing with frequent subgraph matching only, we avoided the heavy computation of direct graph matching or similarity assessment. For future work, we are interested in incorporating relevance feedback function, which will allow semi-supervised learning from reused results and/or re-ranked feedback from readers.

References

1. In, Y., Oie, T., Higuchi, M., Kawasaki, S., Koike, A., Murakami, H.: Fast frame decomposition and sorting by contour tracing for mobile phone comic images. In: Proceedings of the 3rd WSEAS International Conference on Visualization, Imaging and Simulation, VIS 2010, Stevens Point, Wisconsin, USA, pp. 23–28 (2010)
2. Chan, C.H., Leung, H., Komura, T.: Automatic panel extraction of color comic images. In: Ip, H.H.-S., Au, O.C., Leung, H., Sun, M.-T., Ma, W.-Y., Hu, S.-M. (eds.) PCM 2007. LNCS, vol. 4810, pp. 775–784. Springer, Heidelberg (2007)
3. Ishii, D., Watanabe, H.: A study on frame position detection of digitized comics images. In: Workshop on Picture Coding and Image Processing (PCSJ), Nagoya, vol. 2010, pp. 124–125 (2010)

4. Ho, A.K.N., Burie, J.C., Ogier, J.: Panel and speech balloon extraction from comic books. In: 2012 10th IAPR International Workshop on Document Analysis Systems (DAS), pp. 424–428 (March 2012)
5. Arai, K., Tolle, H.: Automatic e-comic content adaptation. International Journal of Ubiquitous Computing 1(1), 1–11 (2010)
6. Rigaud, C., Karatzas, D., de Weijer, J.V., Burie, J.C., Ogier, J.M.: Automatic text localisation in scanned comic books. In: 9th International Conference on Computer Vision Theory and Applications, Barcelona (2013)
7. Rigaud, C., Burie, J.C., Ogier, J.M., Karatzas, D.: Color descriptor for content-based drawing retrieval. In: 2014 11th IAPR International Workshop on Document Analysis Systems (DAS), pp. 267–271 (April 2014)
8. Luqman, M.M., Ho, H.N., Burie, J.C., Ogier, J.M.: Automatic indexing of comic page images for query by example based focused content retrieval. In: 10th IAPR International Workshop on Graphics Recognition, United States (August 2013)
9. Yanagisawa, H., Ishii, D., Watanabe, H.: Face detection for comic images with deformable part model. In: 4th IIEEJ International Workshop on Image Electronics and Visual Computing (October 2014)
10. Liu, Y., Zhang, D., Lu, G., Ma, W.Y.: A survey of content-based image retrieval with high-level semantics. Pattern Recognition 40(1), 262–282 (2007)
11. Smeulders, A.W., Worring, M., Santini, S., Gupta, A., Jain, R.: Content-based image retrieval at the end of the early years. IEEE Transactions on Pattern Analysis and Machine Intelligence 22(12), 1349–1380 (2000)
12. Bunke, H., Riesen, K.: Recent advances in graph-based pattern recognition with applications in document analysis. Pattern Recognition 44(5), 1057–1067 (2011)
13. Rigaud, C., Tsopze, N., Burie, J.-C., Ogier, J.-M.: Robust frame and text extraction from comic books. In: Kwon, Y.-B., Ogier, J.-M. (eds.) GREC 2011. LNCS, vol. 7423, pp. 129–138. Springer, Heidelberg (2013)
14. Matas, J., Chum, O., Urban, M., Pajdla, T.: Robust wide-baseline stereo from maximally stable extremal regions. Image and Vision Computing 22(10), 761–767 (2004)
15. Mikolajczyk, K., Tuytelaars, T., Schmid, C., Zisserman, A., Matas, J., Schaffalitzky, F., Kadir, T., Van Gool, L.: A comparison of affine region detectors. International Journal of Computer Vision 65(1-2), 43–72 (2005)
16. Hu, M.-K.: Visual pattern recognition by moment invariants. IRE Transactions on Information Theory 8(2), 179–187 (1962)
17. Jain, A.K., Vailaya, A.: Image retrieval using color and shape. Pattern Recognition 29, 1233–1244 (1996)
18. Jain, A.K.: Data clustering: 50 years beyond k-means. Pattern Recognition Letters 31(8), 651–666 (2010)
19. Yan, X., Han, J.: gspan: Graph-based substructure pattern mining. In: Proceedings of 2002 IEEE International Conference on Data Mining, ICDM 2003, pp. 721–724. IEEE (2002)
20. Cordella, L., Foggia, P., Sansone, C., Vento, M.: A (sub)graph isomorphism algorithm for matching large graphs. IEEE Transactions on Pattern Analysis and Machine Intelligence 26(10), 1367–1372 (2004)
21. Guerin, C., Rigaud, C., Mercier, A., Ammar-Boudjelal, F., Bertet, K., Bouju, A., Burie, J.C., Louis, G., Ogier, J.M., Revel, A.: ebdtheque: A representative database of comics. In: 2013 12th International Conference on Document Analysis and Recognition (ICDAR), pp. 1145–1149 (August 2013)

Learning High-Order Structures for Texture Retrieval

Ni Liu[✉], Georgy Gimel'farb, and Patrice Delmas

Department of Computer Science, The University of Auckland
Auckland 1142, New Zealand
nliu866@aucklanduni.ac.nz, {g.gimelfarb,p.delmas}@auckland.ac.nz

Abstract. Learning interaction structures of graphical models, such as a high-order Markov-Gibbs random field (MGRF), is an open challenge. For a translation and contrast/offset invariant MGRF model of image texture, we sequentially construct higher-order cliques from the lower-order ones, starting from characteristic 2nd-order interactions (i.e. edges of the interaction graph). Every next-order clique is built by adding the available required edges to one of the current cliques. Experiments on texture databases resulted in the interpretable and intuitive learned cliques of up to 20th order. The learned high-order cliques consistently outperform in texture retrieval the multiple 2nd-order ones and the state-of-the-art local binary (LBP) and ternary patterns (LTP).

1 Introduction

Undirected graphical models, a.k.a. MGRFs, are natural and powerful tools for data modeling and problems solving in image processing and analysis [3,10,23]. However, learning from one or more training images an interaction graph, which cliques support factors defining an unknown MGRF of the images, is a hard problem due to combinatorial complexity of a space of all possible structures [21]. Today's structural learning searches for an interaction graph by scoring, constraints testing, and regression [11,24]. These learning strategies have been successful in multiple applications. A $L1$-regularised regressive learning [21] estimated an interpretable and intuitive structure of a 28-nodal Markov chain model of rain data, as well as structures on the news data set under specific setting of parameters. An efficient Gaussian model of a gene regulatory 957-nodal network with a biologically informative and comprehensible structure was learned in [11].

Contrastingly, the interaction structure of the MGRF image models is often assumed fixed. The earlier models mostly have only 2nd-order interactions between either the nearest 4- or 8-neighbours neighbours, like in the Ising/Potts models [2,19], the classical Gaussian and auto-regressive Gauss-Markov models [12]. A subsequent generic 2nd-order MGRF model with multiple arbitrary close- and long-range interactions [5,6] allowed for learning its structure to more accurately represent a training texture. The low-order cliques in principle cannot capture rich statistics of natural scenes and thus severely restrict descriptive abilities of the MGRFs. Long-standing efforts to improve their expressiveness focused on enhancing the interaction structures and developing relevant learning and inference algorithms.

The well-known FRAME [8,25,26] and Fields-of-Experts (FoE) [20] models involve the high-order interactions implicitly, via "whitening" the MGRF to be learned, i.e. transforming it into a stack of independent random fields with a bank of either selected from a fixed collection (FRAME), or optimised (FoE) square fixed-size linear and non-linear filters, their supports in the lattice acting as cliques. Our previous work [14,16] extended the learned structures of the multiple 2nd-order interactions to ordinal interactions in up to 4th-order cliques of the arbitrary learned shapes. A higher-order generic MGRF [15], which inherited ordinal interactions from generalised LBPs or local ternary patterns (LTPs), allows for learning most characteristic cliques of different shapes, adapted to visual appearance of a training texture. For computational simplicity, these learning schemes took into account only partial clique-wise ordinal interactions between a guiding vertex and all other vertices of an LBP/LTP, instead of all the conditional dependencies over the clique. However, our previous experiments with natural textures confirmed that these simplified learned higher-order cliques can consistently improve the texture retrieval accuracy, comparing with the lower-order ones and the state-of-the-art LBPs.

For modelling translation and contrast/offset invariant image textures, this paper explores possibilities to build characteristic high-order cliques sequentially, starting from an initial interaction graph and increasing their order step-by-step towards the largest arbitrary-shaped cliques. This learning process relies on the same conjecture as the aforementioned constraints testing: the interaction graph is completely determined by the learned 2nd-order interactions, i.e. no edge between any two conditionally independent vertices. After the characteristic edges are selected, each higher-order clique is built by adding a new vertex to a lower-order clique only if all the new edges are characteristic. These steps continue until no higher-order clique can be built or a prescribed top order is reached. Experiments with natural textures confirmed the learned up to 20th-order models outperform the multiple 2nd-order cliques and the state-of-the-art LBPs in recognition and retrieval of individual textures from a large database.

The paper is organized as follows. Section 2 presents a generic MGRF with multiple ordinal interactions. The sequential learning of the 2nd- and higher-order cliques for this model is described in Section 3. Section 4 discusses experiments and outlines conclusions and future work.

2   Generic Low- and High-Order MGRF Models

Let $\mathbb{Q}$ and $\mathbb{R}$ denote, respectively, a finite set of grey levels: $\mathbb{Q} = \{0, 1, \ldots, Q-1\}$ and a finite arithmetic 2D or 3D lattice, supporting greyscale images $g : \mathbb{R} \to \mathbb{Q}$. Probabilistic dependencies between the lattice sites (pixels or voxels), $\mathbf{r} \in \mathbb{R}$, are described by an interaction graph $\Gamma = (\mathbb{R}, \mathbf{E} \subset \mathbb{R}^2)$ with set $\mathbb{R}$ of vertices (points) and a set of arcs, or edges $\mathbf{E}$. Each edge $(\mathbf{r}, \mathbf{r}')$ connects an interacting pair of points, called neighbours and considered a second-order clique in Γ.

The MGRF modelling assumes that an interaction structure over $\mathbb{R}$ is formed by a system of cliques, or complete subgraphs of Γ. For a spatially repetitive texture, the structure is reduced to N; $N \geq 1$, families $\mathbb{C} = \{\mathbb{C}_n : n = 1, \ldots, N\}$ of translation-invariant cliques of order K_n; $K_n \geq 2$. Each individual family can be specified by $K_n - 1$ fixed coordinate offsets of points of each its clique with respect to a "guiding" point $\mathbf{r}$ of the clique: $(\mathbf{s}_{n:k} = \mathbf{r}'_k - \mathbf{r} : k = 1, \ldots, K_n - 1)$.

An MGRF model of images is defined by its Gibbs probability distribution (GPD), $\mathbf{P} = \left[P(\mathbf{g}) : \mathbf{g} \in \mathbb{G}; \sum_{\mathbf{g} \in \mathbb{G}} P(\mathbf{g}) = 1 \right]$, ($\mathbb{G}$ is the set of all images) such that each probability, $P(\mathbf{g})$, is factored over the maximal cliques of an interaction graph, $\Gamma = (\mathbb{R}, \mathbb{E})$. The cliques support non-trivial (non-constant) factors and potentials (logarithms of the factors). Let, for simplicity, all the cliques are of the same order K (i.e., $K_n = K$ for $n = 1, \ldots, N$. A translation-invariant K-order interaction structure on $\mathbb{R}$ is a system, $\mathbb{C}$, of N, $N \geq 1$, K-order clique families, $\mathbb{C}_n$; $\bigcup_{n=1}^{N} \mathbb{C}_n = \mathbb{C}$, such that each family has its own geometrical shape of cliques, $\mathbf{c}_{n:\mathbf{r}} \in \mathbb{C}_a$, with origins at the lattice sites, $\mathbf{r} \in \mathbb{R}$, and a K-variate potential function, $\mathbf{V}(q_1, \ldots, q_K) : q_k = g(\mathbf{r}'_k) \in \mathbb{Q}; \mathbf{r}'_k \in \mathbf{c}_{a:\mathbf{r}}$, depending on ordinal relationships between the origin, $\mathbf{r}$, and its $K - 1$ neighbours, $\mathbf{r}' \in \mathbf{c}_{n:\mathbf{r}}; \mathbf{r}' \neq \mathbf{r}$, in each clique. The GPD for such translation- and contrast/offset invariant MGRF is:

$$P(\mathbf{g}) = \frac{1}{Z} \exp\left(-\sum_{n=1}^{N} \sum_{\mathbf{c}_{n:\mathbf{r}} \in \mathbb{C}_a} \mathbf{V}_n \left(g(\mathbf{r}') : \mathbf{r}' \in \mathbf{c}_{a:\mathbf{r}} \right) \right) \equiv \frac{1}{Z} \exp\left(-E(\mathbf{g}) \right) \quad (1)$$

where $E(\mathbf{g}) = \sum_{\mathbf{c}_{n:\mathbf{r}} \in \mathbb{C}_a} \mathbf{V}_a \left(g(\mathbf{r}') : \mathbf{r}' \in \mathbf{c}_{a:\mathbf{r}} \right)$ is the Gibbs energy and the partition function Z normalises the distribution over the entire parent population $\mathbb{G} = \mathbb{Q}^{|\mathbb{R}|}$ of the images: $Z = \sum_{\mathbf{g} \in \mathbb{G}} \exp(-E(\mathbf{g}))$.

Both the shapes of the cliques and potentials, quantifying the ordinal interactions in Eq. (1) can be learned approximately from a given training image $\mathbf{g}^\circ$, rather than chosen by hand, much in the same way as for a generic 2nd-order MGRFs accounting for signal co-occurrences [7].

$$\mathbf{V}_n = -\lambda_n \left(\mathbf{F}_n(\mathbf{g}^\circ) - \mathbf{F}_{n:\text{irf}} \right); \ n = 1, \ldots, N$$

where $\mathbf{F}_n(\mathbf{g}^\circ)$ is the empirical marginal probability distribution, or the normalised histogram of the ordinal signal configurations over the clique family $\mathbb{C}_n$ for the training image, $\mathbf{g}^\circ$, and $\mathbf{F}_{n:\text{irf}}$ is the like probability disturbution for the independent random field (IRF) with probabilities of pixel- or voxel-wise signal magnitudes, which are equal to the corresponding empirical pixel/voxel-wise marginals for the training image $\mathbf{g}^\circ$. In principle, the values $\mathbf{F}_{n:\text{irf}}$ can be derived from the signal marginals or evaluated empirically from a generated sample of this IRF. The factor λ_a is also computed analytically [6,7]. To keep learning such a K-order model feasible for K at most 20 , only partial clique-wise ordinal relations in the K-order arbitrary-shaped LTPs are taken into account below.

3   Sequential Structural Learning

Learning the interaction graph. The characteristic clique families are found like in [6,7] by analysing the family-wise partial interaction energies for a large search pool of the candidate families. In our experiments (Section 4), the 2nd-order candidates for modelling 2D images are limited to the intra-clique pixel x- and y-offsets in the ranges $[-50, 50]$ and $[0, 50]$, respectively, giving in total the $5,150$ candidates. The most characteristic (least-energetic) families were selected by representing the empirical distribution of the family-wise interaction energies as a mixture of Gaussians, and the Akaike information criterion (AIC) was used to estimate the number of mixture components. This process was implemented by employing the "gmdistribution" class from the MATLAB statistics toolbox.

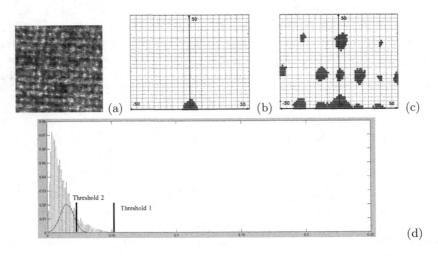

Fig. 1. (a) Texture "blanket1" and its 2nd-order characteristic clique families learned for Threshold 1 (b) and Threshold 2 (c); "0" – the origin vertex; the blue dots – the offsets $\boldsymbol{\delta}$ from the origin for each 2nd-order clique; (d) the empirical energy distribution for the 2nd-order candidates. The first three components of the mixture (with the means of 0.17, 0.03, and 0.02) are plotted: Thresholds 1 and 2 are at the intersections of the components 1 and 2 and 2 and 3, respectively

Figure 1 shows the distribution of the negated Gibbs energies (negenergies) for the $5,150$ 2nd-order candidates and its Gaussian mixture approximation for the texture "blanket1". The estimated five mixture components have the means, variances, and weights of $(0.17, 0.03, 0.01)$; $(0.03, 10^{-4}, 0.14)$; $(0.02, 2.8 \cdot 10^{-5}, 0.26)$, $(0.001, 1.1 \cdot 10^{-5}, 0.26)$, and $(0.004, 4.8 \cdot 10^{-6}, 0.33)$, respectively. Only the first three mixture components are plotted in Fig. 1.

The intersection of the first and second components (Threshold 1) selects mostly the short-range cliques, as shown in Fig. 1 where only 58 families with small intra-clique coordinate offsets were chosen. The intersection of the second

Algorithm 1. Learning the interaction graph (2nd- order clique families)

Input: Training image $\mathbf{g}^{\circ}$
Output: List $\boldsymbol{\Delta}_2 = \{\boldsymbol{\delta}_j : j = 1, \dots, N_2\}$ of coordinate offsets $\boldsymbol{\delta}_j = (\delta_{\mathrm{x}:j}, \delta_{\mathrm{y}:j})$ specifying the learned characteristic 2nd-order clique families, $\{\mathbf{C}_{2:j}; j = 1, \dots, N_2\}$.

1. Compute the normalized histogram $\mathbf{F}_{\mathrm{irf}}$ of the IRF signal configurations for all candidate families.
2. Compute the normalized histograms $\{\mathbf{F}_{2:j}; j = 1, \dots, 5150\}$ of signal configurations in the training image for all candidate families.
3. Estimate the relative (scaled) potentials: $\{\mathbf{V}_{2:j} = \mathbf{F}_{2:j} - \mathbf{F}_{\mathrm{irf}}; j = 1, \dots, 5150\}$
4. Compute the relative Gibbs energy for each clique family.
5. Represent the Gibbs energy distribution as a mixture of Gaussians and estimate the number of its components.
6. Find the threshold at the intersection of the second and third Gaussians to select the characteristic cliques with the lower Gibbs energy, $\{\mathbf{C}_{2:j}; j = 1, \dots, n_2\}$.

and third components (Threshold 2) yields in this example the 538 families with both the short- and long-range offsets. Although the selection of the characteristic edges deserves a more in-depth study (especially, to separate the true primary dependencies from the false secondary ones produced by statistical interplays of the primary ones), for simplicity, our experiments below applied to all the training textures the latter threshold. Algorithm 1 presents the basic steps of learning the characteristic 2nd-order clique families specifying the interaction graph.

Constructing the Higher-Order Cliques. Since all pairs of vertices in a clique are connected by edges, a naïve, but still workable way to construct the higher-order cliques is to start from the lowest-order ones, i.e., the learned edges $\boldsymbol{\Delta}_2$, and check whether adding a new vertex to an existing clique meets this constraint. If the new vertex is connected to each node of the original lower-order clique, then the high-order clique exists, too, so that all cliques of the next order are sequentially learned from the current-order ones (with simultaneous exclusion of possible duplicates). Algorithm 2 starts from the learned translation-invariant edges specified by the coordinate offsets $\boldsymbol{\Delta}_2$ of the size N_2 and builds sequentially the cliques of the orders $k = 3, \dots, 20$. Given a new offset, $\boldsymbol{\delta}$; $\boldsymbol{\delta} \in \boldsymbol{\Delta}_2$, and an offset $\boldsymbol{\delta}_{k-1:i:\kappa}$ for the node κ of the i-th clique of the order $k - 1$, the relative offset between both the nodes, $\boldsymbol{\rho} = \boldsymbol{\delta}_{k-1:i:\kappa} - \boldsymbol{\delta}$ is checked against the list $\boldsymbol{\Delta}_2$. If either $\boldsymbol{\rho}$ or $-\boldsymbol{\rho}$ is in the list, the next clique's node $\kappa + 1$ is checked, otherwise the higher-order clique with the candidate node does not exist.

Algorithm 2 becomes computationally infeasible when the number of the higher-order cliques is too big. Thus in the experiments below some of the constructed high-order cliques are discarded after their number becomes greater than 5000. To circumvent such cases, the number of edges $\boldsymbol{\Delta}_2$ should be minimised by excluding low-energetic secondary interactions produced by interplays of the primary ones and similar long-range offsets with too small angular and

---

Algorithm 2. Constructing high-order cliques

Input: List of the characteristic offsets Δ_2 from Algorithm 1.
Output: Lists $\Delta_k = [\Delta_{k:j} = \{\delta_{k:j:1}, \ldots, \delta_{k:j:k-1}\} : j = 1, \ldots, N_k]; \ k = 3, \ldots, 20$ of
descriptors for the k-order translation invariant families.

1: **while** $k = 2 \ldots 11$ **do**
2:     $k \leftarrow k + 1; \ j \leftarrow 0$
3:     **for** each offset $\delta \in \Delta_2$ **do**
4:         **for** each $(k-1)$-order descriptor $\Delta_{k-1:i} \in \Delta_{k-1}$ such that $\delta \notin \Delta_{k-1:i}$ **do**
5:             **for** each offset $\delta_{k-1:i:\kappa} \in \Delta_{k-1:i}; \ \kappa = 1, \ldots, k-1$ **do**
6:                 $\rho = \delta_{k-1:i:\kappa} - \delta$
7:                 **if** $\rho \in \Delta_2$ OR $-\rho \in \Delta_2$ for all $\kappa = 1, \ldots, k-1$ **then**
8:                     $j \leftarrow j + 1$
9:                     Add δ to $\Delta_{k-1:i}$ to form the k-order clique descriptor $\Delta_{k:j}$.
10:    **if** $(N_k \leftarrow j) > 0$ **then**
11:        Exclude duplicate offsets from each descriptor $\Delta_{k:j}$.
12:        Exclude duplicate descriptors from the list Δ_k.
13:        Exclude from Δ_2 all the offsets, being absent in the list Δ_k.
14:    **if** $N_k = 0$ OR $k = 20$ **then break while**

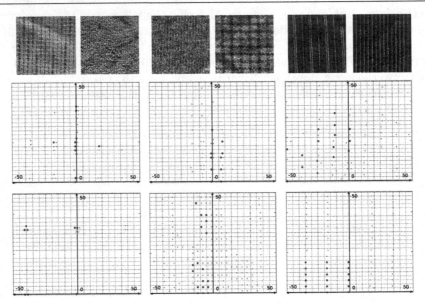

Fig. 2. High-order cliques for the query textures 1–6 from the DB1 (red and blue dots – the offsets in Δ_2 and a high-order clique, respectively)

radial differences. Figure 2 presents the high-order cliques learned for the six natural textures from the test database DB1.

Our present mixed-code (Matlab and C) implementation on a 3.33 GHz Intel Core i5-2500 processor with 8 GB memory takes about 5 sec for selecting 2nd-order characteristic clique families from the 5150 candidates for a training

256×256 image. The complexity for constructing the higher-order families depends on the number of their candidates. An empirically chosen upper limit of 5000 candidates made such sequential learning (constructing the higher-order families from the loweer-order ones) in Algorithms 1 and 2 tractable. In the average, it takes about 11.5 min per the 256×256 training image to build the models at most 20th order.

4 Experimental Results and Conclusions

The proposed sequential structural learning was tested experimentally on texture retrieval based on up to 20th-order partial ordinal relations in the learned multiple translation invariant cliques.

Fig. 3. Query textures: 7 – 24 from the DB 1 (top 2 rows) and DB2 (other 5 rows)

The goal interaction graph is built in such a way, as to exclude possible overlaps between the cliques from the different families. For all clique families from high to low order, the first highest-order clique family is always chosen, and every next family is selected from only the candidates having no common interactions with any already chosen family, until all the candidates are checked. Because the

goal was to evaluate the performance of the highest-order clique families, the top
three high-order families in the interaction graph for the query image were used
for retrieving each class of textures.

Table 1. Class-wise and total accuracy of the χ^2-based retrieval on the DB 1 (D_k;
D_{to}, CLB; HF; SLB; LTP$_j$ – the learned least-energetic k-order clique and 3 top-order
cliques vs. the classic LBP; LBP-HF; shift LBP, and j LTPs, respectively)

#	Texture	D_2	D_8	D_{to}	CLB	HF	SLB	LTP$_1$	LTP$_3$
				DB 1					
1	canvas001	0.50	1.00	1.00	0.50	0.50	0.88	1.00	1.00
2	canvas002	1.00	1.00	1.00	1.00	1.00	1.00	1.00	1.00
3	canvas003	0.63	1.00	1.00	0.13	0.13	0.88	1.00	1.00
4	canvas005	0.88	1.00	1.00	0.75	0.88	0.88	1.00	1.00
5	canvas006	0.88	1.00	1.00	0.63	1.00	1.00	1.00	1.00
6	canvas009	0.75	0.88	0.88	0.75	0.88	0.88	0.75	0.75
7	canvas011	1.00	1.00	1.00	1.00	1.00	1.00	1.00	1.00
8	canvas021	0.75	0.88	1.00	1.00	1.00	1.00	1.00	1.00
9	canvas022	0.63	1.00	1.00	0.50	1.00	1.00	1.00	1.00
10	canvas023	0.50	0.63	1.00	0.38	0.38	0.63	0.63	1.00
11	canvas025	0.50	0.75	1.00	0.75	1.00	0.63	0.75	1.00
12	canvas026	0.38	1.00	0.88	1.00	1.00	1.00	1.00	1.00
13	canvas031	0.63	1.00	1.00	0.88	0.88	0.75	1.00	1.00
14	canvas032	0.75	1.00	1.00	0.63	0.63	0.63	0.63	0.63
15	canvas033	0.13	0.63	1.00	0.13	0.13	0.25	0.38	1.00
16	canvas035	0.63	1.00	1.00	0.50	0.63	0.50	0.63	1.00
17	canvas038	0.63	0.88	1.00	0.75	0.75	0.63	0.75	1.00
18	canvas039	0.75	0.75	0.88	0.63	0.75	0.63	0.75	1.00
19	carpet002	0.75	1.00	1.00	1.00	1.00	1.00	1.00	1.00
20	carpet004	1.00	1.00	1.00	1.00	1.00	1.00	1.00	1.00
21	carpet005	0.88	1.00	1.00	1.00	1.00	1.00	1.00	1.00
22	carpet009	0.63	0.75	1.00	0.38	0.38	0.75	1.00	1.00
23	tile005	0.75	1.00	1.00	0.50	0.75	0.88	0.88	1.00
24	tile006	0.88	1.00	1.00	0.88	0.88	0.88	1.00	1.00
	Mean	0.70	0.92	**0.99**	0.70	0.77	0.82	0.88	0.97
	St.dev.	0.20	0.13	**0.04**	0.20	0.28	0.20	0.17	0.08
				DB 2					
	Mean	0.74	0.94	**0.97**	0.79	0.86	0.87	0.93	0.96
	St.dev.	0.22	0.14	**0.08**	0.25	0.24	0.17	0.14	0.12

Two test databases: DB 1 and DB 2 – for our experiments are exemplified
in Fig. 3. The DB 1, having been already used in [13], contains 192 textures
from the Outex_TC_00000 test suit [17]: 24 classes and 8 samples 256×256
per class, randomly cropped from the 746×538 source images. The DB 2
from [15] contains 320 Outex images (40 classes; 8 samples 256×256 per class,
randomly cropped from 512×512 source images). Only one sample, randomly
selected from each class, was used as a query or training texture (Figs. 2 and 3).

All other samples from this class, together with all samples of the remaining classes served as test images for retrieval. The χ^2 distance between the query and test statistics (empirical distributions of the chosen ordinal relations) was used to measure the dissimilarity between two textures.

The performance of each descriptive model was evaluated using the same metric as in [4]. Let k be the number of samples for one class in a database. Let one sample of each class serve as a query, all other samples belonging to this class and other classes in the database form the test images. After the top k images are retrieved, the percentage of the correctly retrieved images is stored for each query, and the average percentage over all the queries specifies the total performance of the model on this database.

Table 1 compares the retrieval results on the DB 1 and DB 2 using the classic LBP [18], Fourier features of the LBP histogram (LBP-HF) [1], one single-scale ($\nu = 8$ points at radius $r = 1$) or three multiple-scale ($r = 1, \nu = 8$; $r = 2, \nu = 12$; $r = 4, \nu = 20$) symmetric LTPs [22], and the shift LBP [9] as the controls. These pilot experiments have shown that the learned highest-order interaction structures outperform all the control methods using heuristic structures. The straightforward sequential learning of characteristic high-order cliques for translation- and contrast/offset-invariant MGRFs is computationally feasible and useful for solving applied problems. The structures learned result in richer and more comprehensive texture descriptions. Future work will focus on effective elimination of redundancies in the interaction graphs and more diverse texture recognition and retrieval experiments.

References

1. Ahonen, T., Matas, J., He, C., Pietikäinen, M.: Rotation Invariant Image Description with Local Binary Pattern Histogram Fourier Features. In: Salberg, A.-B., Hardeberg, J.Y., Jenssen, R. (eds.) SCIA 2009. LNCS, vol. 5575, pp. 61–70. Springer, Heidelberg (2009)
2. Baxter, R.J.: Exactly Solved Models in Statistical Mechanics. Academic Press (1982)
3. Blake, A., Kohli, P., Rother, C.: Markov Random Fields for Vision and Image Processing. MIT Press (2011)
4. Doshi, N.P., Schaefer, G.: A comprehensive benchmark of local binary pattern algorithms for texture retrieval. In: Proc. 21st IAPR Intern. Conf. on Pattern Recognition (ICPR 2012), pp. 2760–2763 (2012)
5. Gimel'farb, G.: Texture modeling with multiple pairwise pixel interactions. IEEE Trans. on Pattern Analysis and Machine Intelligence 18, 1110–1114 (1996)
6. Gimel'farb, G.: Image Textures and Gibbs Random Fields. Kluwer Academic, Dordrecht (1999)
7. Gimel'farb, G., Zhou, D.: Texture analysis by accurate identification of a generic Markov-Gibbs model. In: Kandel, A., Bunke, H., Last, M. (eds.) Applied Pattern Recognition. SCI, vol. 91, pp. 221–245. Springer, Berlin (2008)
8. Guo, C.-E., Zhu, S.-C., Wu, Y.N.: Modeling visual patterns by integrating descriptive and generative methods. Intern. J. of Computer Vision 53, 5–29 (2003)

9. Kylberg, G., Sintorn, I.-M.: Evaluation of noise robustness for local binary pattern descriptors in texture classification. EURASIP Journal on Image and Video Processing 2013(1), 1–20 (2013)
10. Lézoray, O., Grady, L.: Image Processing and Analysis with Graphs: Theory and Practice. CRC Press (2012)
11. Li, F.: Structure learning with large sparse undirected graphs and its applications. PhD thesis, Citeseer (2007)
12. Li, S.: Markov Random Field Modeling in Image Analysis. Springer, Berlin (2009)
13. Liu, L., Zhao, L., Long, Y., Kuang, G., Fieguth, P.: Extended local binary patterns for texture classification. Image and Vision Computing 30(2), 86–99 (2012)
14. Liu, N., Gimel'farb, G., Delmas, P.: Texture modelling with generic translation- and contrast/offset-invariant 2nd–4th-order MGRFs. In: Proc. 28th Intern. Conf. on Image and Vision Computing New Zealand (IVCNZ 2013), pp. 370–375 (2013)
15. Liu, N., Gimel'farb, G., Delmas, P.: High-order mgrf models for contrast/offset invariant texture retrieval. In: Proc. 29th Intern. Conf. on Image and Vision Computing New Zealand (IVCNZ 2014), pp. 96–101 (2014)
16. Liu, N., Gimel'farb, G., Delmas, P., Chan, Y.H.: Contrast/offset-invariant generic low-order MGRF models of uniform textures. In: AIP Conf. Prioceedings, vol. 1559(1), pp. 145–154 (2013)
17. Ojala, T., Mäenpää, T., Pietikäinen, M., Viertola, J., Kyllönen, J., Huovinen, S.: Outex – New framework for empirical evaluation of texture analysis algorithms. In: Proc. 16th Intern. Conf. on Pattern Recognition, vol. 1, pp. 701–706 (2002)
18. Ojala, T., Pietikainen, M., Maenpaa, T.: Multiresolution gray-scale and rotation invariant texture classification with local binary patterns. IEEE Transactions on Pattern Analysis and Machine Intelligence 24(7), 971–987 (2002)
19. Potts, R.B.: Some generalized order-disorder transformations. Mathematical Proceedings of the Cambridge Philosophical Society 48(1), 106–109 (1952)
20. Roth, S., Black, M.J.: Fields of Experts. Intern. Journal of Computer Vision 82(2), 205–229 (2009)
21. Schmidt, M.: Graphical Model Structure Learning with l1-regularization. PhD thesis, University of British Columbia (2010)
22. Tan, X., Triggs, B.: Enhanced local texture feature sets for face recognition under difficult lighting conditions. IEEE Transactions on Image Processing 19(6), 1635–1650 (2010)
23. Tappen, M.F., Russell, B.C., Freeman, W.T.: Efficient graphical models for processing images. In: Proc. 2004 IEEE Computer Society Conf. on Computer Vision and Pattern Recognition (CVPR 2004), vol. 2, p. II–673. IEEE (2004)
24. Zhou, Y.: Structure learning of probabilistic graphical models: a comprehensive survey. arXiv preprint arXiv:1111.6925 (2011)
25. Zhu, S.C., Wu, Y., Mumford, D.: Minimax entropy principle and its application to texture modeling. Neural Computation 9(8), 1627–1660 (1997)
26. Zhu, S.C., Wu, Y., Mumford, D.: Filters, random fields and maximum entropy (FRAME): Towards a unified theory for texture modeling. Intern. Journal of Computer Vision 27(2), 107–126 (1998)

Author Index